Farnesyltransferase Inhibitors in Cancer Therapy

Cancer Drug Discovery and Development

Beverly A. Teicher, Series Editor

10. **Matrix Metalloproteinase Inhibitors in Cancer Therapy,** edited by *Neil J. Clendeninn and Krzysztof Appelt,* 2000
9. **Tumor Suppresor Genes in Human Cancer,** edited by *David E. Fisher,* 2000
8. **Farnesyltransferase Inhibitors in Cancer Therapy,** edited by *Saïd M. Sebti and Andrew D. Hamilton,* 2000
7. **Platinum-Based Drugs in Cancer Therapy,** edited by *Lloyd R. Kelland and Nicholas P. Farrell,* 2000
6. **Signaling Networks and Cell Cycle Control: *The Molecular Basis of Cancer and Other Diseases,*** edited by *J. Silvio Gutkind,* 2000
5. **Apoptosis and Cancer Chemotherapy**, edited by *John A. Hickman and Caroline Dive,* 1999
4. **Antifolate Drugs in Cancer Therapy,** edited by *Ann L. Jackman,* 1999
3. **Antiangiogenic Agents in Cancer Therapy,** edited by *Beverly A. Teicher,* 1999
2. **Anticancer Drug Development Guide: *Preclinical Screening, Clinical Trials, and Approval,*** edited by *Beverly A. Teicher,* 1997
1. **Cancer Therapeutics: *Experimental and Clinical Agents,*** edited by *Beverly A. Teicher,* 1997

Farnesyltransferase Inhibitors in Cancer Therapy

Edited by

Saïd M. Sebti

H. Lee Moffitt Cancer Center and Research Institute, Tampa, FL

and

Andrew D. Hamilton

Yale University, New Haven, CT

Humana Press

Totowa, New Jersey

999 Riverview Drive, Suite 208
Totowa, New Jersey 07512

For additional copies, pricing for bulk purchases, and/or information about other Humana titles, contact Humana at the above address or at any of the following numbers: Tel.: 973-256-1699; Fax: 973-256-8341; E-mail: humana@humanapr.com or visit our Website: http://humanapress.com

The crystal structure of protein farnesyltransferase has provided a structural framework for understanding how inhibitors and substrates interact with this therapeutically important enzyme. Shown here, the beta subunit of this heterodimeric enzyme is lined with aromatic residues that interact with both substrates in the reaction. For more detail please see Chapter 3. Cover artwork courtesy of Drs. Lorena Beese and Stephen Long.

Cover design by Patricia F. Cleary.

This publication is printed on acid-free paper. ∞
ANSI Z39.48-1984 (American National Standards Institute) Permanence of Paper for Printed Library Materials.

Printed in the United States of America. 10 9 8 7 6 5 4 3 2 1

Library of Congress Cataloging-in-Publication Data

Farnesyltransferase inhibitors in cancer therapy/edited by Saïd M. Sebti and Andrew D. Hamilton.
p.;cm.—(Cancer drug discovery and development)
Includes bibliographical references and index.
ISBN 0-89603-629-4 (alk. paper)
1. Dimethylalyltranstransferase—Inhibitors—Therapeutic use—Testing. 2. Cancer—Chemotherapy. I. Sebti, Saïd M. II. Hamilton, Andrew D. III. Series.
[DNLM: 1. Dimethylalyltranstransferase—antagonists & inhibitors. 2. Antineoplastic Agents—chemical synthesis. 3. Drug design. 4. Enzyme Inhibitors—therapeutic use. 5. Genes, ras. QU 141 P927 2000]
RC271.D534 P74 2000
616.99'4061—dc21

00-035038

PREFACE

Most presently used anticancer drugs were developed based on their antiproliferative rather than antioncogenic properties and consequently suffer from two major limitations. Many are cytotoxic and cause major thwarted effects owing to their ability to inhibit indiscriminately the growth of fast dividing cells. Drug resistance, the second major limitation of these drugs, arises primarily from the lack of activity against the more slowly growing solid tumors.

The recent explosion of knowledge gained from genes capable of causing cancer, and the pivotal role they play in growth factor signal transduction, have opened up new avenues for rationally designing novel anticancer drugs. One of the best studied signal transduction pathways, which contains a gold mine of anticancer drug discovery targets, is that of receptor tyrosine kinase signaling. A key molecular switch within this pathway is a small GTPase called Ras. Ras mediated the transfer of biological information from extracellular signals to the nucleus and is a major regulator of cell division. Oncogenic mutations in the ras gene are found in about 30% of all human cancers and result in a constitutively activated protein that sends uninterrupted signals to the nucleus. Over the last two decades several approaches have failed to reverse the constitutive activation of the Ras protein. Recently, however, the realization that farnesylation, a lipid posttranslational modification, of Ras is required for its cancer-causing activity, prompted an intense search for farnesyltransferase inhibitors as novel anticancer agents.

Farnesyltransferase Inhibitors in Cancer Therapy describes the efforts of several groups to design, synthesize, and evaluate the biological activities of farnesyltransferase inhibitors. Rational design of small organic molecules that mimic the carboxyl terminal tetrapeptide farnesylation site of Ras resulted in pharmacological agents capable of inhibiting Ras processing and selectively antagonizing oncogenic signaling and suppressing human tumor growth in mouse models without side effects. These agents are presently undergoing advanced preclinical studies. Several important issues, such as the mechanism of action of farnesyltransferase inhibitors and the potential mechanisms of resistance to inhibition of K-Ras farnesylation, are also discussed. Furthermore, the recent observation that K-Ras 4B, the most frequently mutated form of Ras in human tumors, can be geranylgeranylated and that, in addition to Ras, there are other geranylgeranylated small G-proteins that play an important role in smooth muscle proliferation and apoptosis, stimulated the search for inhibitors of a closely related enzyme, geranylgeranyltransferase I. Thus, the current volume also discusses geranylgeranyltransferase I inhibitors as modulators of cell cycle and apoptosis, and as potential therapeutic agents for cardiovascular disease.

Saïd M. Sebti
Andrew D. Hamilton

CONTENTS

CONTRIBUTORS

LYDIA ARMSTRONG, PHD • *Schering-Plough Research Institute, Kenilworth, NJ*
LORENA S. BEESE, PHD • *Department of Biochemistry, Duke University Medical Center, Durham NC*
ERIC J. BERNHARD, PHD • *Department of Radiation Oncology, University of Pennsylvania Medical Center, Philadelphia, PA*
W. ROBERT BISHOP, PHD • *Schering-Plough Research Institute, Kenilworth, NJ*
MARIE-CHRISTINE BISSERY, PHD • *Rhone-Poulenc-Rorer S.A., Vitry sur Seine, France*
JEAN-DOMINIQUE BOURZAT, PHD • *Rhone-Poulenc-Rorer S.A., Vitry sur Seine, France*
CLIVE BREALEY, PHD • *Rhone-Poulenc-Rorer S.A., Vitry sur Seine, France*
FREDERICK S. BUCKNER, PHD • *Department of Medicine, University of Washington, Seattle, WA*
MARC CAPET, PHD • *Rhone-Poulenc-Rorer S.A., Vitry sur Seine, France*
PATRICK J. CASEY, PHD • *Department of Pharmacology and Cancer Biology, Duke University Medical Center, Durham, NC*
MICHEL CHEVÉ, PHD • *Rhone-Poulenc-Rorer S.A., Vitry sur Seine, France*
ELIZABETH COHEN-JONATHAN, MD, PHD • *Department of Radiation Oncology, University of Pennsylvania Medical Center, Philadelphia, PA*
ALAIN COMMERÇON, PHD • *Rhone-Poulenc-Rorer S.A., Vitry sur Seine, France*
ADRIENNE D. COX, PHD • *Departments of Radiation Oncology, Pharmacology, and Curriculum in Genetics and Molecular Biology, Lineberger Comprehensive Cancer Center, University of North Carolina, Chapel Hill, NC*
MARGUERITE B. DALTON, PHD • *Department of Biochemistry and Molecular Biology, East Tennessee State University, Johnson City, TN*
KEITH DEL VILLAR, PHD • *Department of Microbiology and Molecular Genetics, Jonsson Comprehensive Cancer Center, University of California, Los Angeles, CA*
CHANNING J. DER, PHD • *Department of Pharmacology and Curriculum in Genetics and Molecular Biology, Curriculum in Toxicology, Lineberger Comprehensive Cancer Center, University of North Carolina, Chapel Hill, NC*
NORBERT DEREU, PHD • *Rhone-Poulenc-Rorer S.A., Vitry sur Seine, France*
RONALD J. DOLL, PHD • *Schering-Plough Research Institute, Kenilworth, NJ*
MARC DUCHESNE, PHD • *Rhone-Poulenc-Rorer S.A., Vitry sur Seine, France*
CRAIG FAIRCHILD, PHD • *Oncology Drug Discovery, Bristol-Myers Squibb Pharmaceutical Research Institute, Princeton, NJ*
GILLES FAVRE, PHD • *Centre Claudius Regaud, Toulouse, France*
CAROL A. FIERKE, PHD • *Department of Biochemistry, Duke University Medical Center, Durham, NC*
JAMES J. FIORDALISI, PHD • *Department of Radiation Oncology, Lineberger Comprehensive Cancer Center, University of North Carolina, Chapel Hill, NC*
MICHAEL H. GELB, PHD • *Departments of Chemistry and Biochemistry, University of Washington, Seattle, WA*
JACKSON B. GIBBS, PHD • *Department of Cancer Research, Merck Research Laboratories, West Point, PA*
SAMUEL L. GRAHAM, PHD • *Department of Medicinal Chemistry, Merck Research Laboratories, West Point, PA*

Jean-Dominique Guitton, PhD • *Rhone-Poulenc-Rorer S.A., Vitry sur Seine, France*
Andrew D. Hamilton, PhD • *Department of Chemistry, Yale University, New Haven, CT*
George D. Hartman, PhD • *Department of Medicinal Chemistry, Merck Research Laboratories, West Point, PA*
Chih-Chin Huang, PhD • *Department of Biochemistry, Duke University Medical Center, Durham NC*
Gary R. Hudes, MD • *Department of Medical Oncology, Fox Chase Cancer Center, Philadelphia, PA*
MeeRhan Kim, PhD • *Department of Microbiology and Molecular Genetics, Jonsson Comprehensive Cancer Center, University of California, Los Angeles, CA*
Paul Kirschmeier, PhD • *Schering-Plough Research Institute, Kenilworth, NJ*
Kenneth S. Koblan, PhD • *Department of Cancer Research, Merck Research Laboratories, West Point, PA*
Nancy E. Kohl, PhD • *Department of Cancer Research, Merck Research Laboratories, West Point, PA*
David Knowles, PhD • *Department of Chemistry, University of Pittsburgh, PA*
Abdel Laoui, PhD • *Rhone-Poulenc-Rorer S.A., Vitry sur Seine, France*
Jacques Lavayre, PhD • *Rhone-Poulenc-Rorer S.A., Vitry sur Seine, France*
François Lavelle, PhD • *Rhone-Poulenc-Rorer S.A., Vitry sur Seine, France*
Alain LeBrun, PhD • *Rhone-Poulenc-Rorer S.A., Vitry sur Seine, France*
Frank Lee, PhD • *Oncology Drug Discovery, Bristol-Myers Squibb Pharmaceutical Research Institute, Princeton, NJ*
Yves Lelièvre, PhD • *Rhone-Poulenc-Rorer S.A., Vitry sur Seine, France*
Daniele M. Leonard, PhD • *Department of Chemistry, Parke-Davis Pharmaceutical Research, Division of Warner-Lambert Co., Ann Arbor, MI*
W. R. Leopold, PhD • *Department of Cancer Research, Parke-Davis Pharmaceutical Research, Division of Warner-Lambert Co., Ann Arbor, MI*
Ming Liu, PhD • *Schering-Plough Research Institute, Kenilworth, NJ*
Stephen B. Long, PhD • *Department of Biochemistry, Duke University Medical Center, Durham, NC*
Patrick Mailliet, MD • *Rhone-Poulenc-Rorer S.A., Vitry sur Seine, France*
Veeraswamy Manne, PhD • *Oncology Drug Discovery, Bristol-Myers Squibb Pharmaceutical Research Institute, Princeton, NJ*
Jean-Paul Martin, MD • *Rhone-Poulenc-Rorer S.A., Vitry sur Seine, France*
W. Gillies McKenna, MD, PhD • *Department of Radiation Oncology, University of Pennsylvania Medical Center, Philadelphia, PA*
Ruth J. Muschel, MD, PhD • *Department of Pathology and Laboratory Medicine, University of Pennsylvania Medical Center, Philadelphia, PA*
Lisa Nguyen, PhD • *Department of Medicine, University of Washington, Seattle, WA*
George Njoroge, PhD • *Schering-Plough Research Institute, Kenilworth, NJ*
Junko Ohkanda, PhD • *Department of Chemistry, Yale University, New Haven, CT*
Allen Oliff, MD • *Department of Cancer Research, Merck Research Laboratories, West Point, PA*
Charles Omer, PhD • *Department of Cancer Research, Merck Research Laboratories, West Point, PA*
James J.-K. Pai, PhD • *Schering-Plough Research Institute, Kenilworth, NJ*
Guy Pantel, MD • *Rhone-Poulenc-Rorer S.A., Vitry sur Seine, France*

AMITA PATNAIK, MD • *Institute for Drug Development, Cancer Therapy and Research Center, San Antonio, TX*
ANGEL PELLICIER, MD, PHD • *Department of Pathology and Kaplan Cancer Center, NYU Medical Center, New York, NY*
JEAN-FRANÇOIS PEYRONEL, PHD • *Rhone-Poulenc-Rorer S.A., Vitry sur Seine, France*
GEORGE C. PRENDERGAST, PHD • *Dupont Pharmaceuticals, Glenolden, PA*
JEAN-FRANÇOIS RIOU, PHD • *Rhone-Poulenc-Rorer S.A., Vitry sur Seine, France*
NICOLE ROBINSON, PHD • *Department of Microbiology and Molecular Genetics, Jonsson Comprehensive Cancer Center, University of California, Los Angeles, CA*
KELLEY ROGERS-GRAHAM, BS • *Department of Pharmacology, Lineberger Comprehensive Cancer Center, University of North Carolina, Chapel Hill, NC*
WILLIAM C. ROSE, PHD • *Oncology Drug Discovery, Bristol-Myers Squibb Pharmaceutical Research Institute, Princeton, NJ*
SAUL ROSENBERG, PHD • *Department of Cancer Research, Abbott Laboratories, Abbott Park, IL*
BARTIRA ROSSI-BERGMANN, PHD • *Instituto de Biofisica Carlos Chagas Filho, Universidade Federal do Rio de Janeiro, Brazil*
ERIC K. ROWINSKY, MD • *Institute for Drug Development, Cancer Therapy and Research Center, San Antonio, TX*
CHRISTOPHE SALAGNAD, MD • *Rhone-Poulenc-Rorer S.A., Vitry sur Seine, France*
JESSIE SCHOL, RN • *Department of Medical Oncology, Fox Chase Cancer Center, Philadelphia, PA*
JUDITH S. SEBOLT-LEOPOLD, PHD • *Department of Cell Biology, Parke-Davis Pharmaceutical Research, Division of Warner-Lambert Co., Ann Arbor, MI*
SAÏD M. SEBTI, PHD • *Drug Discovery Program, H. Lee Moffitt Cancer Center and Research Institute, University of South Florida, Tampa, FL*
MICHAEL SINENSKY, PHD • *Department of Biochemistry and Molecular Biology, East Tennessee State University, Johnson City, TN*
KENNETH D. STUART, PHD • *Department of Pathobiology, University of Washington, Seattle, WA; and Seattle Biomedical Research Institute, Seattle, WA*
JIAHZI SUN, PHD • *Drug Discovery Program, H. Lee Moffitt Cancer Center, University of South Florida, Tampa, FL*
FUYUHIKO TAMANOI, PHD • *Department of Microbiology and Molecular Genetics, Jonsson Comprehensive Cancer Center, University of California, Los Angeles, CA*
ARTHUR TAVERAS, PHD • *Schering-Plough Research Institute, Kenilworth, NJ*
FABIENNE THOMPSON, MD • *Rhone-Poulenc-Rorer S.A., Vitry sur Seine, France*
L. GERARD TOUSSAINT III, MD • *Departments of Radiation Oncology and Pharmacology, Lineberger Comprehensive Cancer Center, University of North Carolina, Chapel Hill, NC*
JUN URANO, PHD • *Department of Microbiology and Molecular Genetics, Jonsson Comprehensive Cancer Center, University of California, Los Angeles, CA*
WESLEY C. VAN VOORHIS, PHD • *Department of Medicine, University of Washington, Seattle, WA*
PATRICIA VRIGNAUD, PHD • *Rhone-Poulenc-Rorer S.A., Vitry sur Seine, France*
JOLENE WINDLE, PHD • *Cancer Therapy and Research Center, San Antonio, TX*
PAUL WORKMAN, PHD • *CRC Center for Cancer Therapeutics at the Institute for Cancer Research, Surrey, UK*

NING YAN, PHD • *Oncology Drug Discovery, Bristol-Myers Squibb Pharmaceutical Research Institute, Princeton, NJ*
WENLI YANG, PHD • *Department of Microbiology and Molecular Genetics, Jonsson Comprehensive Cancer Center, University of California, Los Angeles, CA*
KOHEI YOKOYAMA, PHD • *Departments of Chemistry and Biochemistry, University of Washington, Seattle, WA*
FANG ZHANG, PHD • *Schering-Plough Research Institute, Kenilworth, NJ*
MARTINE ZUCCO, PHD • *Rhone-Poulenc-Rorer S.A., Vitry sur Seine, France*

1 Signal Transduction Pathways

A Goldmine for Therapeutic Targets

Paul Workman, PHD

Contents

It is by testing we discern fine gold.—*Leonardo da Vinci* (1452–1519)

1. INTRODUCTION

Farnesyltransferase and geranylgeranyltransferase inhibitors have remarkable potential for the treatment of cancer, cardiovascular disease, and a wide range of other disorders. The objective of this introductory chapter is to set the development of these prenylation inhibitors within the general context of contemporary drug discovery, covering the major drivers of unmet medical need, commercial imperatives, and intellectual challenge, and focusing on the issues surrounding development of signal transduction inhibitors in particular. The emphasis is on small molecule drugs, but much of the chapter is relevant to approaches with higher molecular weight agents, such as antibodies, other therapeutic proteins, antisense oligonucleotides, and gene therapy.

2. THE NEED FOR INNOVATIVE THERAPEUTIC AGENTS IN THE NEW MILLENNIUM

Current drug therapies are remarkably effective across a range of therapeutic areas. The use of H2-receptor antagonists, and more recently proton pump inhibitors, has

From: *Farnesyltransferase Inhibitors in Cancer Therapy*
Edited by: S. M. Sebti and A. D. Hamilton © Humana Press Inc., Totowa, NJ

revolutionized the management of gastric ulcers. β-adrenergic receptor blockers and inhibitors of angiotensin-converting enzyme (ACE) have had a major impact in the area of cardiovascular disease. However, for many diseases cure or even good management cannot be achieved. The treatment of infectious diseases, often cited as a success story of drug development, remains suboptimal, as exemplified by the emergence of HIV/AIDS and of drug-resistant strains. Moreover, in view of changing global demographics, the management of the chronic diseases associated with the aging population in developed countries represents a major unmet medical need. As examples we can cite cardiovascular disease, diabetes, inflammatory diseases (such as arthritis), asthma, neurological disorders, and cancer.

Let us look more closely at the case of cancer, as we shall do for illustrative purposes throughout this introductory chapter. The statistics are stark. More than 1 in 4 of the population will suffer from cancer in the developed world. In 1997, cancer caused 2.5 million, or 21%, of deaths in the developed world and a further 3.6 million in developing countries *(1–3)*. The World Health Organization (WHO) predicts a rise in the annual worldwide incidence of cancer from 10 million new cases in 1997 to 20 million in 2020. Deaths from cancer will increase from 6 million to 10 million. For society this represents a major concern and cost. For the pharmaceutical industry, it represents a major market opportunity. For science it represents a major intellectual challenge.

Excellent results can be obtained with chemotherapy in a small group of cancers. For example, cures can be achieved in several childhood tumors and some adult malignancies, including testicular cancer, lymphoma, and leukemia. Incremental survival benefits are being achieved in certain adult tumors, for example by adjuvant treatment of breast and colon cancer *(4)*. The incidence and morbidity rate for cancer in the United States has fallen over the period 1990–1995 *(5)*. Useful palliation can be achieved in all patients. But we have a long way to go to achieve a significant gain in the curability of the major solid tumors, especially in their disseminated (metastatic or spread) stage, for which systemic drug therapy is the only realistic treatment choice.

The figures for 1999 show that a total of 92 cancer drugs had been approved by FDA for marketing in the United States *(6)*. A recent WHO consultation process resulted in the categorization of 17 anticancers (plus two antiemetics) as justifying widespread availability and an additional 12 were seen as having some advantages in particular clinical settings. A further 13 drugs were viewed as not essential currently for the delivery of effective cancer care.

A great many of our current cancer drugs work by inhibiting DNA synthesis or the mechanics of cell replication. As a result of such nonspecific actions, biochemical selectivity is poor, side effects are frequently very severe, and drug resistance is the norm.

A very considerable amount of effort and research dollars are going into improving cancer treatment. For example, the Pharmaceutical Research and Manufacturers Association of America (PhRMA) reported that 316 cancer medicines were in development in 1997 and the research and development spending for the research-based pharmaceutical industry was estimated at almost $19 billion *(7)*. In May 1999 it was reported that 1422 projects were in development, making cancer the leading developmental therapeutic area *(8)*. This is perhaps not so surprising as the worldwide market for cancer drugs was $11.7 billion in 1997, about 4% of the worldwide pharmaceuticals market, and this figure is projected to hit $14.7 billion by the year 2000 *(3)*. In view of the demographics and the scientific and recent technological advances (*see later*) the potential market is much greater.

Preclinical	**6.5 years**
↓	↓
Phase 1	**1.5 years**
↓	↓
Phase 2	**2 years**
↓	↓
Phase 3	**3.5 years**
↓	↓
FDA approval	**1.5 years**
Total	**15 years**

Fig. 1. Timescale for the average drug discovery and development project in the 1990s.

Moreover, cancer drugs have the potential to make a major contribution to the so-called "new chemical entity (NCE) gap" *(9)*. This is the shortfall in productivity when the current output from the 50 or so top-tier pharmaceutical companies, currently running at an average of one new product per company every two years, is compared to the figure of two per company per year that is necessary to sustain the desired growth requirement of 10% per annum *(10)*.

So what is going wrong? A review of data for drug development across all therapeutic areas in the 1990s reveals that it has most often taken 15 yr to progress from the initial work in the lab to the point of marketing approval, which allows widespread patient availability *(6,7)*. The breakdown of the timescale, in terms of the various stages of discovery and development, is shown in Fig. 1.

If we look at a topical example, the taxane antitubulin cancer drug palitaxel was discovered as a crude extract activity as early as 1963 but did not gain approval by the regulatory authorities until 1992 *(11)*. A common aim is to reduce discovery and development timescales dramatically, down to 5–7.5 yr or less.

In addition, the attrition rate at all stages must be reduced. For example, the figures show that out of every 5,000–10,000 chemical compounds evaluated in the preclinical phase, only five of these enter early clinical trials. Furthermore, of these only one at the most gains regulatory approval, giving a success of 20% at best and more likely 10%, at a cost of around $500 million per drug *(6,7)*.

Therefore, the process of drug discovery and development is too slow, too inefficient, and too expensive and is failing to meet the unmet medical need in many important disease areas. How can it be improved? The contemporary view is that major advances in disease therapy will come from two principal sources *(12)*. The first is an improvement in the efficiency of the drug development process by the implementation of a number of modern technological advances. The second is to enhance innovation by focusing on novel molecular targets for drug action, as revealed by new gene discovery and the elucidation

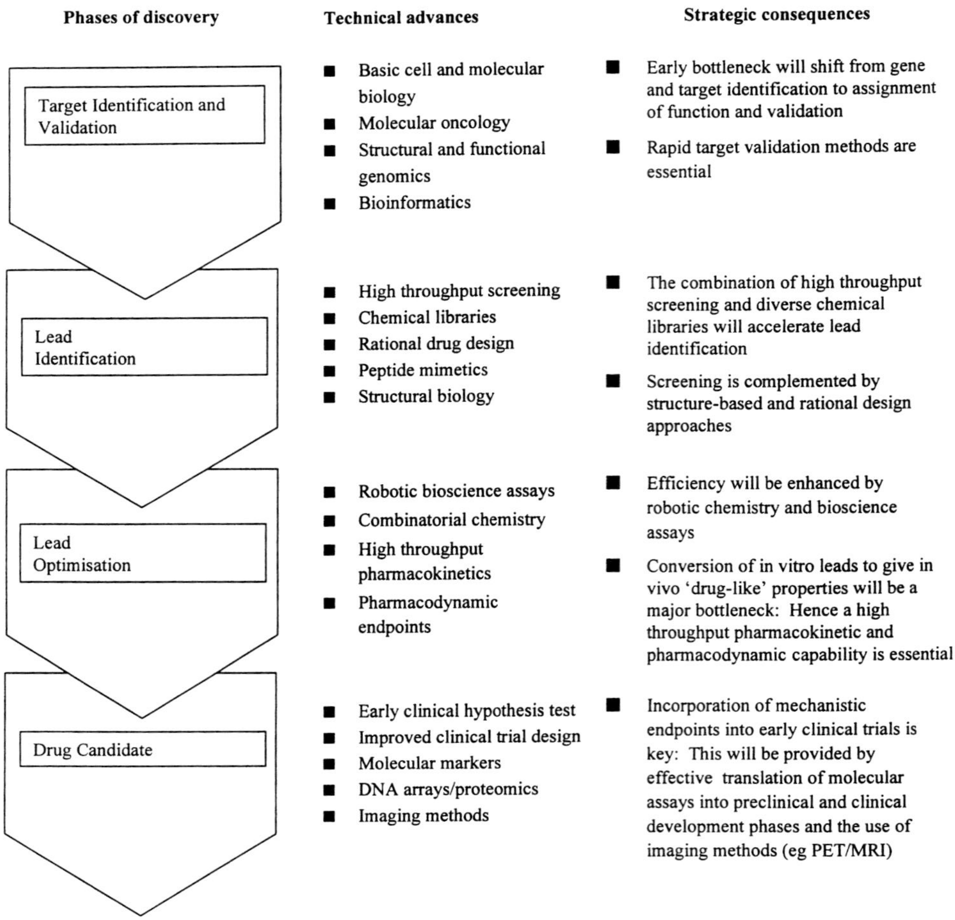

Fig. 2. Schema showing the phases of contemporary small molecule drug discovery, the technical advances that are impacting these phases, and the consequences arising from their implementation, particularly in terms of overcoming hurdles and removing roadblocks in the process.

of the signal transduction pathways that the newly discovered genes participate in and control. Let us examine these developments in more detail.

3. OVERCOMING THE HURDLES WITH NEW TECHNOLOGIES

The modern drug discovery process in operation today is frequently described in terms of the sequential elements of target identification and target validation, lead identification and lead optimization, followed by the selection and development of a clinical candidate. This process is illustrated in Fig. 2. Also shown are the technological advances that are being implemented to improve efficiency, as well as the strategic impact of these advances. The discovery of new genes is being accelerated enormously by the new science of genomics *(9,13,14)*. Through the use of robotic high throughput sequencing methods all the potential 100,000 genes or more in the human genome will be identified and sequenced in a working draft version by spring 2000 and in fully accurate, completed form by 2003 *(15)*. As gene discovery gets faster and is eventually completed, the major

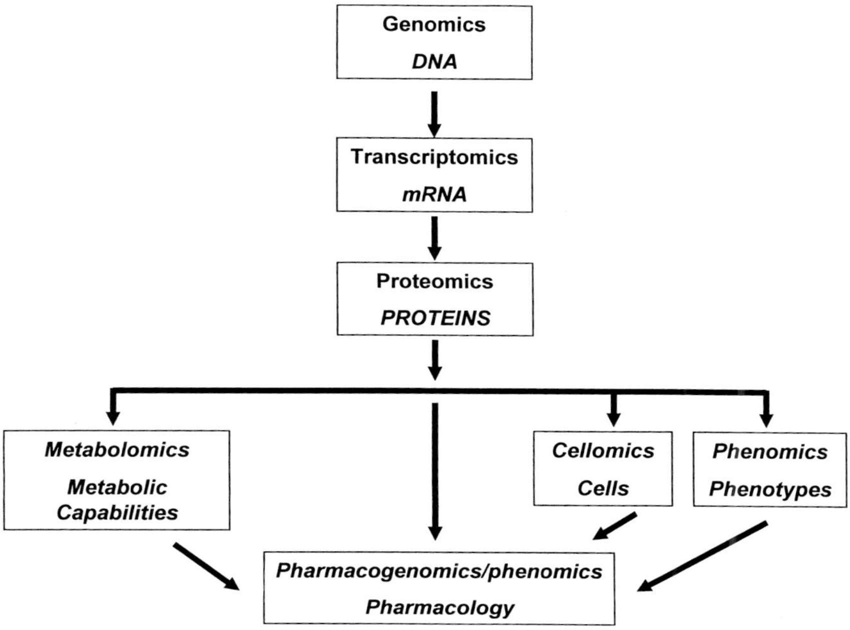

Fig. 3. The proliferation of "omics" research.

bottlenecks become the assignment of function to the encoded proteins, their positioning within cellular complexes and pathways, and their validation as feasible and attractive targets for pharmacological intervention. This process is being greatly facilitated by the high throughput, global biology methodologies of nucleic acid microarrays *(16)* and proteomics *(17)*.

There is currently a proliferation of "omics" research (Fig. 3) from genomics (DNA level), transcriptomics (mRNA), and proteomics (proteins), through to the more complex, higher organizational level omics of cellomics (cells), phenomics (phenotypes), metabolomics (metabolic capabilities), and finally pharmacogenomics/ phenomics in humans. The determination of gene expression patterns in normal and disease tissues, using expressed sequence tag (EST) libraries or gene microarrays, is proving to be exceptionally useful. As a case in point, cathepsin K was identified as a novel cysteine protease in a database of EST sequences derived from an osteoclast library *(13)*. This led to the rapid development of cathepsin K inhibitors, which should block bone resorption by osteoclasts and thereby prevent osteoporosis.

In cancer, target validation is helped considerably by studying the genetics and gene expression patterns of tumors. Cancer is now frequently referred to as a genetic disease. It is, in fact, a series of around 100 diseases in which sets of genes undergo germline or, more commonly, somatic mutation, or are subject to aberrant expression *(18,19)*. With regard to the validation of potential therapeutic targets and the selection of those likely to be more promising for therapeutic intervention, it seems logical to propose that those genes and pathways that are most commonly subject to mutation or deregulated expression are likely to be the most fruitful to pursue *(12,20)*. The receptor tyrosine kinase → Ras → Raf → MEK → MAP kinase pathway regulating proliferation and the control of the cell cycle by the cyclin-dependent kinase (CDK)-retinoblastoma gene product axis

(21) are excellent examples. A further recent example of genetic validation is that of PI3 kinase. The *PIK3CA* gene that encodes the p110α catalytic subunit of PI3 kinase is amplified and overexpressed in ovarian cancer *(22)*. Activation of p110α PI3 kinase signal transduction may also be achieved by genetic deletion of the *PTEN* tumor suppressor gene; the product of the *PTEN* gene acts as a lipid phosphatase that reverses the reaction catalysed by p110α PI3 kinase *(23,24)*. Enzymes, such as kinases, are very good examples of pharmacologically tractable targets, because the technical feasibility of discovering small molecule enzyme inhibitors is very high as a result of the presence of a small-molecule binding site. This is not the case for blocking protein–protein interactions, e.g., as exemplified by the search for nonpeptidic, "drug-like" small molecule SH2 domain inhibitors which has not been successful so far.

Data are now beginning to emerge on the differences in global gene expression between normal and cancer cells. For example, in one of the first papers in this area, more than 300,000 transcripts derived from at least 45,000 different genes were analyzed in gastrointestinal tumors and corresponding normal cells, using the technique of serial analysis of gene expression (SAGE; *25*). Although there was extensive similarity, 548 transcripts (1.5%) were differentially expressed. Perhaps surprisingly, the main differences were seen in differentiation markers, genes associated with protein synthesis, ribosomal proteins, elongation factors, and glycolysis rather than oncogenes.

Following identification, validation, and selection of a target, the next phase is lead identification. This phase is now mainly carried out by high throughput screening of large chemical libraries against recombinant protein targets *(12,26,27)*. Compound libraries must be chemically diverse. Computational chemistry methods can be used to maximize diversity in an efficient way *(28)*. There are advantages to removing chemicals that are generally poor starting points for a medicinal chemistry program, are unlikely from past experience to result in drugs, or are highly toxic. Such nondrug-friendly chemical types include highly charged compounds, heavy metals, alkylating agents, and Michael acceptors. Chemically reactive compounds cause considerable problems *(29)*. The screening approach is complemented by molecular design, often aided by the use of peptide mimetic chemistry and the structural biology techniques of X-ray crystallography and NMR *(30,31)*.

A novel strategy is to combine organic synthesis, screening of libraries of natural product-like substances, site-directed mutagenesis, and X-ray crystallography in a creative approach known as "chemical genetics" or "chemical biology" *(32)*. This approach uses chemical compounds to probe protein function, in an analogous way to the genetic mutation method, and also to develop synthetic derivatives of natural products as potential drug candidates. This approach has been applied to signal transduction targets. Examples include the use of rapamycin and trapoxin to study FRAP and histone deacetylase, respectively, and the design of potent peptide ligands for SH3 domains *(32,33)*.

Leads discovered by screening or design are then optimized by iterative cycles of medicinal chemistry refinement and rapid feedback from biological evaluation, building up an understanding of structure–activity relationships for the desirable and undesirable features of the lead molecule *(34)*. This process is aided today by the revolutionary method of combinatorial chemistry *(35,36)*, which also generates chemical diversity for primary screening. By these means, identification of leads and their optimization in terms of potency, selectivity, and activity in intact cells by the desired mechanism has become much more readily achievable.

A major hurdle is, however, the transition from activity at the level of in vitro cell culture to activity in the intact animal. Poor pharmacokinetics is a major bottleneck *(12,37)*. Problems can arise in all aspects of pharmacokinetic behavior, i.e., absorption, distribution, metabolism, and excretion (ADME). Oral absorption (usually required for chronic administration schedules) can be problematic, as can tissue uptake. Elimination from the body may be too fast because of overly rapid metabolism or renal excretion. As a result of such problems, pharmacologically and therapeutically active drug levels of lead compounds may not be achieved. At this stage of the project some degree of target potency and selectivity may have to be sacrificed in order to gain the necessary improvement in ADME properties. Progress can be made by modifying the physicochemical properties of compounds, e.g., introducing solubilizing functions, and there are valuable guidelines, such as Lipinski's "rule of five", to help optimize bioavailability *(38)*. Although useful rules of thumb can be employed to improve ADME within a particular lead series (e.g., *39,40*), the development of structure-pharmacokinetic relationships is fairly primitive *(41,42)* and it is extremely difficult to predict the pharmacokinetic properties of a given chemical compound.

In modern drug discovery projects, this important issue is currently being addressed in a pragmatic way by increasing the throughput of pharmacokinetic and metabolic analysis, for example using "cassettes" or mixtures of compounds dosed together in low amounts with very sensitive HPLC-MS-MS detection *(43,44)*. The most promising structures can then be selected for further rounds of optimization. Nevertheless, the development of chemoinformatics algorithms to predict structures having good pharmacokinetic properties is an important future objective. However, the rational design of robust, drug-like character looks set to remain a major bottleneck for the next several years.

With sufficient bioavailability achieved, the next hurdle is to demonstrate some mechanism-based pharmacodynamic activity in the animal. Better animal models are needed to give a faster readout of mechanism-related pharmacodynamic activity. These may involve the use of reporter genes in transgenic animals or other genetically engineered models. A relatively high throughput model for the effects of anticancer agents in vivo is the hollow fiber assay *(45)*. This has some advantages in terms of speed and cost over solid tumor xenografts, but has not been validated for the new generation of signal transduction inhibitors and it is unsuitable for antiangiogenics.

Current issues surrounding late preclinical and clinical development of cancer drugs have been discussed recently *(12,46)*. Of particular importance in cancer and most therapeutic areas is the need for pharmacodynamic endpoints that will enable us to judge whether the molecular target is being modulated in the intact animal and patient. Early clinical trials of agents affecting novel molecular targets must contain a strong component of hypothesis testing. Is the intended molecular target being affected (e.g., kinase or farnesyltransferase inhibition)? Is the biochemical pathway being modulated (e.g., MAP kinase activation)? And is the desired biological effect being achieved (e.g., inhibition of proliferation, cell cycle transit, survival, or angiogenesis)? If the answer to those questions is yes this provides confidence to move forward to the more expensive phases of clinical development. A structured, logically based approach to clinical development can be a major aid to decision making. If problems are seen at any level then these can be addressed or resources reallocated to other more promising projects. For example, if the target is failing to be appropriately modulated, this might suggest a limitation with the drug candidate, and indicate that an appropriately designed back-up compound could be more effective. If the target is being modulated to the required degree but the desired biological effect is not seen, this suggests that the target is not valid but modulation of other targets in the pathway to achieve the biological effect may be worth pursuing. If, however, both the target and the biochemical pathway or biological effect are being suitably affected (i.e., to an extent defined in a preclinical model) but there is no impact on the disease process, then this would indicate that the pathway or biological effect is not linked to the disease in humans, and further approaches to the whole biochemical pathway and biological effect may not be worthwhile.

In terms of the development of pharmacodynamic endpoints, the use of modern molecular techniques will be crucial, and it seems likely that nucleic acid microarray and proteomics technology in particular are poised to play a major role. Noninvasive imaging technologies can provide valuable information, especially positron emission tomography and magnetic resonance imaging *(47)*.

Creative trial design will be important *(46)*. Trial designs aimed at demonstrating a slowing of disease progression, as in Alzheimer's disease (with tacrine), amyotrophic lateral sclerosis (with riluzole), and rheumatoid arthritis (with prednisolone) find parallels in other therapeutic areas, including cancer, where prolonged disease control, rather than cure, would have significant value.

Toxicology is essential to ensure acceptable safety in humans. However, since excessive toxicology can cause delay and in oncology it is often poorly predictive with respect to the qualitative nature of particular organ toxicities, nonprofit organizations in Europe (the Cancer Research Campaign [CRC] and European Organization for Research and Treatment of Cancer [EORTC]) utilize a system with a relatively simple program of rodent-only toxicology *(48)*. This has proved safe and effective *(49,50)*.

Regulatory review is also speeding up. FDA approval for Herceptin (triatuzumab), a humanized monoclonal antibody against the tyrosine kinase receptor erbB2, was obtained in a record 4.5 mo *(51)*. Arguably the first molecular target therapy based on cancer genomics, this agent shows promising activity in breast cancer.

4. THE CONTEMPORARY PARADIGM: HOW MANY NEW TARGETS?

The contemporary paradigm of new drug discovery can be summarized as:

New Genes → Novel Targets → Innovative Medicines

This paradigm is based on the premise that the discovery of innovative agents having a high degree of selectivity for a given molecular target involved in disease causation and progression will lead to drugs that have markedly improved efficacy and tolerability in humans. Thus, much will depend on the correct identification and selection of the disease target. What is our expectation of the likely numbers of targets arising now that we have entered the genomic era of drug discovery? It has been estimated that genomics has the potential to deliver 3000–10,000 interesting new targets for therapeutic intervention out of approx 100,000 genes in the human genome *(9)*. How is this figure computed?

The calculation is based on the proposal that there are likely to be 5–10 disease-related genes for each of the 100 or so, at a conservative estimate, really important human diseases with major unmet medical need: hence there could be 500–1000 key disease-related genes. Each gene product interacts in biochemical pathways with, say, 3–10

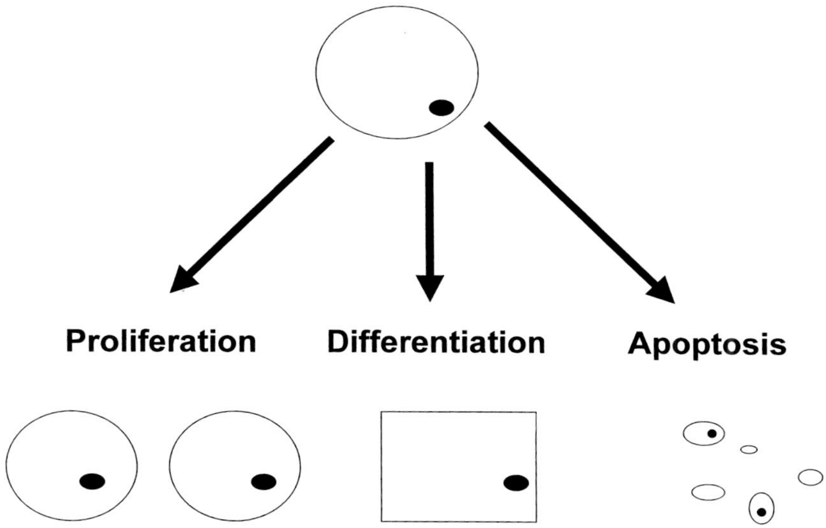

Fig. 4. Cell fate is controlled by signal transduction pathways, e.g., the decision to proliferate, differentiate, or undergo programmed cell death.

upstream or downstream partners, which when multiplied up gives a total of 1500–10,000 potentially interesting drug targets. This number of targets represents a very significant increase over today. As shown in a 1996 survey *(9)*, today's drugs, across all therapeutic areas, act on only 417 or so targets (enzymes, receptors, ion channels, and so forth, excluding anti-infectives). These drugs were mainly discovered by classical, empirical methods, usually without detailed knowledge of the molecular target. Not all of the potential 10,000 new targets will prove pharmacologically tractable (e.g., some will be structural proteins that are difficult to modulate), but it is clear that there should be a major opportunity to increase the number of therapeutic targets. Moreover, these targets will be genetically and biochemically validated and ideal for highly focused, mechanism-based drug discovery and development.

5. THE IMPORTANCE OF SIGNAL TRANSDUCTION TARGETS

Why are signal transduction pathways seen as so attractive for pharmacological and therapeutic intervention? The first reason is because the whole of biology and physiology is controlled by a network of biochemical interconnections, globally referred to as signal transduction. Signal transduction proteins control cell fate, for example, regulating decisions to proliferate, differentiate, or undergo apoptosis (Fig. 4). Malfunction of signal transduction processes caused by mutation or abnormal gene expression leads to incorrect decisions being made about cell fate and function, resulting in disease. Abnormal signaling can bring about hyperproliferative disorders, such as cancer, atherosclerosis, restinosis, and psoriasis, and also inflammatory diseases, such as rheumatoid arthritis *(52,53)*.

In cancer, oncogenes and tumor suppressor genes are positioned at critical points on signal transduction pathways, or oncoprotein networks, that control cell fate *(54)*. Mutation or abnormal expression of genes involving signal transduction proteins leads to neoplastic transformation and malignant progression.

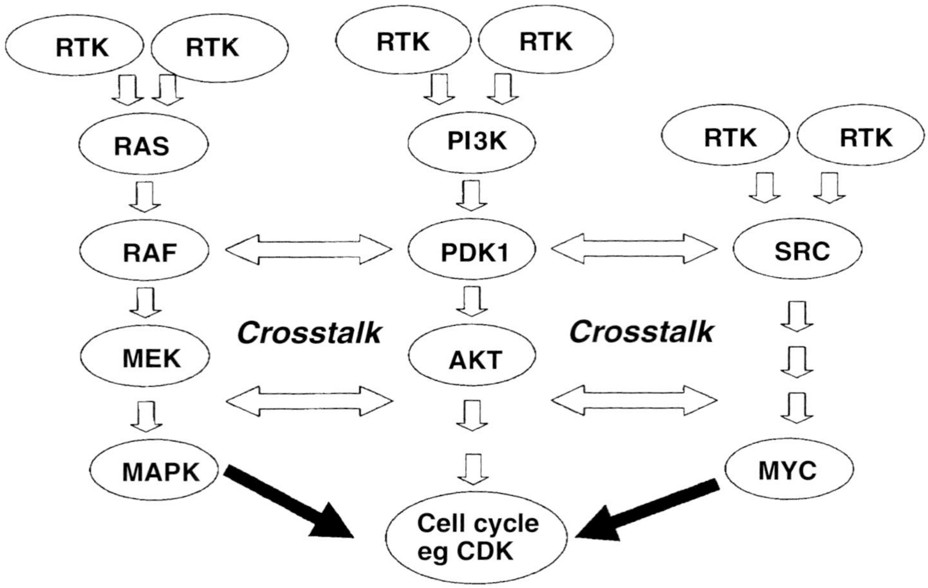

Fig. 5. A simplified view of signal transduction from receptor tyrosine kinases (RTKs) at the cell membrane to the control of the cell cycle and gene expression in the nucleus.

Figure 5 shows a very simplified view of the signal transduction pathways leading from receptor tyrosine kinases located at the cell membrane and connecting with the cell-cycle machinery and the control of gene expression in the nucleus. Cancer cells activate these pathways by, for example, overexpression of receptors, such as epidermal growth factor (EGF) receptor and erbB2; mutational activation of the Ras oncogene; biochemical activation of the Src tyrosine kinase; deregulation of the transcription factor Myc; and various alterations in the cyclins, cyclin-dependent kinases, their regulatory proteins, or loss of the retinoblastoma gene product *(12,55)*.

6. THE RAS SIGNALING PATHWAY

An enormous amount of research across many species has gone into elucidating the genetics and biochemistry of the Ras → Raf → MEK → MAP kinase signaling pathway. This is because this signal transduction cascade plays such a pivotal role in a large number of cellular functions. It is involved not only in signaling downstream from membrane receptor tyrosine kinases, but also plays a key role in T-cell receptor signaling, which utilizes nonreceptor tyrosine kinases, and also in signaling from G protein-coupled receptors *(56)*. As shown by the early, pioneering experiments of Stacey and colleagues *(57, 58)*, Ras function is essential for stimulation of cell proliferation by all growth factors tested, as well as for oncogenic tyrosine kinases to cause cell transformation.

The Ras → MAP kinase pathway is often depicted as a linear one, connecting the initiating event (e.g., activation of receptor tyrosine kinase by ligand binding, leading to receptor dimerization) to the control of gene expression in the nucleus (Fig. 6). This can be a useful simplification and may capture the predominant flow of information in certain circumstances. However, signal transduction pathways have many potential branchpoints that provide opportunities for signaling crosstalk (Figs. 5 and 7).

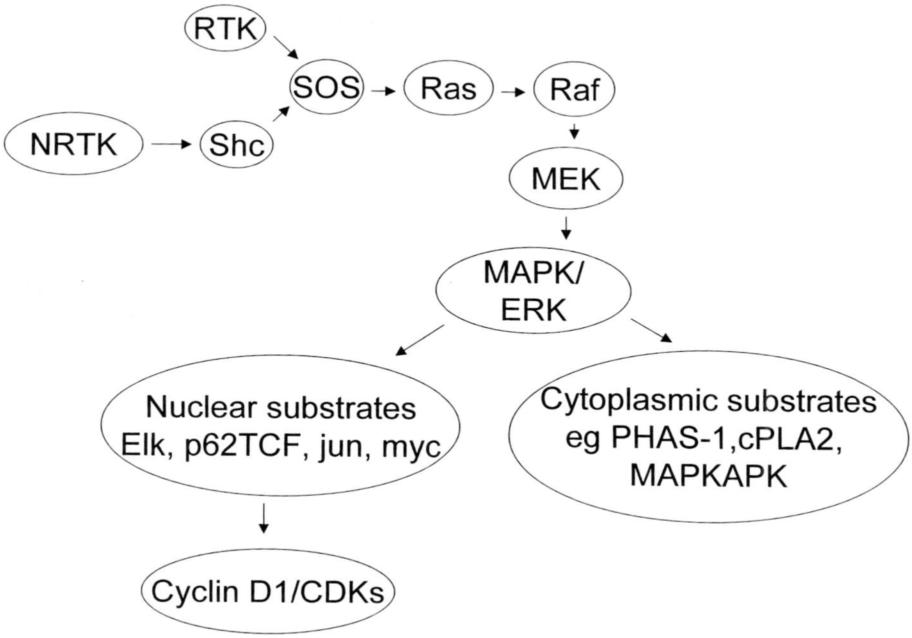

Fig. 6. An alternative depiction of the Ras signal transduction pathway from membrane to nucleus. The pathway is initiated by the activation of receptor-tyrosine kinases (RTK) or nonreceptor tyrosine kinase (NRTK) at the cell membrane. A series of phosphorylation events and protein–protein interactions then lead to activation of cytoplasmic and nuclear substrates, the latter being transcription factors leading to expression of genes required for cell cycle progression. Of these cyclin D1 provides the best understood connection between the Ras signaling pathway and cell cycle control.

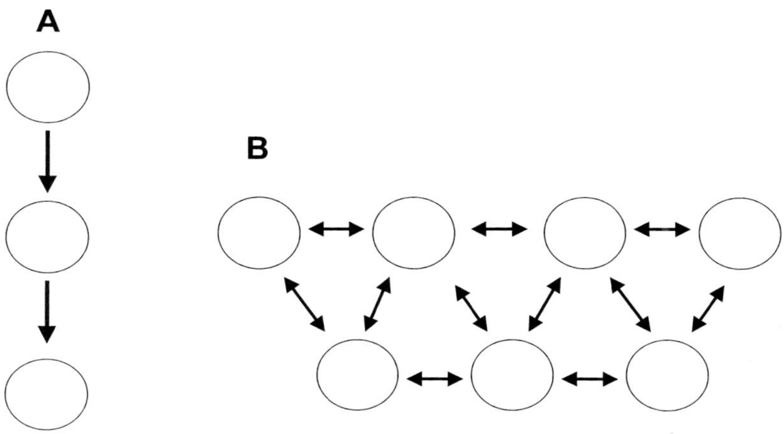

Fig. 7. Schematic representation of a simple linear pathway (A) and a more complex, branched pathway with multiple points for crosstalk.

A key feature of the control of the Ras pathway is its ability to connect events external to the cell (e.g., growth factor binding) to intracellular events, such as gene transcription and the control of the cell cycle. By this means, the cell can make an appropriate response to the external environment. The upstream components of the pathway from the activation of membrane tyrosine kinases to the phosphorylation of ERK/MAP kinase are reasonably well understood. In particular, the binding of the Raf-1 protein kinase to the active GTP-bound form of Ras serves to recruit cytoplasmic Raf-1 to the plasma membrane, where it undergoes activation *(59,60)*. In addition to inhibitory antibodies, the use of interfering mutants has proved valuable in determining the order of components in the pathway. Overexpression of the Raf-1 N-terminal regulatory domain prevents activation of MAP kinase by various stimuli *(61)*. This "dominant negative" effect of the N-terminal domain of Raf-1 probably results from the formation of a complex with Ras-GTP, thereby soaking up activated Ras present in the cell and preventing it from signaling to the endogenous Raf-1. Thus, Raf-1 is positioned upstream of MAP kinase. Similarly, overexpression of MEK containing a blocking mutation in its activating phosphorylation sites prevents the activation of endogenous MEK in the cell and at the same time blocks transformation by the Ras and Src oncogenes *(62)*. Hence, MEK is placed downstream of Ras and Src in oncogenic transformation.

The precise details of the mechanism by which ERK/MAP kinase cascade connects to gene expression and in particular cell cycle control are not so clear. ERK/MAP kinase phosphorylates a variety of cytoplasmic and nuclear substrates (Fig. 6), including the transcription factor Elk-1 *(63)*. An important consequence of the activation of the Ras → Raf → MEK → MAP kinase cascade is the induction of cyclin D1 synthesis *(64,65)*. Cyclin levels in turn orchestrate the activities of CDKs, which are also activated by the action of cdc 25A phosphatase. Expression of cdc 25A is increased by the action of Myc, an oncoprotein transcription factor that is also activated by mitogens *(66)*.

Activation of CDKs is crucial for cell cycle progression. In collaboration with their G1 cyclin partners (cyclins D and E) CDKs 2, 4, and 6 are responsible for phosphorylation of the hypophosphorylated form of Rb, the retinoblastoma gene product. Rb phosphorylation is initiated by D-type cyclins (D1, 2, and 3) in association with CDK4 or CDK6, following which hyperphosphorylation is completed by cyclin E-CDK2 complexes *(21)*. In turn, hyperphosphorylation of Rb displaces it from its repressive binding to the E2F family of transcription factors. Thus released, E2F family members activate a plethora of genes involved in the S phase of cell cycle progression.

Consistent with the view that Rb phosphorylation is a critically important target for cell-cycle control downstream of the Ras → Raf → MEK → MAP kinase cascade, inhibition of this pathway fails to prevent G1 progression in certain cells lacking Rb *(67–69)*, as is the case for a significant proportion of cancers.

In addition to processing growth stimulatory signals, the CDK–Rb control axis also integrates growth inhibitory signals. Serum withdrawal, which removes mitogenic growth factors and causes cell-cycle arrest, is associated with a decrease in cyclin D levels and increased expression of the CDK inhibitory protein p27 *(70)*. Another CDK inhibitory protein, $p15^{INK4B}$, is induced by the growth inhibitory factor TGFβ *(71)*.

Recent results indicate that Rb phosphorylation and the Rb–E2F interaction are involved not only in the G1 cell-cycle transition in response to mitogens and antimitogens, but also in the control of apoptosis *(21)*.

7. INHIBITORS OF THE RAS PATHWAY

All points in the RTK → Ras → Raf → MEK → MAP kinase pathway have potential for therapeutic intervention *(56)*. The entire pathway is well-validated by basic cell and molecular biology research, particularly the use of inhibitory antibodies and interfering mutants (dominant negatives), as described earlier. Validation by linkage of the pathway to a particular disease is strong in the case of cancer. Ras genes (H-Ras, K-Ras, and N-Ras) are mutated in about one-third of all cancers *(72,73)*. The mutations generate forms of Ras with impaired GTPase activity, resulting in persistent activation of the downstream signaling pathway *(74)*. The Ras molecular switch is permanently on. In colorectal cancer, a correlation has been seen between mutation and clinical outcome *(75)*. RTKs, such as EGF receptor and erbB2, are frequently overexpressed or in some cases subject to activating mutation. Again, in some examples such RTK deregulation is linked to clinical outcome *(55)*. Further downstream, the cyclin-CDK-Rb axis is frequently deregulated in cancer at the level of one (but not more in any one tumor) of the individual molecular players discussed in the previous section *(21)*. Defects include overexpression of cyclins, loss of CDK regulation, loss of Rb, and loss of p16.

Numerous experiments have shown that recapitulation of such changes in model systems can lead to conversion of normal cells (usually rodent fibroblasts) into the transformed phenotype, including tumor formation in nude mice. Recent findings show that human epithelial cells can be made cancerous by the introduction of an oncogenic allele of the H-Ras gene, in collaboration with expression of another oncogene (the simian virus 40 large-T oncoprotein, which inactivates Rb and p53) and the catalytic subunit of telomerase *(76)*. The need for more than one genetic abnormality is consistent with laboratory and epidemiological data that cancer is a multistep process. Downregulation of various points in the pathway by various means (genetic constructs, antibodies, interfering mutants, antisense oligonucleotides, and more recently chemical inhibitors) has been shown to reverse the transformation process. For example, a recent study has shown that continued expression of oncogenic H-Ras is essential for the genesis and maintenance of melanomas in a transgenic mouse model *(77)*.

What, then, is the preferred point of intervention in the RTK → Ras → Raf → MEK → ERK/MAP kinase → CDK/Rb pathway? A number of factors influence this selection, including issues of technical feasibility, extent of validation and proof principle, and the potential for therapeutic selectivity. One argument claims we should seek to interfere as closely as possible to the point of molecular deregulation, e.g., at the overexpressed RTK, the mutated Ras, or the deregulated CDK. This may provide maximum pharmacological selectivity, since the therapy would be tailored to the particular genetic makeup of the individual tumor.

Another argument is in favor of targeting the pathway as far upstream as possible, since this would block all the signals downstream of that point. On the other hand, it might be expected that it would be inappropriate to block at a point upstream of a major deregulation locus. However, as mentioned earlier, signal transduction pathways, including those involving Ras, are rarely linear but involve complex branchpoints, crosstalk, and feedback loops. For example, at least part of the oncogenic effect of Ras is likely to involve the production of transforming growth factors, which then act in an autocrine fashion on cell membrane receptors to activate other Ras-dependent signaling pathways *(78)*. A potential concern about intervening well downstream in a signal transduction cascade is that alternative pathways may feed in at or above that point. Hence, blockade of a downstream locus could take out signals that we do not wish to inhibit since they might result in toxic side effects. Equally well, however, interfering upstream in a cascade that branches to form multiple downstream pathways could result in the ablation of those that are not driving the particular cellular outcome that we seek to inhibit. Once again, toxic side effects could follow from this.

These issues associated with selecting targets are easily understood by reference to Fig. 7. The simple linear pathway on the left can be blocked at various points in the cascade. The more complex pathway on the right provides a much more difficult challenge. Given the ubiquitous participation of the major signal transduction pathways in many functions of normal tissues, there are concerns about potential side effects of using inhibitors. These issues and the potential sources of therapeutic selectivity in cancer have been discussed in detail previously *(53,55,79)*. It can, for example, be reasoned that tumor cells may be excessively dependent on activation of particular signaling pathways and are therefore more vulnerable to inhibition than are normal tissues. The latter might be spared to some degree by redundancy in cell signaling, i.e., the same result may be achieved by various means. These explanations are, however, rather lacking in hard data. It is, in fact, extremely difficult to predict how inhibitors will affect complex pathways, although computer models and lessons from metabolic engineering may be valuable *(80–82)*. In the light of these complexities many drug discovery groups favor the pragmatic view of screening for small molecule inhibitors of several of the more attractive points within a validated pathway (from a technical feasibility point of view) and then evaluating the efficacy and toxicity of these inhibitors.

Figure 8 shows the potential sites for pharmacological intervention in the Ras → ERK → CDK/Rb pathway. Protein–protein interaction inhibitors are possible at various points *(56)*. Options would be to block the SH2 domain of Grb2 binding to phosphotyrosines in the activated RTK, to prevent the Ras-Raf interaction, and to interfere with the interaction between cyclins and their cognate CDKs. However, as mentioned earlier, experience suggests that such inhibitors are hard to find by screening and peptide-based approaches to block protein–protein interactions have generally not delivered the goods in terms of achieving mimetics with drug-like properties.

By contrast, the development of kinase inhibitors has been very successful. Various inhibitors of RTKs, including EGF receptor, erbB2, bcr-abl, and Src are in preclinical and clinical development *(83,84)*. Excellent responses are seen in animal models and clinical activity has been seen with, for example, the EGF tyrosine kinase inhibitor ZD1839 and the bcr-abl inhibitor CGP57148B/STI-571. Inhibitors of Raf-1 kinase have been identified, although those disclosed have the drawback of causing a paradoxical activation of Raf-1, which becomes manifest when the inhibitor is removed *(85,86)*. MEK inhibitors appear more promising and one agent has been disclosed that shows promising activity in animal models *(87)*. Considerable efforts are going into the development of CDK inhibitors. A range of these are showing promising activity in model systems *(88)* and flavopiridol is already in clinical trial. An interesting new approach to blocking oncogenic kinases is the use of Hsp90 chaperone inhibitors, such as the geldanamycins. These deplete oncogenic kinases at the protein level by increasing proteosomal degradation *(89,90)*. A geldanamycin analog (17AAG) that inhibits Hsp90 is already in clinical trial.

At the level of Ras itself, options have included blockers of guanine nucleotide binding, Ras-effector interaction antagonists, and farnesylation inhibitors *(56)*. Of these, the

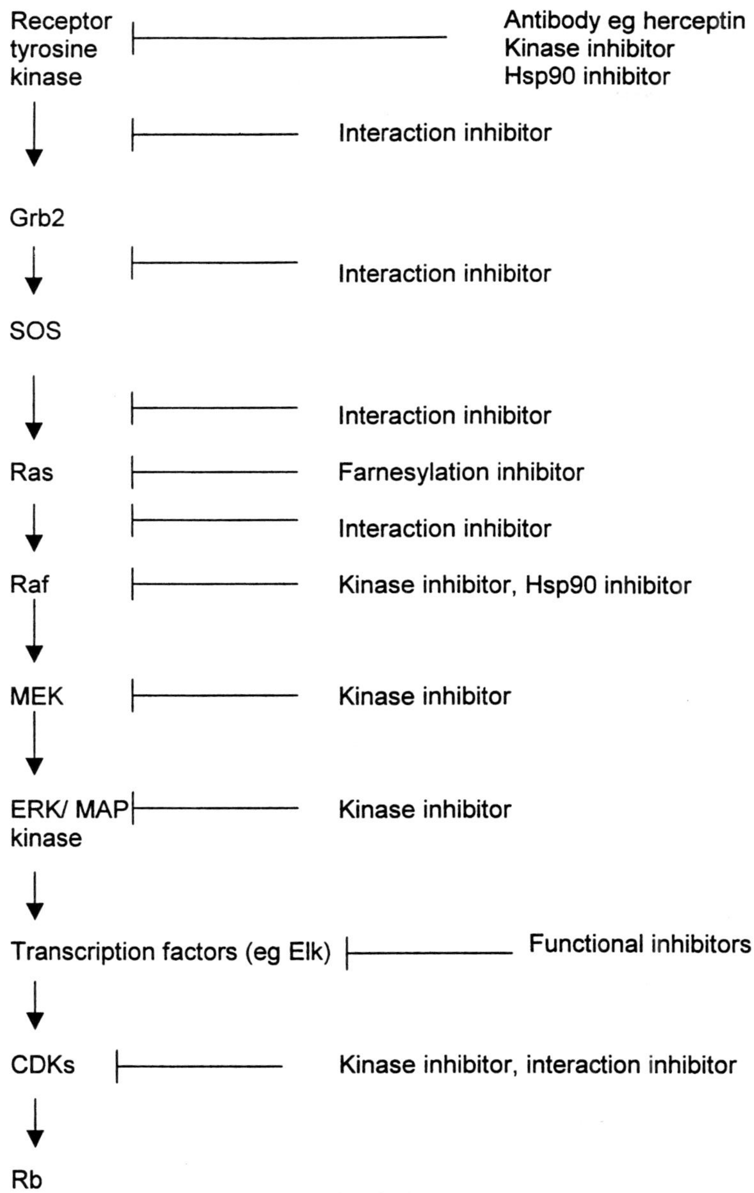

Fig. 8. Cell membrane to nucleus signaling via the Ras pathway and sites for pharmacological intervention.

farnesylation inhibitor approach has been the most successful, with several agents now in clinical trials. Since farnesylation of Ras is required for membrane docking, farnesyl transferase inhibitors block Ras function. It is inappropriate to review the development of farnesylation inhibitors here, since these are dealt with in detail elsewhere in this book. A couple of points are, however, worthy of note. First, it has taken almost 20 years from the initial cloning of the Ras oncogene to the first clinical trials of drugs targeted to this locus. Second, the discovery of farnesyltransferase inhibitors has illustrated the value of both screening and peptide mimetic rational design approaches, but has also confirmed the difficulties of optimizing highly potent and selective analogs to achieve in vivo efficacy. For example, lead CAAX box mimetics retained both a C-terminal amino acid containing a carboxylate function together with a cysteine residue *(91)*, and these have the potential to adversely affect bioavailability. Finally, the uncertainties regarding the precise mechanism of action of prenylation inhibitors illustrate the complexities and challenges of drug discovery and development. The more commonly mutated K-Ras is more difficult to inhibit than H-Ras. Also, Ras is not alone in being prenylated by farnesyl-transferase. Hence inhibition of farnesylation of other protein substrates could contribute to therapeutic and toxic effects. Furthermore, geranylgeranylation of Ras and other proteins can occur when their normal farnesylation is blocked *(92)*, and this could also have an impact on signaling, disease response, and side effects. Moreover, it must be remembered that Raf-1 is not the sole target for Ras signaling. Other candidate effectors include NF1, PI3K, RalGDS, and p120 GAP *(93)*. Oncogenic Ras does not cause transformation solely through activation of the Raf → MAP kinase pathway and there is evidence that Rho proteins (RhoA, Rac1, CDC42) are involved in cytoskeletal organization downstream of Ras *(93)*. Signals from Ras and Rho interact to regulate expression of the p21/Waf1/Cip1 CDK inhibitor *(94)*. Thus, much remains to be learned about Ras signaling in normal and tumor cells in the unperturbed state, even without the complications of adding in pharmacological inhibitors. Farnesyl transferase inhibitors may exert their effects in part by affecting RhoA, RhoB, and the expression of peptide CDK inhibitors. An interesting proposal for the mode of action of farnesyl transferase involves the geranylgeranylation of RhoB. The reasons for therapeutic selectivity for tumor versus normal cells remain unclear. One hypothesis is based on the observation that the dominant negative in Ras mutants selectively inhibit the activity of either normal or oncogenic Ras *(58)*. However, tumor cell line activity does not appear to be dependent on the presence or absence of Ras mutations. At the in vivo level, effects on angiogenesis are quite likely involved.

The observation that Ras is geranylgeranylated when farnesyltransferase is blocked, together with the participation of geranylgeranylated proteins like Rac and Rho in malignant transformation, cell cycle control, and apoptosis, supports a role for geranylgeranylase I as a target for therapy *(92)*. There may well be applications in cardiovascular and other diseases as well as cancer.

8. CONCLUDING REMARKS

There is enormous potential for the development of mechanism-based inhibitors of signal transduction to meet the unmet medical need of cancer, cardiovascular disease, and many other disorders. Development of these agents is benefiting greatly from a range of technological innovations. Roadblocks remain, however, and careful clinical trials with appropriate pharmacodynamic endpoints are critical for success.

The recent experience with kinase inhibitors and farnesylation blockers in the cancer clinic suggests that signal transduction inhibitors can be very well tolerated. Activity in animal models suggests promise for good therapeutic indices to be achieved in patients, despite the concerns about the potential side effects of blocking signal transduction in normal cells. This will be essential, since such agents are likely to require chronic administration over prolonged periods. Encouragingly, early clinical trials with emerging signal transduction inhibitors show that side effects are manageable and tumor responses are achievable.

A large number of signal transduction targets remain to be discovered and exploited in many disease areas. The potential to target signal transduction therapies to particular conditions or individual patients based on gene expression profiles or proteomic patterns is especially appealing. Many barriers remain to be overcome, but we are beginning to mine the gold that has remained buried for so long.

ACKNOWLEDGMENT

The author is a Life Fellow of the UK Cancer Research Campaign (CRC), Director of the CRC Centre for Cancer Therapeutics at the Institute of Cancer Research (Sutton), and Harrap Professor of Pharmacology and Therapeutics in the University of London.

REFERENCES

1. World Health Organization, World Health Report, Life in the 21st century. A vision for all. World Health Organisation, 1998, Geneva, Switzerland.
2. Sikora K, Advani S, Koroltchouk V, Magrath I, Levy L, Pinedo H, Schwartsmann G, Tattersall M, Yan S. Essential drugs for cancer therapy: A World Health Organisation consultation. *Ann Oncol* 1999; 10:385–390.
3. Bonney R. (ed). The Complete Guide to Cancer: 2nd ed. PJB Publications Ltd, Richmond, Surrey, UK, 1998.
4. Tannock IF. Conventional cancer therapy: promise broken or promise delayed? *Lancet* 1998; 351 (Suppl 2):S116–S119.
5. Cimons M. Cancer rates decline as funding increases. *Nature Med* 1998; 4:544.
6. Food and Drug Administration Web Site. Available at: http://www.fda.gov/10ashi/cancer/cdrug.html.
7. PhRMA Web Site. Available at http://www.phrma.org.
8. Pharmaprojects, PJB Publications, Richmond, 1999.
9. Drews J. Genomic sciences and the medicine of tomorrow. *Nature Biotechnol* 1996; 14:1516–1518.
10. Goodfellow P. Cited in: Investigational Drugs Weekly Highlights. Current Drugs Ltd, London, UK, 1999;8.
11. Rowinsky EK, Donehower RC. Drug therapy: paclitaxel (Taxol). *N Engl J Med* 1995; 332:1004–1114.
12. Garrett MD, Workman P. Discovering novel chemotherapeutic drugs for the third millenium. *Eur J Cancer* 1999; 35:2010–2030.
13. Debouck C, Goodfellow PN. DNA microassays in drug discovery and development. *Nature Genet* 1999; 21:48–51.
14. Anderson WF, Field C, Venter JC. Mammalian gene studies: editorial overview. *Curr Opin Biotechnol* 1994; 5:577,578.
15. Marshall E. A high-stakes gamble on genome sequencing. *Science* 1999; 284:1906–1909.
16. Nature Genetics. 1999; vol. 21 (Suppl).
17. Page MJ, Amess B, Rohlff C, Stubberfield C, Parekh R. Proteomics: a major new technology for the drug discovery process. *Drug Discovery Today* 1999; 4:55–62.
18. Bishop JM. Molecular themes in oncogenesis. *Cell* 1991; 64:235–248.
19. Haber DA, Fearon ER. The promise of cancer genetics. *Lancet* 1998; 351 (Suppl II):1–8.
20. Gibbs JB, Oliff A. Pharmaceutical research in molecular oncology. *Cell* 1994; 78:193–198.
21. Lundberg AS, Weinberg RA. Control of the cell cycle and apoptosis. *Eur J Cancer* 1999; 35:531–539.
22. Shayesteh L, Lu Y, Kuo W-L, Baldocchi R, Godfrey T, Collins C, Pinkel D, Powell B, Mills GB, Gray JW. PI3KCA is implicated as an oncogene in ovarian cancer. *Nature Genet* 1999; 21:99–102.
23. Furnari FB, Lin H, Huang H-JS, Cavanee WK. Growth suppression of glioma cells requires a functional phosphatase domain. *Proc Natl Acad Sci USA* 1997; 94;12,479–12,484.
24. Stambolic V, Suzuki A, de la Pompa JL, Brothers GM, Mirtros C, Sasaki T, Rutland J, Penninger JM, Siderovski DP, Mak TW. Negative regulation of PKB/Akt-dependent cell survival by the tumour suppressor gene PTEN. *Cell* 1998; 95:29–39.
25. Zhang L, Zhou W, Velculescu VE, Kern SE, Hruban RH, Hamilton SR, Vogelstein B, Kinzler KW. Gene expression profiles in normal and cancer cells. *Science* 1997; 276:1268–1272.

26. Bevan P, Ryder H, Shaw I. Identiftying small-molecule lead compounds: the screening approach to drug discovery. *Trends Biotechnol* 1995; 13:115–121.
27. Workman P. Towards intelligent anticancer drug screening in the post-genome era? *Anti-Cancer Drug Design* 1997; 12:525–531.
28. Willett P (ed). Computational methods for the analysis of molecular diversity. In: Perspectives in Drug Design, vols. 7 and 8. Kluwer, Dordrecht, The Netherlands, 1997.
29. Rishton GM. Reactive compounds and in vitro false positives in HTS. *Drug Discovery Today* 1997; 2:382–384.
30. Blundell TL. Structure-based drug design. *Nature* 1996; 384 6604(Suppl):23–25.
31. Kubinyi H. Structure-based design of enzyme inhibitors and receptor ligands. *Curr Opin Drug Discovery Develop* 1998; 1(1):4–15.
32. Schreiber S. http://www-schreiber.chem.harvard.edu/home/research/html. Visited 5/3/99.
33. Feng S, Kasahara C, Rickles RJ, Schreiber SL. Specific interactions outside the proline-rich core of two classes of Src homology 3 ligands. *Proc Natl Acad Sci USA* 1995; 92;2,408–12,415.
34. King FD. (ed). Medicinal Chemistry: Principles and Practice. Royal Society of Chemistry, Cambridge, UK, 1994.
35. Terrett NK, Gardner M, Gordon DW, Kobylecki RJ, Steele J. Combinatorial synthesis—the design of compound libraries and the application to drug discovery. *Tetrahedron* 1995; 51:8135–8173.
36. Hogan JC. Combinatorial chemistry in drug discovery. *Nature Biotechnol* 1997; 15:328–340.
37. Workman P. Pharmacokinetics and cancer: successes, failures and future prospects. In: Workman P, Graham MA, eds. Pharmacokinetics and Cancer Chemotherapy. Cancer Surveys, vol 17, Cold Spring Harbor Laboratory, Cold Spring Harbor, New York, 1993; 1–26.
38. Lipinski CA, Lombardo F, Dominy BW, Feeney PJ. Experimental and computational approaches to estimate solubility and permeability in drug discovery and development settings. *Adv Drug Delivery Rev* 1996; 23:3–25.
39. Brown JM, Workman P. Partition coefficient as a guide to the development of radiosensitizers that are less toxic than misonidazole. *Radiation Res* 1990; 82:171–190.
40. Workman P, Brown JM. Structure-pharmacokinetic relationships for misonidazole analogues in mice. *Cancer Chemother Pharmacol* 1981; 6:39–49.
41. Rowland M. Pharmacokinetics-QSAR: definitions concepts and models. In: Dearden JC, ed. Quantitative Approaches to Drug Design, Elsevier, Amsterdam, The Netherlands, 1993; 155–161.
42. Mayer JM, van de Waterbeem D. Development of structure-pharmacokinetic relationships. *Environ Health Perspect* 1985; 61:295–306.
43. Olah TV, McLoughlin DA, Gilbert JD. The simultaneous determination of mixtures of drug candidates by liquid chromatography/atmospheric pressure chemical ionization mass spectrometry as an *in vivo* drug screening procedure. *Rapid Commun Mass Spectrom* 1997; 11:17–23.
44. Rodrigues AK. Preclinical drug metabolism in the age of high throughput screening: an industrial perspective. *Pharmaceutical Res* 1997; 14:1504–1515,
45. Hollingshead MG, Alley MC, Camelerier RF, Abbot BJ, Mayo JG, Malspeis L, Grever MR. In vivo cultivation of tumour cells in hollow-fibers. *Life Sci* 1995; 57:131–141.
46. Gelman KA, Eisenhauer EA, Harris AL, Ratain MJ, Workman P. Anticancer agents targeting signaling molecules and cancer cell environment: challenges for drug development. *J Natl Cancer Inst* 1999; 91: 1281–1287.
47. Weissleder R. Molecular imaging: exploring the next frontier. *Radiology* 1999; 212:609–614.
48. Burtles SS, Newell DR, Henrar REC, Connors TA. Revisions of general guidelines for the preclinical toxicology of new cytotoxic anticancer agents in Europe. *Eur J Cancer* 1995; 31A:408–410.
49. Burtles SS, Joddrell DI, Newell DR. Evaluation of "rodent only" preclinical toxicology for Phase I trials of new cancer treatments—the Cancer Research Campaign (CRC) experience. *Proc Am Assoc Cancer Res* 1998; 39:363.
50. Newell DR, Burtles SS, Fox BW, Jodrell DI, Connors TA. Evaluation of rodent-only toxicology for early clinical trials with novel cancer therapeutics. *Br J Cancer* 1999; 81:760–768.
51. Scrip No. 2374 September 30th, 1998; 20.
52. Levitzki A. Targeting signal transduction for disease therapy. *Med Oncol* 1997; 14:83–89.
53. Brunton VG, Workman P. Cell-signalling targets for antitumour drug development. *Cancer Chemother Pharmacol* 1993; 32:1–19.
54. Hunter T. Oncoprotein networks. *Cell* 1997; 88:333–346.

55. Workman P. The potential for molecular oncology to define new drug targets. In: Kerr DR, Workman P, eds. New Molecular Targets for Cancer Chemotherapy. CRC, Boca Raton, FL, 1994; 1–29.
56. Marshall CJ. Opportunities for pharmacological intervention in the ras pathway. *Ann Oncol* 1995; 6(Suppl 1):S63–S67.
57. Mulcahy LS, Smith MR, Stacey DW. Requirement for ras proto-oncogene function during serum-stimulated growth of NIH 3T3 cells. *Nature* 1985; 313:241,242.
58. Stacey DW, Feig LA, Gibbs JB. Dominant inhibitory Ras mutants selectively inhibit the activity of either cellular or oncogenic Ras. *Mol Cell Biol* 1991; 11:4053–4064.
59. Leevers SJ, Paterson HF, Marshall CJ. Requirement for ras in raf activation is overcome by targeting raf to the plasma membrane. *Nature* 1994; 369:411–414.
60. Stokoe D, Macdonald SG, Cadwallader G, et al. Activation of Raf as a result of recruitment to the plasma membrane. *Science* 1994; 264:1463–1467.
61. Schaap D, Van der Wal J, Howe LR, et al. A dominant-negative mutant of raf blocks mitogen-activated protein kinase activation by growth factors and oncogenic p21 (ras). *J Biol Chem* 1993; 68:20,232–20,236.
62. Cowley S, Paterson H, Kemp P, Marshall CJ. Activation of MAP kinase kinase is necessary and sufficient for PC12 differentiation and for transformation of NIH cells. *Cell* 1994; 77:841–852.
63. Marais R, Wynne J, Treisman R. The SRF accessory protein Elk-1 contains a growth factor-regulated transcriptional activation domain. *Cell* 1993; 73:381–383.
64. Lui JJ, Chao JR, Jiang MC, Ng SY, Yen JJ, Yang-Yen HF. Ras transformation results in an elevated level of cyclin D1 and acceleration of G1 progression in NIH 3T3 cells. *Mol Cell Biol* 1995; 15:3654–3663.
65. Filmus J, Robles A, Shi W, Wong MJ, Colombo LL, Conti CJ. Induction of cyclin D1 overexpression by activated ras. *Oncogene* 1994; 9:3627–3633.
66. Galactionov K, Chen X, Beach D. Cdc25 cell-cycle phosphatase as a target for c-myc. *Nature* 1996; 382: 511–517.
67. Leone G, DeGregori J, Sears R, Jakoi L, Neving JR. Myc and Ras collaborate in inducing accumulation of active cyclin E/Cdk2 and E2F. *Nature* 1997; 387:422–426.
68. Mittnacht S, Paterson H, Olson MF, Marshall CJ. Ras signalling is required for inactivation of the tumour suppressors pRb cell-cycle control protein. *Curr Biol* 1997; 7:219–221.
69. Peeper DS, Upton TM, Ladha MH, et al. Ras signalling linked to the cell-cycle machinery by the retinoblastoma protein. *Nature* 1997; 386:177–181.
70. Polyak K, Lee MH, Erdjument-Bromage H, et al. Cloning of p27 Kip1, a cyclin-dependent kinase inhibitor and a potential mediator of extracellular antimitogenic signals. *Cell* 1994; 78:59–66.
71. Hannon GJ, Beach D. $p15^{INK4B}$ is a potential effector of TGF-beta-induced cell cycle arrest. *Nature* 1994; 3715:257–261.
72. Bos JL. The ras gene family and human carcinogenesis. *Mutat Res* 1988; 195:255–271.
73. Kiaris H, Spandidos DA. Mutations of *ras* genes in human tumours (review). *Int J Oncol* 1995; 7: 413–421.
74. Wittinghofer A, Pai EF. The structure of Ras protein: a model for a universal switch. *Trends Biochem Sci* 1991; 16:382–387.
75. Andreyev JHNA, Norman AR, Cunningham D, Oates JR, Clarke PA. Kirsten ras mutations in patients with colorectal cancer: the multicenter "RASCAL" study. *J Natl Cancer Inst* 1998; 90(9):675–684.
76. Hahn WC, Counter CM, Lundberg AS, Beijersbergen RL, Brooks MW, Weinberg RA. Creation of human tumour cells with defined genetic elements. *Nature* 1999; 400:464–468.
77. Chin L, Tam A, Pomerantz J, Wong M, Holash J, Bardeesy N, Shen Q, O'Hagan R, Pantginis J, Zhou H, Horner JW II, Cordon-Cardo C, Yancopoulos GD, DePinho RA. Essential role for oncogenic Ras in tumour maintenance. *Nature* 1999; 400:468–472.
78. Marshall CJ. Ras effectors. *Curr Opin Cell Biol* 1996; 8:197–204.
79. Powis G. Signalling targets for anticancer drug development. *Trends Pharm Sci* 1991; 12:188–194.
80. Jackson RC. The kinetic properties of switch antimetabolites. *J Natl Cancer Inst* 1993; 85:539–545.
81. Morrison PF, Dedrick RL. Discovering pharmaceuticals from multistationary biochemical networks. *J Natl Cancer Inst* 1993; 85:518–519.
82. Bailey JE. Lessons from metabolic engineering for functional genomics and drug discovery. *Nature Biotechnol* 1999; 17:616–618.
83. Strawn LM, Schawver LK. Tyrosine kinase in disease. *Exp Opin Invest Drugs* 1998; 7:553–573.
84. Workman P. Towards genomic cancer pharmacology: innovative drugs for the new millenium. *Curr Opin Oncol Endo Metab Investig Drugs* 2000; 2:21–25.

85. Hall-Jackson CA, Goedert M, Hedge P, Cohen P. Effect of SB 203580 on the activity of c-Raf in vitro and in vivo. *Oncogene* 1999; 18:2047–2054.
86. Hall-Jackson CA, Eyers PA, Cohen P, Goedert M, Boyle FT, Hewitt N, Hedge P. Paradoxical activation of Raf by a novel Raf inhibitor. *Chem Biol* 1999; 6:559–568.
87. Sebolt-Leopold JS, Dudley DT, Herrera R, van Becelaere K, Wiland A, Gowan RC, Tecle H, Barrett SD, Bridges A, Przybranowski S, Leopold WR, Saltiel AR. Blockade of the MAP kinase pathway suppresses growth of colon tumors in vivo. *Nature Med* 1999; 5(7):810–816.
88. Gray NS, Wodicka L, Thunnissen A-MWH, Norman TC, Kwon S, Espinoza FH, Morgan DO, Barnes G, LeClerc S, Meijer L, Kim S-H, Lockhart DJ, Schultz PG. Exploiting chemical libraries, structure and genomics in the search for kinase inhibitors. Science 1998; 281:533–538.
89. Whitesell L, Mimnaugh EG, Decosta B, Myers CE, Neckers LM. Inhibition of heat-shock protein Hsp90-*p60(v-Src)* heteroprotein complex-formation by benzoquinone ansamycins—essential role for stress proteins in oncogenic transformation. *Proc Natl Acad Sci USA* 1994; 91:8324–8328.
90. Kelland LR, Sharp SY, Rogers PM, Myers TG, Workman P. DT-diaphorase expression and tumor cell sensitivity to 17-oallylamino, 17-demethoxygeldanamycin, an inhibitor of heat shock protein 90. *J Natl Cancer Inst* 1999; 91:1940–1949.
91. Lerner EC, Hamilton AD, Sebti SM. Inhibition of Ras prenylation:a signalling target for novel anti-cancer drug design. *Anti-Cancer Drug Design* 1997; 12:229–238.
92. Sebti S, Hamilton A. Inhibitors of prenyl transferases. *Curr Opin Oncol* 1997; 9:557–561.
93. Der CJ. American Association of Clinical Oncology Educational Book, 1996; 108–110.
94. Olson MF, Paterson HF, Marshall CJ. Signals from Ras and Rho GTPases interact to regulate expression of $p21^{Waf1/Gip1}$. *Nature* 1998; 394:295–299.

2 The Biochemistry of Farnesyltransferase and Geranylgeranyltransferase I

Chih-Chin Huang, PHD, *Carol A. Fierke,* PHD, *and Patrick J. Casey,* PHD

CONTENTS

1. INTRODUCTION TO PROTEIN PRENYLATION

Protein prenylation refers to a type of covalent posttranslational modification by lipids at cysteine residues near the C-terminus of a protein; either a 15-carbon farnesyl or a 20-carbon geranylgeranyl isoprenoid is attached to the protein via a thioether linkage *(1–3)* (Fig. 1). Protein prenylation is ubiquitous in the eukaryotic world, and most prenylated proteins are membrane-associated for at least part of their lifetime. The majority of prenylated proteins are involved in cellular signaling and/or regulatory events that occur at or near the cytoplasmic surfaces of cellular membranes *(4,5)*.

The first evidence for protein prenylation came from studies in Japan in the late 1970s from structural analysis of specific fungal mating factor peptides *(6,7)*. On some such mating factors, a 15-carbon farnesyl isoprenoid was found linked to the peptide through a thioether bond from a cysteine sulfhydryl to the C-1 carbon of the farnesyl. Biochemical data suggested that this modification was an important element of the mating factor peptide. The discovery of mammalian prenylated proteins arose from studies concerning the effects of inhibiting isoprenoid biosynthesis on cell growth. Inhibitors of HMG-CoA reductase, the rate-limiting enzyme in isoprenoid biosynthesis (Fig. 1), were found to block cell growth in a fashion that could not be reversed by adding exogenous sterols (the major end products of the isoprenoid pathway) to the media *(8,9)*. However, the effects could be reversed by addition of small amounts of mevalonate, suggesting that a nonsterol metabolite of mevalonate was involved in this cell growth control. When ^{3}H-mevalonate

From: *Farnesyltransferase Inhibitors in Cancer Therapy*
Edited by: S. M. Sebti and A. D. Hamilton © Humana Press Inc., Totowa, NJ

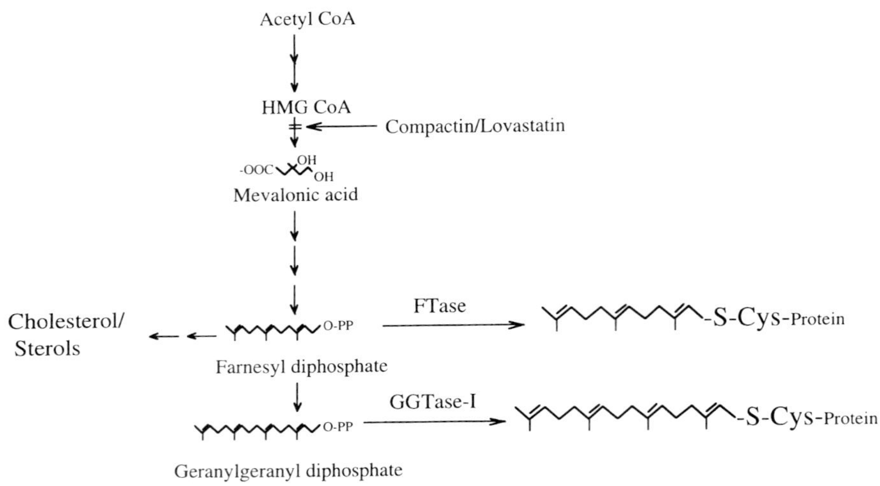

Fig. 1. Overview of the isoprenoid biosynthetic pathway and the structures of the prenyl groups attached to proteins by FTase and GGTase I.

was used in these types of experiments, the label was found to be incorporated into a number of cellular proteins, dubbed "prenylated proteins" *(10,11)*.

The first prenylated mammalian protein identified was the nuclear protein lamin B *(12,13)*. At about the same time, independent studies on the a-factor mating peptide of *Saccharomyces cerevisiae* revealed that this peptide was modified by a farnesyl isoprenoid *(14)*. The realization that both lamin B and the a-factor peptide contained a so-called "CAAX motif" at their carboxyl terminus (where "C" was the cysteine residue that served as the isoprenoid attachment site, "A" signified an aliphatic amino acid, and "X" denoted an undefined amino acid) prompted examination of other proteins containing the motif to determine if they too were prenylated. Foremost among the CAAX-containing proteins examined were the products of the Ras family of protooncogenes. The discoveries that Ras proteins were modified by farnesylation and that the modification was required for the oncogenic forms of these proteins to transform cells triggered an immediate and widespread interest in this form of lipid modification *(15–17)*. Subsequent studies have identified almost a hundred prenylated proteins in mammalian cells *(1,5,18)*, and revealed that, in addition to the 15-carbon farnesyl moiety, the 20-carbon geranylgeranyl isoprenoid could also be attached to proteins *(19,20)* (Fig. 1).

For CAAX-containing proteins, prenylation is but the first step in a series of three posttranslational modifications that occur at the C-terminus of most of these proteins. The three C-terminal amino acids (i.e., the -AAX) are subsequently removed by a membrane-bound protease, and finally a membrane-bound enzyme methylates the newly-formed carboxyl group to produce a methylester at the C-terminus *(21,22)*. Furthermore, in addition to the modifications at the CAAX motif, in some prenylated proteins other modifications such as palmitoylation and phosphorylation can occur in the C-terminal region just upstream of the CAAX motif *(5)*.

Monomeric guanine nucleotide (GTP)-binding proteins (G proteins) such as Ras, Rap, Rho, and Rab comprise the largest set of prenylated proteins *(5,23)*. Among these G proteins, Ras proteins have attracted particular attention because of the important role of Ras in carcinogenesis *(24,25)*. The normal functions of Ras proteins are in cellular signal transduction pathways that are essential for cell growth and differentiation *(25–27)*. Moreover, specific mutations in Ras proteins render them oncogenic, and such mutations are found in about 30% of all human tumors, including over 90% of human pancreatic cancer and 50% of human colon cancer *(18,24)*. The dependence of the transformed phenotype on the constitutive activity of Ras prompted considerable speculation that blocking the Ras signaling pathway could provide a way to treat such cancers *(24)*. Hence, the finding that farnesylation is absolutely required for oncogenic Ras function identified a specific point in the process, i.e., the attachment of the isoprenoid, for which development of specific inhibitors might provide an approach to this type of cancer chemotherapy *(18,28–30)*.

2. FTASE AND GGTASE-I: THE CAAX PRENYLTRANSFERASES

2.1. General Features of the Enzymes

The first identified protein prenyltransferase was protein farnesyltransferase (FTase), originally isolated from rat brain cytosol using an assay that followed the incorporation of radiolabel from ^{3}H-FPP into a recombinant Ras protein *(31)*. The finding that CAAX proteins containing methionine or serine at their C-terminus were farnesylated, whereas those ending in leucine were modified by a geranylgeranyl moiety *(32–34)*, provided the initial evidence for the existence of a distinct enzyme that would catalyze the addition of geranylgeranyl to certain proteins in the CAAX class. Using an approach similar to that which led to the identification of FTase, an enzymatic activity capable of transferring the geranylgeranyl group from geranylgeranyl diphosphate to candidate proteins was identified *(35,36)*. This enzyme, protein geranylgeranyltransferase type I (GGTase I), exhibited properties similar to those of FTase (*see below*). The C-terminal leucine residue was shown to be responsible for the specific recognition of substrate proteins by GGTase I by producing a Ras protein with a leucine-for-serine switch at the COOH-terminal position, a switch that converted the Ras protein from a FTase to a GGTase I substrate *(35)*.

Mammalian FTase and GGTase I share many properties *(37)*. Both enzymes are heterodimers that contain a common subunit (designated the α-subunit) of 48 kDa and distinct β-subunits of 46 kDa (FTase) and 43 kDa (GGTase I) *(31,38,39)*. Both proteins are zinc metalloenzymes that operate through apparently quite similar kinetic and chemical mechanisms (*see below*). Both enzymes have been cloned, and sequence analysis has 1) confirmed that the α-subunits are the products of the same gene and 2) revealed that the β-subunits had ~35% sequence identity at the amino acid level *(40–42)*. The significance of the two enzymes sharing a common subunit is not yet clear, but the existence of an identical α-subunit and a highly homologous β-subunit for these two protein prenyltransferases provided the initial evidence that discrete segment(s) of the β-subunit would be responsible for the remarkable substrate specificities of the enzymes.

Structural information just recently has begun to emerge on the CAAX prenyltransferases. Data to date have come from analysis of mammalian FTase, whose X-ray crystal structure was determined at 2.2 Å resolution *(43)*. In this structure, which was of the free (i.e., unliganded) enzyme, the α-subunit was found to be folded into a crescent-shaped domain composed of seven successive pairs of coiled coils termed "helical hairpins," which contact a significant portion of the β-subunit. The existence of repeat motifs in this subunit was first predicted from sequence alignments of mammalian and fungal α-subunits *(44)*. The β-subunit was also found to consist largely of helical domains, with the majority of the helices arranged into an α–α barrel structure. One end of the barrel was open to the solvent, while the other end was blocked by a short stretch of residues near the C-terminus of the β-subunit. This arrangement results in a structure containing a deep cleft in the center of the barrel that possesses all of the features expected for the active site of the enzyme, including the aforementioned bound zinc ion (*see* Subheading 2.4.1.). Quite recently, crystal structures of the complex of FTase with its isoprenoid substrate FPP have been reported that provide a quite detailed snapshot of the binding site for this substrate on the enzyme *(45,46)* (*see also* Chapter 3).

2.2. Recognition of Substrates

2.2.1. Recognition of Isoprenoid Substrates

Binding of isoprenoid substrates by both CAAX prenyltransferases is of very high affinity with K_D values being in the low n*M* range *(47–49)*; the initial realization of this property came from findings that the enzyme-isoprenoid complexes could be isolated by gel filtration *(50,51)*. The use of photoactivatable analogs of both FPP and GGPP revealed that the analogs could be specifically crosslinked to the β-subunits of FTase and GGTase I, respectively, upon activation *(52–54)*; these and related findings with peptide substrates (*see* Subheading 2.2.2.) provided the initial evidence that the active sites for the enzymes were, as expected, predominately associated with the β-subunits. The recent crystal structures determined for FTase-FPP complexes have provided the formal proof of this hypothesis *(45,46)*.

An early observation made with FTase was that the enzyme could bind both FPP and GGPP with relatively high affinity, although only FPP could serve as a substrate in the reaction *(50)*. A more detailed study of the FPP binding properties of FTase revealed that there is in fact a significant difference in the binding of the two isoprenoids, with GGPP binding being some 15-fold weaker than that of FPP *(49)*, although this still translates to an apparent affinity of ~100 n*M* for GGPP binding to FTase. A structural-based hypothesis why FTase exhibits such high affinity binding of GGPP to form a complex that is essentially catalytically inactive has been advanced *(46)*. Briefly, this hypothesis—discussed in detail in Chapter 3—is that the depth of a hydrophobic binding cavity in the β-subunit acts as a ruler that discriminates between the two isoprenoids based on their chain length. No structural information is yet available for GGTase I, although this enzyme exhibits much higher selectivity; binding of GGPP to the enzyme is several hundred-fold tighter than that of FPP *(48,49)*.

Analogs of FPP have been identified that bind to FTase with high affinity but cannot participate in catalysis *(55,56)*. These analogs have been quite useful in mechanistic studies of FTase, because they allow formation of an inactive FTase-isoprenoid binary complex (*see* Subheading 2.4.). Analogs of GGPP that should allow similar studies with GGTase I have also been described *(57,58)*.

2.2.2. Recognition of Protein Substrates

As noted in Subheading 1, mammalian cells contain a wide variety of proteins that are processed by CAAX prenyltransferases. Substrates of FTase include all four Ras proteins, nuclear lamins A and B, the γ-subunit of the retinal trimeric G protein transducin, and a variety of kinases and phosphatases *(18,59–63)*. Known targets of GGTase I include most identified γ-subunits of heterotrimeric G proteins and a multitude of Ras-related monomeric G proteins, including most members of the Rac, Rho, and Rap subfamilies

(1,5). All these protein substrates contain a Cys residue precisely four amino acids from the C-terminus. Furthermore, as noted in Subheading 2.1., the identity of the C-terminal residue (i.e., the "X" of the CAAX motif) determines which of the two enzymes will act on the protein. FTase prefers proteins containing Ser, Met, Ala or Gln, whereas Leu at this position directs modification by GGTase I *(1,21)*. This property of the enzymes make it possible to predict with reasonable accuracy from its primary sequence which prenyl modification will be on a protein.

An important property of both FTase and GGTase I is that they can recognize short peptides containing appropriate CAAX motifs as substrates *(31,36,64)*. A quite detailed analysis of specificity in recognition of Ca_1a_2X sequences by FTase indicates that the a_1 position has a relaxed amino acid specificity, while variability at a_2 and X are more restricted. Basic and aromatic side chains are tolerated at a_1 but much less so at a_2, whereas acidic residues are not well-tolerated at either position *(64,65)*. Moreover, substitution at the a_2 position by an aromatic residue in the context of a tetrapeptide creates a molecule that has been reported to be not a substrate for FTase but rather a competitive inhibitor *(66)*. One such peptide, CVFM, has served as the basis for design of peptidomimetic inhibitors of FTase *(67–69)*.

Binding of peptide substrates to FTase has been examined by nuclear magnetic resonance (NMR) using a resonance transfer approach. One such study reported that the CAAX sequence of a peptide substrate adopts a Type I β-turn conformation when bound to the enzyme *(70)*. However, a similar study of binding of a peptidomimetic inhibitor of FTase termed L-739,787 revealed a slightly different conformation most closely approximating a Type III β-turn *(71)*. A note of caution here is that, in both cases, the binding of the peptide/peptidomimetic was examined in the absence of bound isoprenoid on the enzyme. The recent realization that the kinetic mechanism is most likely an ordered one in which isoprenoid binding precedes that of the peptide/protein substrate *(47)*, and that the binding of the isoprenoid markedly increases the affinity for the peptide substrate *(72)* (*see* Subheading 2.3.) may have profound implications for this data, as the binding of the peptide/protein substrate to the enzyme-isoprenoid complex may be very different than its binding to free enzyme.

The zinc ion in both FTase and GGTase I is essential for the high affinity binding of the protein substrate (but not, however, for binding of the isoprenoid substrate) *(48,73)*, and recent studies indicate a direct coordination of the thiolate of the Cys residue of the protein substrate with the metal ion during catalysis *(72,74)*. Further evidence supporting a metal-substrate interaction in the enzymes comes from studies showing that the zinc ion can be replaced by cadmium, and the cadmium-substituted enzymes retain steady-state activity but have somewhat altered protein substrate specificities *(48,75)*. The location of the zinc ion was determined in the crystal structure of FTase to be in the β-subunit near the interface with the α-subunit *(43)*, consistent with the findings that both protein and peptide substrates can be crosslinked to the β-subunit of FTase *(50,76)*, and that short peptide substrates containing divalent affinity groups label both the α- and β-subunits upon photoactivation *(76)*.

2.2.3. Cross Prenylation by CAAX Prenyltransferases: Is it Important?

Although FTase and GGTase I seem to be quite selective for their substrates, cross-specificity (i.e., modification of a protein by either enzyme) has been observed *(36,65)*. However, whether this ability to modify alternate substrates is of physiological significance is still somewhat unclear. The most compelling evidence in this regard comes from studies in fungal systems. Yeast lacking FTase are viable, although they exhibit growth defects *(77)*. Overexpression of GGTase I in these mutants can partially suppress the growth defects, suggesting that GGTase I can at some level prenylate substrates of FTase *(78)*. Although yeast lacking GGTase I are not viable, the phenotype can be rescued by overexpression of two essential G protein substrates of the enzyme *(78)*, suggesting that FTase can prenylate some substrates of GGTase I if the substrate proteins are overproduced.

In terms of mammalian CAAX proteins, two Ras isoforms—K-Ras4B and N-Ras—can serve as substrates for both FTase and GGTase I in vitro, although they are much better substrates for FTase *(79,80)*. Although under normal conditions these two Ras isoforms seem to be modified solely by the farnesyl group, geranylgeranylation of the proteins can be detected in cells if FTase is inhibited *(81,82)*. The primary determinant for this type of cross-prenylation appears to be the existence of a Met as the C-terminal residue of these proteins *(80)*. Although these studies do not provide evidence to support the notion that cross-prenylation has significance under normal physiological conditions, it is certainly a concern in terms of the biology associated with FTase inhibition. The discussion of these concerns can be found in Chapters 5,13, and 15.

There do appear to be some mammalian proteins that can be normally modified by either isoprenoid. One such example is the Ras-related small G protein TC21, where the presence of Phe as the C-terminal residue apparently allows modification by either enzyme *(83)*. Additionally, another Ras-related small G protein, RhoB, has been shown to be farnesylated as well as geranylgeranylated even though its C-terminal residue is Leu *(84,85)*; farnesylation of this protein is most likely due to an ability to be processed by FTase, rather than an alternate activity of GGTase I *(86)*. How RhoB gets recognized and farnesylated by FTase is not yet clear, although a Lys residue in the second position of the CAAX motif may be partly responsible. Whatever property is responsible for this "dual prenylation," the differently prenylated forms of RhoB apparently have unique functions, as suppressing the farnesylated population by treatment of cells with a FTase inhibitor suppresses RhoB-dependent cell growth *(86)*.

2.3. Kinetic Mechanism

Mammalian CAAX prenyltransferases are quite slow enzymes, with k_{cat} values in the range of 0.05 s^{-1}*(31,87)*. Steady-state kinetics of mammalian FTase were initially interpreted as indicating a random-order binding mechanism in which either substrate could bind first *(87)*. However, the failure to trap enzyme-bound protein or peptide substrate in transient kinetic experiments suggested that either substrate binding is actually ordered or that the dissociation rate constant of the protein/peptide substrate is so fast and the affinity is so weak that farnesyl diphosphate (FPP) binding first is the kinetically preferred pathway *(47,87)* (Fig. 2). Consistent with this functionally ordered mechanism, the affinity of FPP for FTase is in the low n*M* range; whereas the affinity of the peptide substrate in the absence of bound FPP is relatively weak but this affinity is increased several hundred-fold by the binding of FPP analogs *(72)*. The aforementioned pre-steady-state kinetic studies also revealed that the association of the peptide substrate with the FTase · FPP binary complex was effectively irreversible with a k_{assoc} of 2×10^5 $M^{-1}s^{-1}$ *(47)*. While the rate constant for product formation could not be accurately determined in these studies, a lower limit of > 12 s^{-1} was established using protein fluorescence *(47)*. A more precise determination of the rate constant for product formation has come from

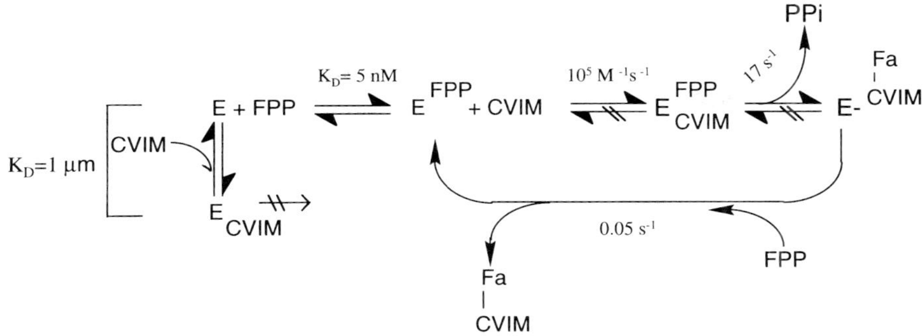

Fig. 2. Kinetic scheme for FTase. The overall kinetic scheme for the FTase reaction is shown; the available data indicates that the kinetic pathway for GGTase I will be quite similar. The abbreviations used are: E, FTase; CVIM, the C-terminal peptide of protein substrates (in the case shown this is K-Ras); FPP, farnesyl diphosphate; Fa, the 15-carbon farnesyl group attached to the Cys residue of the protein/peptide substrate. (*See text* for details.)

measurements of changes in the absorption spectrum of cobalt-substituted FTase during the catalytic process *(74)*, in which a value of 17 s^{-1} was obtained. The finding that the rate constant for the product formation was much greater than the steady-state k_{cat} revealed that the rate-limiting step under steady-state conditions was a step after formation of the thioether product, most likely product dissociation. In fact, in the absence of excess substrates, the product dissociation rate constant is so slow that the adduct of FTase with bound product can be isolated *(88)*. However, product dissociation can be enhanced by the addition of either substrate, with FPP being the most efficient in this regard (Fig. 2). Surprisingly, the affinity of FTase for the thioether product is weak (>1 m*M*), suggesting that product dissociation is kinetically controlled by an associated step that is triggered by substrate binding *(88)*.

The kinetics of yeast FTase differ from mammalian FTase in several aspects. For the yeast enzyme, steady-state kinetics clearly show that the mechanism is ordered with FPP binding first, because yeast FTase is inhibited by the peptide substrate *(89)*. In addition, the dissociation rate constant of the peptide substrate (33 s^{-1}) is faster than the chemical step (10.5 s^{-1}) *(90)*. Most importantly, the k_{cat} (4.5 s^{-1}) is much faster than that of the mammalian enzyme, such that product dissociation is not the sole rate-limiting step for steady-state turnover at saturating substrate concentrations *(89)*.

GGTase I has been less well studied than FTase, because only recently has there been much interest in this enzyme as a drug target (*see* Chapters 15 and 16). The catalytic activity of mammalian GGTase I is very similar to that of mammalian FTase, with the exception that this enzyme does not require added magnesium for optimal turnover *(48)*. The steady-state kinetic parameters of GGTase I are also comparable to those of FTase *(53,57)*. Yeast GGTase I, similar to yeast FTase, follows an ordered binding mechanism; however, steady-state turnover at saturating substrate (k_{cat} = 0.34 s^{-1}) is 10-fold slower than that of yeast FTase *(91)*.

2.4. Chemical Mechanism

2.4.1. Role of Zinc in Catalysis

In zinc proteins, the major role of the zinc ion can be either catalytic or structural. A catalytic zinc is involved in the chemical reaction directly, whereas a structural zinc is only required for the structural stability of the protein. A catalytic zinc ion is located at

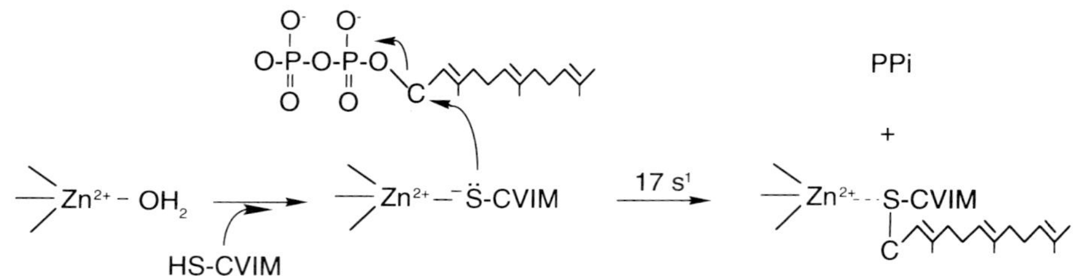

Fig. 3. The basic mechanism of FTase. Shown is the active site zinc ion with an open coordination sphere (i.e., with a water molecule providing one of the metal ligands). The peptide/protein substrate is designated HS-CVIM, although the thiolate anion form is the zinc-bound species that participates in the catalytic step. The net result is the formation of the enzyme-bound thioether product. (*See text* for details.)

the active site of an enzyme, where it participates directly in the catalytic mechanism. A unique feature of a catalytic zinc site is the existence of an open coordination sphere; that is, the zinc binding polyhedron contains at least one water molecule in addition to three or four protein ligands *(92)*; such an open coordination sphere for the bound zinc in FTase was revealed in the initial crystal structure of the enzyme *(43)*. The accessibility of the open coordination sphere to solvent and substrates implies that the zinc site can function in a catalytic manner. Unlike other first-row transition metals, the zinc ion (Zn^{2+}) contains a filled d orbital (d^{10}) and therefore does not participate in redox reactions, but rather functions as a Lewis acid to accept a pair of electrons *(93)*. The zinc-bound water is a critical component for a catalytic zinc site, because it can be either ionized to zinc-bound hydroxide (e.g., in carbonic anhydrase), polarized by a general base (e.g., in carboxypeptidase A) to generate a nucleophile for catalysis, or displaced by the substrate (e.g., in alkaline phosphatase) *(94)*.

Zinc is a metal of borderline "softness;" therefore, it can coordinate ligands comprised of either oxygen, nitrogen, or sulfur atoms *(95)*. In most catalytic zinc sites, the zinc ion is coordinated by different combinations of protein ligands, including the nitrogen of histidine, the oxygen of aspartate or glutamate, and the sulfur of cysteine; among these, histidine is most commonly observed *(96)*; in FTase, the three protein-derived zinc ligands are Asp297, Cys299, and His362 *(43,97)*. This spacing between the protein ligands is also characteristic of many catalytic zinc sites, which show a regular pattern that is not observed for structural zinc sites. This pattern consists of a short spacer (1–3 amino acids) between the first and the second ligands, and a long spacer (20–120 amino acids) between the second and the third ligands. The short spacer, with a rigid arrangement, may constitute a nucleus for the zinc binding site; whereas the third ligand, distant from the other two ligands in the linear sequence, may be responsible for the spatial formation of the active site and increase the stability of zinc coordination. The long spacer may also imply that flexibility is essential for the change of geometry and number of ligands that occurs in the zinc polyhedron during catalysis *(96)*.

The CAAX prenyltransferases are members of a new class of zinc metalloproteins that possess a previously-unappreciated catalytic function of the zinc ion: to enhance the nucleophilicity of a thiol group at neutral pH *(98)* (Fig. 3). The "founding member" of this family is a DNA repair protein termed "Ada." The function of Ada is to remove irreversibly the methyl group from the S_p diastereomer of DNA methylphosphotriesters *(99)*. Recently, the bound zinc ion found at the N-terminal domain of Ada has been proposed

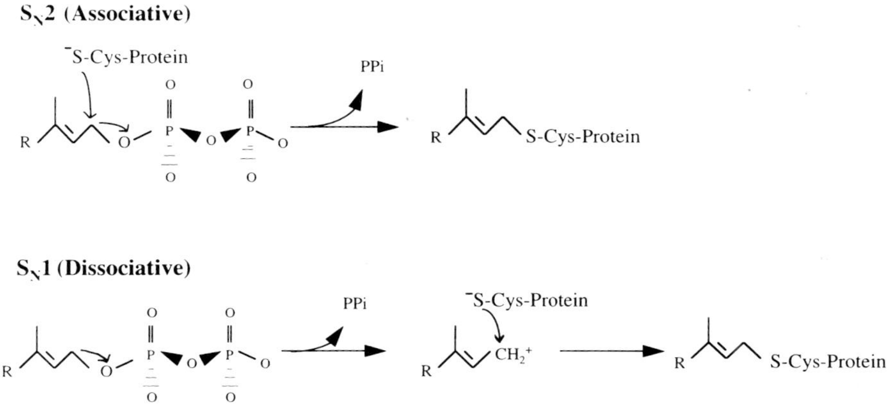

Fig. 4. Possible chemical mechanisms of CAAX prenyltransferases. Shown are the two extremes for potential mechanisms of catalysis, either a purely nucleophilic attack of the thiolate anion at C-1 of the isoprenoid substrate (the S_N2 mechanism) or an electrophilic reaction that involves preliminary formation of a carbocation at C-1 of the isoprenoid that then "captures" the appropriately positioned thiolate anion (the S_N1 mechanism). (*See text* for details.)

to catalyze the methyl transfer reaction in addition to stabilizing the structure of Ada *(99)*. The zinc ion is coordinated by the sulfur atoms of four cysteines, and one of the zinc ligands, Cys_{69}, is the residue that is methylated to form a thioether bond during the reaction with the methylphosphotriesters. The primary function proposed for this zinc ion is to coordinate the cysteine thiolate to lower the pK_a and perhaps enhance the reactivity of this group *(99)*. A similar mechanism has been recently proposed for several additional zinc-containing enzymes that catalyze S-alkylation reactions; these include, in addition to the CAAX prenyltransferases, the enzymes cobalamin-dependent methionine synthase, cobalamin-independent methionine synthase, and methanol:coenzyme M *(98,100–103)*.

2.4.2. Mechanism of Catalysis by FTase

Both electrophilic and nucleophilic mechanisms have been proposed for CAAX prenyltransferases *(37,104)* (Fig. 4). The electrophilic mechanism was first suggested by analogy to the somewhat similar enzymes such as prenyl diphosphate synthases and cyclases that use prenyl diphosphate as a substrate *(105–107)*; reactions catalyzed by these enzymes proceed predominately through formation of a carbocation at the site of bond formation. However, a conserved aspartate-rich region in the active sites of these enzymes that is important for catalysis was not observed in FTase, suggesting that the mechanism of the protein prenyltransferases may be somewhat different *(43,107)*. Furthermore, predominately nucleophilic mechanisms have been proposed for the other zinc-containing enzymes and proteins involved in S-alkylation reactions (i.e., Ada, the methionine synthases, and methanol:coenzyme M methyltransferase) noted in Subheading 2.4.1. *(98)*.

A nucleophilic mechanism for FTase was first suggested when it was found that the bound zinc ion in the enzyme was required both for catalysis as well as for the high affinity of peptide and protein substrates *(39,73)*. The initial direct evidence supporting a mechanism with substantial nucleophilic character came from the finding that the metal ion coordinated the thiol(ate) of the peptide substrate in the FTase · isoprenoid · CAAX ternary complex of cobalt-substituted FTase *(74)*; this finding was consistent with a model where the substrate thiol is activated by binding to the metal. Additional evidence in this regard comes from an examination of the pH dependence of peptide substrate binding to FTase. These studies revealed that the peptide thiol coordinates the metal as the thiolate anion, and that the pK_a of the -SH group is shifted from ~8 in the free peptide to ≤6.5 in the enzyme-bound peptide *(72)*. Stereochemical studies using FPP containing chiral deuterium-for-hydrogen substitutions at its C-1 carbon have demonstrated that FTasecarries out the reaction with inversion of configuration at the C-1 farnesyl center *(108,109)*. Although this finding is consistent with a S_N2-type nucleophilic displacement, it could also result from a S_N1-type mechanism in which the configuration of carbocation formed was subject to steric hindrance.

In a recent study, a transient kinetic analysis of FTase in which the zinc ion was substituted with cadmium ion was used to investigate further whether the metal-coordinated substrate thiolate played a nucleophilic role in catalysis by mammalian FTase *(109a)*. Cadmium is a softer metal than zinc, hence it has higher affinity for sulfur atom *(95)*. In model studies using small thiol compounds, Cd^{2+} was found to bind the thiolate 1–2 orders of magnitude more tightly than Zn^{2+} *(110)*. In fact, the affinity of the FTase · isoprenoid binary complex for peptide substrates is increased >10-fold for Cd-FTase compared to Zn-FTase *(109a)*, a finding consistent with the notion that Cd^{2+} enhances the binding of peptide substrate through stronger coordination of the substrate thiolate. Furthermore, in single-turnover experiments, the rate of product formation catalyzed by Cd-FTase was decreased ~sixfold compared to that of Zn-FTase, suggesting that the metal-thiolate nucleophile is important in the transition state *(109a)*.

Evidence has also been obtained for an electrophilic component in the mechanism of FTase. This evidence, initially using the yeast enzyme, came from the use of FPP substrate analogs that contained fluorine substitutions at the C_3 methyl position; fluorine, an electron-withdrawing group, would be expected to destabilize the carbocation in the transition state *(104)*. A decreased activity of yeast FTase with the C_3 fluoromethyl-FPP analogs was in fact observed; furthermore, the decrease in activity paralleled the number of fluorines in the substrate *(104)*. Similar results have recently been obtained for mammalian FTase using the C_3 fluoromethyl FPP analogs (C.C. Huang and C.A. Fierke, unpublished results). In both the yeast and mammalian enzymes, the rate constant for the chemical step of the FTase reaction was significantly decreased with these compounds and the decrease paralleled the number of fluorines in the substrate, strongly suggesting that the transition state has carbocation character. Because the fluoro-substituted FPPs affect a step after the metal-thiolate coordination, the carbocation apparently does not form until the peptide substrate binds. Importantly, the decreases in reactivity caused by fluorine substitution in FPP analogs are significantly smaller than the effects of fluorine substitution on either the solvolysis of dimethylallyl *p*-methoxybenzensulfonates or the reactivity of farnesyl pyrophosphate synthase, both of which proceed via an S_N1 reaction *(104,111)*. In fact, the effects of the fluorine substitution on FTase reactivity more closely parallel the effects on solvolysis reactions in the presence of a potent nucleophile, such as azide, which proceed with significant S_N2 character through what has been termed an open "exploded" transition state *(112)*.

Taken together, the data summarized above suggest that the mechanism of FTase is a carbocation-nucleophile combination reaction (Fig. 5). It seems that neither the nucleophilic nature of the metal-thiolate nor the carbocation character of the C_1 of FPP can be ignored. For such a carbocation-nucleophile combination reaction, whether the reaction

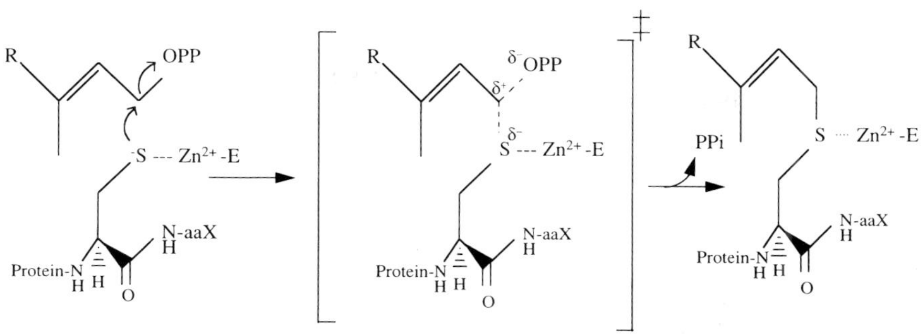

Fig. 5. Proposed mechanism for FTase. Shown is the carbocation-nucleophile combination mechanism proposed for FTase. In this mechanism, the role of the enzyme-bound zinc ion is to both facilitate formation of the substrate protein thiolate anion and position it for attack at the developing carbocation at C-1 of the isoprenoid substrate. (*See text* for details.)

proceeds through an S_N1 or S_N2 mechanism depends on both the stability of the carbocation in the transition state and the strength of the nucleophile *(113)*. When the carbocation intermediate has a long enough lifetime, the reaction can proceed via a stepwise S_N1 mechanism. On the contrary, if the lifetime of the intermediate is short, the reaction may occur through an enforced preassociation mechanism, where the reactants are assembled before the first bond-making or -breaking step occurs. This preassociation mechanism could either be concerted with no intermediate, or stepwise with an intermediate.

3. CONCLUSIONS

The close similarities in both structure and function of GGTase I and FTase, which include conservation of the zinc-binding residues in the β-subunit of GGTase I and the requirement for this metal for protein substrate binding, make it is very likely that both enzymes use a similar catalytic mechanism. Nonetheless, it will still be important to conduct studies similar to those described previously with GGTase I to confirm this hypothesis; furthermore, it could well turn out that there will be some important differences between these enzymes. Moreover, one must always keep in mind with these enzymes that the actual catalytic step has little influence on their steady-state rate and thus extreme caution must be exercised in the use of steady-state data to draw conclusions on such parameters as specificity in recognition of CAAX sequences, and so forth. For example, a peptide substrate that can be prenylated but is poorly released may exhibit the behavior of an inhibitor of this enzyme. To understand fully the mechanism of substrate specificity and other parameters of these enzymes, examination of the individual steps in the catalytic process, i.e., by presteady-state kinetics, is required.

REFERENCES

1. Zhang FL, Casey PJ. Protein prenylation: molecular mechanisms and functional consequences. *Annu Rev Biochem* 1996; 65:241–269.
2. Glomset JA, Gelb MH, Farnsworth CC. Prenyl proteins in eukaryotic cells: a new type of membrane anchor. *Trends Biochem Sci* 1990; 15:139–142.
3. Schafer WR, Rine J. Protein prenylation: genes, enzymes, targets and functions. *Annu Rev Genetics* 1992; 25:209–238.
4. Casey PJ. Protein lipidation in cell signaling. *Science* 1995; 268:221–225.

5. Glomset JA, Farnsworth CC. Role of protein modification reactions in programming interactions between ras-related GTPases and cell membranes. *Annu Rev Cell Biol* 1994; 10:181–205.
6. Kamiya Y, Sakurai A, Tamura S, Takahashi N. Structure of rhodotorucine A, a novel lipopeptide, inducing mating tube formation in Rhodosporidium toruloides. *Biochem Biophys Res Commun* 1978; 83:1077–1083.
7. Ishibashi Y, Sakagami Y, Isogai A, Suzuki A. Structures of tremerogens A-9291-I and A-9291-VII: peptidyl sex hormones of Tremella brasiliensis. *Biochemistry* 1984; 23:1399–1404.
8. Brown MS, Goldstein JL. Multivalent feedback regulation of HMG CoA reductase, a control mechanism coordinating isoprenoid synthesis and cell growth. *J Lipid Res* 1980; 21:505–517.
9. Maltese WA, Sheridan KM. Isoprenylated proteins in cultured cells: subcellular distribution and changes related to altered morphology and growth arrest induced by mevalonate deprivation. *J Cell Physiol* 1987; 133:471–481.
10. Schmidt RA, Schneider CJ, Glomset JA. Evidence for post-translational incorporation of a product of mevalonic acid into Swiss 3T3 cell proteins. *J Biol Chem* 1984; 259:10,175–10,180.
11. Faust J, Krieger M. Expression of specific high capacity mevalonate transport in a Chinese hamster ovary cell variant. *J Biol Chem* 1987; 262:1996–2004.
12. Wolda SL, Glomset JA. Evidence for modification of lamin B by a product of mevalonic acid. *J Biol Chem* 1988; 263:5997–6000.
13. Beck LA, Hosick TJ, Sinensky M. Incorporation of a product of mevalonic acid metabolism into proteins of Chinese hamster ovary cell nuclei. *J Cell Biol* 1988; 107:1307–1316.
14. Anderegg RJ, Betz R, Carr SA, Crabb JW, Duntze W. Structure of *Saccharomyces cerevisiae* mating hormone a-factor. *J Biol Chem* 1988; 263:18,236–18,240.
15. Hancock JF, Magee AI, Childs JE, Marshall CJ. All ras proteins are polyisoprenylated but only some are palmitoylated. *Cell* 1989; 57:1167–1177.
16. Casey PJ, Solski PA, Der CJ, Buss JE. p21ras is modified by a farnesyl isoprenoid. *Proc Natl Acad Sci USA* 1989; 86:8323–8327.
17. Schafer WR, Kim R, Sterne R, Thorner J, Kim S-H, Rine J. Genetic and pharmacological suppression of oncogenic mutations in *RAS* genes of yeast and humans. *Science* 1989; 245:379–385.
18. Cox AD, Der CJ. Farnesyltransferase inhibitors and cancer treatment: targeting simply Ras? *Biochim Biophys Acta* 1997; 1333:F51–F71.
19. Farnsworth CC, Gelb MH, Glomset JA. Identification of geranylgeranyl-modified proteins in HeLa cells. *Science* 1990; 247:320–322.
20. Rilling HC, Breunger E, Epstein WW, Crain PF. Prenylated proteins: the structure of the isoprenoid group. *Science* 1990; 247:318–320.
21. Clarke S. Protein isoprenylation and methylation at carboxyl-terminal cysteine residues. *Annu Rev Biochem* 1992; 61:355–386.
22. Ashby MN. CaaX converting enzymes. *Curr Opinion Lipidol* 1998; 9:99–102.
23. Seabra MC. Membrane association and targeting of prenylated ras-like GTPases. *Cell Signal* 1998; 10: 167–172.
24. Barbacid M. ras Genes. *Ann Rev Biochem* 1987; 56:779–827.
25. Boguski MS, McCormick F. Proteins regulating Ras and its relatives. *Nature* 1993; 366:643–654.
26. Macara IG, Lounsbury KM, Richards SA, McKeirnan C, Bar-Sagi D. The Ras superfamily of GTPases. *FASEB J* 1996; 10:625–630.
27. Johnson L, Greenbaum D, Cichowski K, Mercer K, Murphy E, Schmitt E, et al. K-ras is an essential gene in the mouse with partial function overlap with N-ras. *Genes Dev* 1997; 11:2468–2481.
28. Gibbs JB, Oliff A, Kohl NE. Farnesyltransferase inhibitors: Ras research yields a potential cancer therapeutic. *Cell* 1994; 77:175–178.
29. Lerner EC, Hamilton AD, Sebti SM. Inhibition of Ras prenylation: a signaling target for novel anticancer drug design. *Anti-Cancer Drug Design* 1997; 12:229–238.
30. Gibbs JB, Oliff A. The potential of farnesyltransferase inhibitors as cancer chemotherapeutics. *Annu Rev Pharmacol Toxicol* 1997; 37:143–166.
31. Reiss Y, Goldstein JL, Seabra MC, Casey PJ, Brown MS. Inhibition of purified p21ras farnesyl:protein transferase by Cys-AAX tetrapeptides. *Cell* 1990; 62:81–88.
32. Mumby SM, Casey PJ, Gilman AG, Gutowski S, Sternweis PC. G protein gamma subunits contain a 20-carbon isoprenoid. *Proc Natl Acad Sci USA* 1990; 87:5873–5877.
33. Yamane HK, Farnsworth CC, Xie H, Howald W, Fung BK-K, Clarke S, et al. Brain G protein gamma subunits contain an all-trans-geranylgeranyl-cysteine methyl ester at their carboxyl termini. *Proc Natl Acad Sci USA* 1990; 87:5868–5872.

34. Yamane HK, Farnsworth CC, Xie H, Evans T, Howald WN, Gelb MH, et al. Membrane-binding domain of the small G protein G25K contains an S-(all-trans-geranylgeranyl) cysteine methyl ester at its carboxyl terminus. *Proc Natl Acad Sci USA* 1991; 88:286–290.
35. Casey PJ, Thissen JA, Moomaw JF. Enzymatic modification of proteins with a geranylgeranyl isoprenoid. *Proc Natl Acad Sci USA* 1991; 88:8631–8635.
36. Yokoyama K, Goodwin GW, Ghomashchi F, Glomset JA, Gelb MH. A protein geranylgeranyltransferase from bovine brain: implications for protein prenylation specificity. *Proc Natl Acad Sci USA* 1991; 88:5302–5306.
37. Casey PJ, Seabra MC. Protein prenyltransferases. *J Biol Chem* 1996; 271:5289–5292.
38. Seabra MC, Reiss Y, Casey PJ, Brown MS, Goldstein JL. Protein farnesyltransferase and geranylgeranyltransferase share a common alpha subunit. *Cell* 1991; 65:429–434.
39. Moomaw JF, Casey PJ. Mammalian protein geranylgeranyltransferase: subunit composition and metal requirements. *J Biol Chem* 1992; 267:17,438–17,443.
40. Chen W-J, Andres DA, Goldstein JL, Russell DW, Brown MS. cDNA cloning and expression of the peptide binding beta subunit of rat p21ras farnesyltransferase, the counterpart of yeast *RAM1/DPR1*. *Cell* 1991; 66:327–334.
41. Kohl NE, Diehl RE, Schaber MD, Rands E, Soderman DD, He B, et al. Structural homology among mammalian and Saccharomyces cerevisiae isoprenyl-protein transferases. *J Biol Chem* 1991; 266: 18,884–18,888.
42. Zhang FL, Diehl RE, Kohl NE, Gibbs JB, Giros B, Casey PJ, Omer CA. cDNA cloning and expression of rat and human protein geranylgeranyltransferase Type-I. *J Biol Chem* 1994; 269:3175–3180.
43. Park H-W, Boduluri SR, Moomaw JF, Casey PJ, Beese LS. Crystal structure of protein farnesyltransferase at 2.25 Å resolution. *Science* 1997; 275:1800–1804.
44. Boguski MS, Murray AW, Powers S. Novel repetitive sequence motifs in the alpha and beta-subunits of prenyl-protein transferases and homology of the alpha subunits to the *MAD2* gene product of yeast. *New Biologist* 1992; 4:408–411.
45. Dunten P, Kammlott U, Crowther R, Weber D, Palermo R, Birktoft J. Protein farnesyltransferase: structure and implications for substrate binding. *Biochemistry* 1998; 37:7907–7912.
46. Long SB, Casey PJ, Beese LS. Co-crystal structure of protein farnesyltransferase complexed with a farnesyl diphosphate substrate. *Biochemistry* 1998; 37:9612–9618.
47. Furfine ES, Leban JJ, Landavazo A, Moomaw JF, Casey PJ. Protein farnesyltransferase: kinetics of farnesyl pyrophosphate binding and product release. *Biochemistry* 1995; 34:6857–6862.
48. Zhang FL, Casey PJ. Influence of metals on substrate binding and catalytic activity of mammalian protein geranylgeranyltransferase type-I. *Biochem J* 1996; 320:925–932.
49. Yokoyama K, Zimmerman K, Scholten J, Gelb MH. Differential prenyl pyrophosphate binding to mammalian protein geranylgeranyltransferase-I and protein farnesyltransferase and its consequences on the specificity of protein prenylation. *J Biol Chem* 1997; 272:3944–3952.
50. Reiss Y, Seabra MC, Armstrong SA, Slaughter CA, Goldstein JL, Brown MS. Nonidentical subunits of p21 H-ras farnesyltransferase: peptide binding and farnesyl pyrophosphate carrier functions. *J Biol Chem* 1991; 266:10,672–10,677.
51. Yokoyama K, Gelb MH. Purification of a mammalian protein geranylgeranyltransferase: formation and catalytic properties of an enzyme-geranylgeranyl diphosphate complex. *J Biol Chem* 1993; 268: 4055–4060.
52. Omer CA, Kral AM, Diehl RE, Prendergast GC, Powers S, Allen CM, et al. Characterization of recombinant human farnesyl-protein transferse: cloning, expression, farnesyl diphosphate binding, and functional homology with yeast prenyl-protein transferses. *Biochemistry* 1993; 32:5167–5176.
53. Yokoyama K, McGeady P, Gelb MH. Mammalian protein geranylgeranyltransferase-I: substrate specificity, kinetic mechanism, metal requirements, and affinity labeling. *Biochemistry* 1995; 34:1344–1354.
54. Bukhtiyarov YE, Omer CA, Allen CM. Photoreactive analogues of prenyl diphosphates as inhibitors and probes of human protein farnesyltransferase and geranylgeranyltransferase type I. *J Biol Chem* 1995; 270:19,035–19,040.
55. Gibbs JB, Pompliano DL, Mosser SD, Rands E, Lingham RB, Singh SB, et al. Selective inhibition of farnesyl-protein transferase blocks ras processing in vivo. *J Biol Chem* 1993; 268(11):7617–7620.
56. Patel DV, Schmidt RJ, Biller SA, Gordon EM, Robinson SS, Manne V. Farnesyl diphosphate-based inhibitors of Ras farnesyl protein transferase. *J Med Chem* 1995; 38:2906–2921.
57. Zhang FL, Moomaw JF, Casey PJ. Properties and kinetic mechanism of recombinant mammalian protein geranylgeranyltransferase type I. *J Biol Chem* 1994; 269:23,465–23,470.

58. Macchia M, Jannitti N, Gervasi G, Danesi R. Geranylgeranyl diphosphate-based inhibitors of post-translational geranylgeranylation of cellular proteins. *J Med Chem* 1996; 39:1352–1356.
59. Inglese J, Glickman JF, Lorenz, W, Caron M, Lefkowitz, RJ. Isoprenylation of a protein kinase: requirement of farnesylation/alpha-carboxyl methylation for full enzymatic activity of rhodopsin kinase. *J Biol Chem* 1992; 267:1422–1425.
60. Cox AD, Der CJ. The ras/cholesterol connection: implications for ras oncogenicity. *Crit Rev Oncog* 1992; 3:365–400.
61. James GL, Goldstein JL, Pathak RL, Anderson RGW, Brown MS. PxF, a prenylated protein of peroxisomes. *J Biol Chem* 1994; 269:14,182–14,190.
62. Heilmeyer LMG, Serwe M, Weber C, Metzger J, Hoffmann-Posorske E, Meyer HE. Farnesylcysteine, a constituent of the alpha and beta subunits of rabbit skeletal muscle phosphorylase kinase: localization by conversion to S-ethylcysteine and by tandem mass spectrometry. *Proc Natl Acad Sci USA* 1992; 89:9554–9558.
63. Smed FD, Boom A, Pesesse X, Schiffmann SN, Erneux C. Post-translational modification of human brain type I inositol-1,4,5-triphosphate 5-phosphatase by farnesylation. *J Biol Chem* 1996; 271:10,419–10,424.
64. Reiss Y, Stradley SJ, Gierasch LM, Brown MS, Goldstein JL. Sequence requirements for peptide recognition by rat brain p21ras farnesyl:protein transferase. *Proc Natl Acad Sci USA* 1991; 88:732–736.
65. Moores SL, Schaber MD, Mosser SD, Rands E, O'Hara MB, Garsky VM, et al. Sequence dependence of protein isoprenylation. *J Biol Chem* 1991; 266:14,603–14,610.
66. Goldstein JL, Brown MS, Stradley SJ, Reiss Y, Gierasch LM. Nonfarnesylated tetrapeptide inhibitors of protein farnesyltransferase. *J Biol Chem* 1991; 266:15,575–15,578.
67. Garcia AM, Rowell C, Ackermann K, Kowalczyk JJ, Lewis MD. Peptidomimetic inhibitors of ras farnesylation and function in whole cells. *J Biol Chem* 1993; 268:18,415–18,418.
68. Kohl NE, Mosser SD, deSolms SJ, Giuliani EA, Pompliano DL, Graham SL, et al. Selective inhibition of ras-dependent transformation by a farnesyltransferase inhibitor. *Science* 1993; 260:1934–1937.
69. Nigam M, Seong C-M, Qian Y, Hamilton AD, Sebti SM. Potent inhibition of human tumor p21ras farnesyltransferase by A1A2-lacking p21ras CA1A2X peptidomimetics. *J Biol Chem* 1993; 268: 20,695–20,698.
70. Stradley SJ, Rizo J, Gierasch LM. Conformation of a heptapeptide substrate bound to protein farnesyltransferase. *Biochemistry* 1993; 32:12,586–12,590.
71. Koblan KS, Culberson JC, deSolms SJ, Giuliani EA, Mosser SD, Omer CA, et al. NMR studies of novel inhibitors bound to farnesyl-protein transferase. *Prot Science* 1995; 4:681–688.
72. Hightower KE, Huang C-C, Casey PJ, Fierke CA. H-Ras peptide and protein substrates bind protein farnesyltransferase as an ionized thiolate. *Biochemistry* 1998; 37:15,555–15,562.
73. Reiss Y, Brown MS, Goldstein JL. Divalent cation and prenyl pyrophosphate specificities of the protein farnesyltransferase from rat brain, a zinc metalloenzyme. *J Biol Chem* 1992; 267:6403–6408.
74. Huang C-C, Casey PJ, Fierke CA. Evidence for a catalytic role of zinc in protein farnesyltransferase: spectroscopy of Co^{2+}-FTase indicates metal coordination of the substrate thiolate. *J Biol Chem* 1997; 272:20–23.
75. Zhang FL, Fu HW, Casey PJ, Bishop WR. Substitution of cadmium for zinc in farnesyl:protein transferase alters its substrate specificity. *Biochemistry* 1996; 35:8166–8171.
76. Ying W, Sepp-Lorenzino L, Cai K, Aloise P, Coleman PS. Photoaffinity-labeling peptide substrates for farnesyl-protein transferase and the intersubunit location of the active site. *J Biol Chem* 1994; 269: 470–477.
77. Powers S, Michaelis S, Broek D, Santa-Ana AS, Field J, Herskowitz I, Wigler M. *RAM*, a gene of yeast required for a functional modification of RAS proteins and for production of mating pheromone a-factor. *Cell* 1986; 47:413–422.
78. Trueblood CE, Ohya Y, Rine J. Genetic evidence for in vivo cross-specificity of the CaaX-Box protein prenyltransferases farnesyltransferase and geranylgeranyltransferase-I in Saccharomyces cerevisiae. *Mol Cell Biol* 1993; 13:4260–4275.
79. James GL, Goldstein JL, Brown MS. Polylysine and CVIM sequences of K-RasB dictate specificity of prenylation and confer resistance to benzodiazepine peptidomimetic in vitro. *J Biol Chem* 1995; 270:6221–6226.
80. Zhang FL, Kirschmeier P, Carr D, James L, Bond RW, Wang L, et al. Characterization of Ha-Ras, N-Ras, Ki-Ras4A, Ki-Ras4B as in vitro substrates for farnesyl protein transferase and geranylgeranyl protein transferase type I. *J Biol Chem* 1997; 272:10,232–10,239.

81. Rowell CA, Kowalczyk JJ, Lewis MD, Garcia AM. Direct demonstration of geranylgeranylation and farnesylation of Ki-Ras in vivo. *J Biol Chem* 1997; 272:14,093–14,097.
82. Whyte DB, Kirschmeier P, Hockenberry TN, Nunez-Oliva I, James L, Catino JJ, et al. K- and N-Ras are geranylgeranylated in cells treated with farnesyl protein transferase inhibitors. *J Biol Chem* 1997; 272:14,459–14,464.
83. Carboni JM, Yan N, Cox AD, Bustelo X, Graham SM, Lynch MJ, et al. Farnesyltransferase inhibitors are inhibitors of Ras but not R-Ras2/TC21, transformation. *Oncogene* 1995; 10:1905–1913.
84. Adamson P, Marshall CJ, Hall A, Tilbrook PA. Post-translational modifications of p21rho proteins. *J Biol Chem* 1992; 267:20,033–20,038.
85. Armstrong SA, Hannah VC, Goldstein JL, Brown MS. Caax geranylgeranyl transferase transfers farnesyl as efficiently as geranylgeranyl to RhoB. *J Biol Chem* 1995; 270:7864–7868.
86. Lebowitz, PF, Casey PJ, Prendergast GC, Thissen JA. Farnesyltransferase inhibitors alter the prenylation and growth-stimulating function of RhoB. *J Biol Chem* 1997; 272:15,591–15,594.
87. Pompliano DL, Rands E, Schaber MD, Mosser SD, Anthony NJ, Gibbs JB. Steady-state kinetic mechanism of ras farnesyl:protein transferase. *Biochemistry* 1992; 31:3800–3807.
88. Tschantz WR, Furfine ES, Casey PJ. Substrate binding is required for release of product from mammalian protein farnesyltransferase. *J Biol Chem* 1997; 272:9989–9993.
89. Dolence JM, Cassidy PB, Mathis JR, Poulter CD. Yeast protein farnesyltransferase: steady-state kinetic studies of substrate binding. *Biochemistry* 1995; 34:16,687–16,694.
90. Mathis JR, Poulter CD. Yeast protein farnesyltransferase: a pre-steady-state kinetic analysis. *Biochemistry* 1997; 36:6367–6376.
91. Stirtan WG, Poulter CD. Yeast protein geranylgeranyltransferase Type-I: steady-state kinetics and substrate binding. *Biochemistry* 1997; 36:4552–4557.
92. Vallee BL, Auld DS. Functional zinc-binding motifs in enzymes and DNA-binding proteins. *Faraday Discuss* 1992; 93:47–65.
93. Williams RJP. The biochemistry of zinc. *Polyhedron* 1987; 6:61–69.
94. Vallee BL, Auld DS. New perspective on zinc biochemistry: cocatalytic sites in multi-zinc enzymes. *Biochemistry* 1993; 32:6493–6500.
95. Pearson RG. Hard and soft acids and bases. *J Am Chem Soc* 1963; 85:3533–3539.
96. Vallee BL, Auld DS. Zinc coordination, function, and structure of zinc enzymes and other proteins. *Biochemistry* 1990; 29:5647–5659.
97. Fu H-W, Moomaw JF, Moomaw CR, Casey PJ. Identification of a cysteine residue essential for activity of protein farnesyltransferase: Cys299 is exposed only upon removal of zinc from the enzyme. *J Biol Chem* 1996; 271:28,541–28,548.
98. Matthews RG, Goulding CW. Enzyme-catalyzed methyl transfers to thiols: the role of zinc. *Current Opinion Chem Biol* 1998; 1:332–339.
99. Myers LC, Terranova MP, Ferentz, AE, Wagner G, Verdine GL. Repair of DNA methylphosphotriesters through a metalloactivated cysteine nucleophile. *Science* 1993; 261:1164–1167.
100. Gonzalez, JC, Peariso K, Penner-Hahn JE, Matthews RG. Cobalamin-independent methionine synthase from escherichia coli: a zinc metalloenzyme. *Biochemistry* 1996; 35:12,228–12,234.
101. LeClerc GM, Grahame DA. Methylcobamide:coenzyme M methyltransferase isozymes from methanosarcina barkeri. *J Biol Chem* 1996; 271:18,725–18,731.
102. Hooley R, Yu C-Y, Symons M, Barber DL. Gα13 stimulates Na^+-H^+ exchange through distinct Cdc42-dependent and RhoA-dependent pathways. *J Biol Chem* 1996; 271:6152–6158.
103. Goulding CW, Matthews RG. Cobalamin-dependent methionine synthase from escherichia coli: involvement of zinc in homocysteine activation. *Biochemistry* 1997; 36:15,749–15,757.
104. Dolence JM, Poulter CD. A mechanism for posttranslational modifications of proteins by yeast protein farnesyltransferase. *Proc Natl Acad Sci USA* 1995; 92:5008–5011.
105. Gebler JC, Woodside AB, Poulter CD. Dimethylallytryptophan synthase. An enzyme-catalyzed electrophilic aromatic substitution. *J Am Chem Soc* 1992; 114:7354–7360.
106. Chen A, Kroon PA, Poulter CD. Isoprenyl disphosphate synthases: protein sequence comparisons, a phylogenetic tree, and predictions of secondary structure. *Prot Science* 1994; 3:600–607.
107. Lesburg CA, Zhai G, Cane DE, Christianson DW. Crystal structure of pentalenene synthase: mechanistic insights on terpenoid cyclization reactions in biology. *Science* 1997; 277:1820–1824.
108. Mu Y, Omer CA, Gibbs RA. On the stereochemical course of human protein-farnesyl transferase. *J Am Chem Soc* 1996; 118:1817–1823.

109. Edelstein RL, Weller VA, Distefano MD. Stereochemical analysis of the reaction catalyzed by yeast protein farnesyltransferase. *J Org Chem* 1998; 63:5298,5299.

109a. Huang C-C, Hightower KE, Fierke CA. Mechanistic studies of rat protein farnesyltransferase indicate an associative transition state. *Biochemistry* 2000; 39:2593–2602.

110. Li NC, Manning RA. Some metal complexes of sulfur-containing amino acids. *J Am Chem Soc* 1955; 77:5225–5228.

111. Poulter CD, Wiggins PL, Le AT. Farnesylpyrophosphate synthetase. A new stepwise mechanism for the 1'-4 condensation reaction. *J Am Chem Soc* 1981; 103:3926,3927.

112. Richard JP, Jencks WP. Concerted bimolecular substitution reactions of 1-phenylethyl derivatives. *J Am Chem Soc* 1984; 106:1383–1396.

113. Jencks WP. When is an intermediate not an intermediate? Enforced mechanisms of general acid-base catalyzed, carbocation, carbanion, and liquid exchange reactions. *Acc Chem Res* 1980; 13:161–169.

3 Structures of Protein Farnesyltransferase

Stephen B. Long, PHD *and Lorena S. Beese,* PHD

CONTENTS

INTRODUCTION TO PROTEIN FARNESYLTRANSFERASE STRUCTURE

High resolution three-dimensional crystal structures of protein farnesyltransferase (FTase) complexed with substrates and inhibitors provide a framework for understanding the molecular basis of substrate specificity and mechanism and may facilitate the development of improved chemotherapeutics. The 2.25Å resolution crystal structure of rat FTase provided the first structural information on any protein prenyltransferase enzyme *(1)*. Rat FTase shares 93% sequence identity with the human enzyme and is predicted to be indistinguishable from human FTase in the active site region. Subsequently, a co-crystal structure of rat FTase with bound farnesyl diphosphate (FPP) revealed the location of the isoprenoid binding and gave insight into the molecular basis of isoprenoid substrate specificity *(2)*. Recently, two co-crystal structures of rat FTase with a bound peptide substrate and a nonreactive isoprenoid diphosphate analog have identified the location of both the peptide and isoprenoid binding sites in a ternary enzyme complex *(3,4)*. In this chapter we describe the recent crystal structures of rat FTase, and discuss their implications on understanding substrate specificity, mechanism, and inhibitor design.

1.1. Overall Structure

FTase is an obligatory heterodimer consisting of 48 kDa (α) and 46 kDa (β) subunits *(5–7)*. The α-subunit is also a component of the αβ heterodimeric enzyme, protein geranylgeranyltransferase type I (GGTase I), which adds a 20-carbon isoprenoid group *(8,9)*. The secondary structure of both the α- and β-subunits is largely composed of α helices (Fig. 1). Helices 2 to 15 of the α-subunit are folded into seven successive pairs that form a series of right-handed antiparallel coiled coils. These "helical-hairpins" are arranged

From: *Farnesyltransferase Inhibitors in Cancer Therapy*
Edited by: S. M. Sebti and A. D. Hamilton © Humana Press Inc., Totowa, NJ

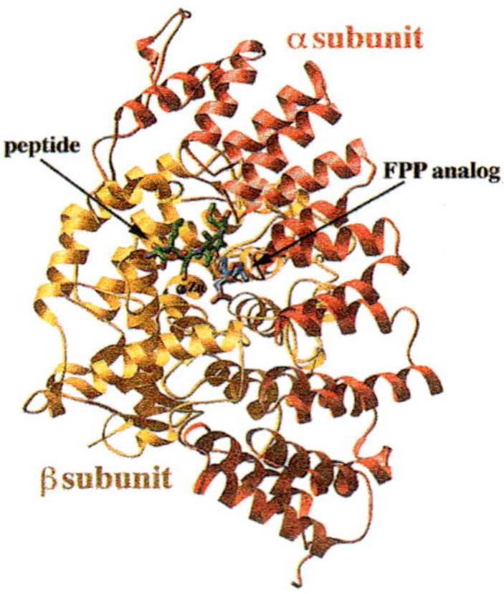

Fig. 1A

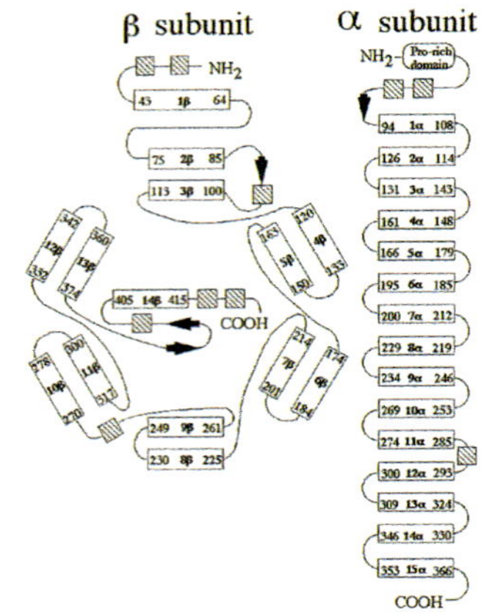

Fig. 1B

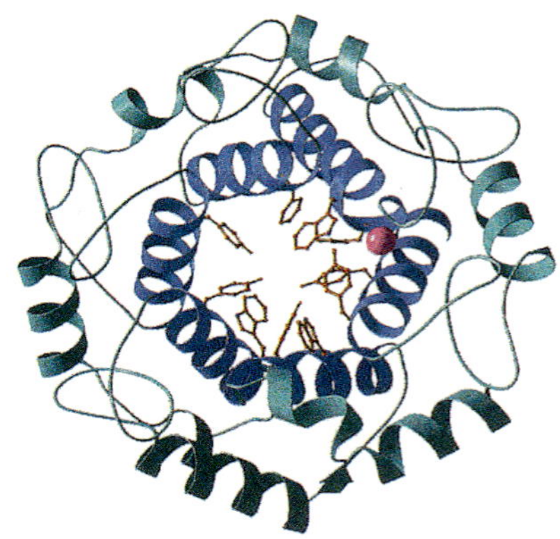

Fig. 2

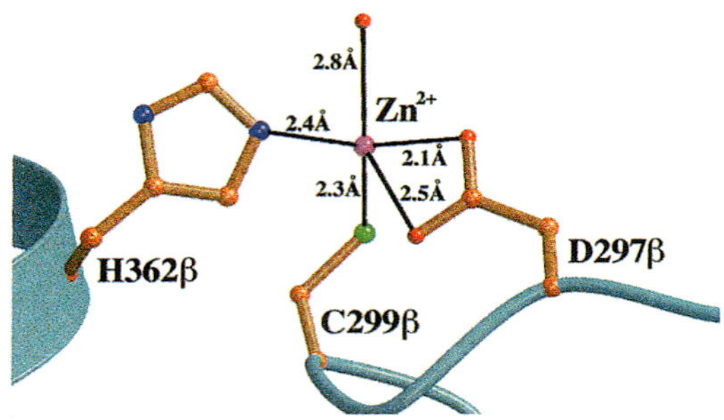

Fig. 3

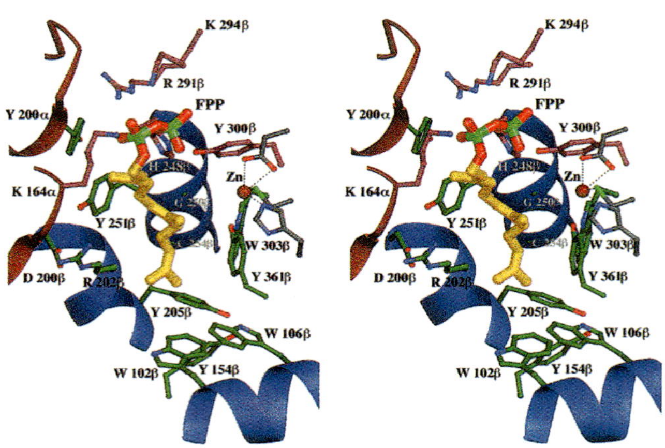

Fig. 4

Fig. 1. (**A**) Overview of the FTase structure with peptide substrate and FPP analog bound. Both the peptide and isoprenoid bind near the subunit interface in the deep hydrophobic cavity of the β-subunit. (**B**) Topology diagram of FTase. Open boxes represent α helices, the striped boxes 3_{10} helices, and the arrows β strands. The residue numbers for α helices are shown.

in a right-handed superhelix resulting in a crescent shaped subunit that envelops part of the β-subunit. This unusual structure has been seen previously in the crystal structures of lipovitellin-phosvitin and bacterial muramidase *(10,11)*. Twelve α helices of the β-subunit are folded into an α–α barrel (Fig. 2), similar to those found in bacterial cellulase, endoglucanase CelA, and glycoamylase *(12–15)*. Six parallel helices (3β, 5β, 7β, 9β, 11β, and 13β) form the core of the barrel. Six additional helices (2β, 4β, 6β, 8β, 10β, and 12β) form the outside of the barrel. These peripheral helices are parallel to each other but anti-parallel to the core helices. One end of the barrel is blocked off by residues 399β to 402β. The opposite end is open to the solvent, forming a deep funnel-shaped cavity in the center of the barrel. This cavity is hydrophobic in nature and is lined with conserved aromatic residues (Fig. 2). The enzymatic active sites of other α–α barrel proteins are located in such cavities.

The N-terminal proline-rich domain of the α-subunit (residues 1–54) is disordered in the crystal structure. Deletion of this domain does not affect the catalytic activity of FTase *(16)*. Taken together, these observations suggest that the proline-rich domain may interact with other factors in the cell, perhaps serving a role in enzyme localization. The crystal structure of a truncated form of rat FTase that lacks the proline-rich domain has been determined to 2.75Å resolution *(17)*. Not surprisingly, the deletion of this domain had no significant effect on the structure of the rest of the protein.

Multiple-sequence alignments of mammalian and yeast α-subunits reveal five tandem sequence repeats *(18)*. Each repeat consists of two highly conserved regions separated by a divergent region of fixed length. These motifs appear in the first five "helical hairpins" of the FTase structure (Fig. 1). The second α helix of each helical pair contains an invariant Trp residue which, together with other hydrophobic residues, forms the hydrophobic core of the hairpin. The conserved sequence motif Pro-X-Asn-Tyr (where X is any amino acid) *(18)* is found in the turns connecting two helices of the coiled-coil. These turns form part of the interface with the β-subunit. Internal repeats of glycine-rich sequences also have been identified in the β-subunits of other protein prenyltransferases *(18)*. These repeats correspond to the loop regions that connect the C-termini of the peripheral helices with the N-termini of the core helices in the barrel.

Fig. 2. The α–α barrel of the β-subunit, showing the aromatic resides (yellow) that line the interior creating a hydrophobic cavity. This view is a 90° clockwise rotation relative to Fig. 1A. The zinc ion is shown as a magenta sphere. Only helices 2β to 13β are shown.

Fig. 3. Zinc binding site. The zinc ion is coordinated by side-chains of Asp297β, Cys299β, His362β, and a well-ordered water molecule. Carbon is shown in coral; oxygen in red; nitrogen in blue; sulfur in green; zinc in magenta; polypeptide chain in cyan. H-bonds are indicated in black.

Fig. 4. Stereoview of the active site with bound FPP. Portions of the α- and β-subunits are drawn as ribbons in red and blue, respectively. The isoprenoid moiety of the FPP molecule binds to the hydrophobic cavity inside the α–α barrel of the β-subunit. Residues colored in green line this hydrophobic cavity. The diphosphate group of the FPP molecule binds in a positively charged cleft near the subunit interface formed by residues colored in pink, and is adjacent to the catalytic zinc ion whose ligands are colored in gray. Confirming the observed location of the FPP molecule, mutation of β-subunit residues His 248β, Arg 291β, Lys 294β, Tyr 300β, or Trp 303β results in an increase in $K_{d(FPP)}$ *(27)*. Lys 164α—which, when mutated to Asn dramatically reduces catalytic turnover *(16)*—is in close proximity to the C1 atom of the FPP molecule and may be directly involved in catalysis.

The FTase subunit interface is quite extensive, burying 3322 $Å^2$ or 19.5% of accessible surface area of the α-subunit and 3220 $Å^2$ or 17.2% of accessible surface area of the β-subunit *(1)*. Although the size of this subunit interface is typical for an oligomeric protein, there are nearly double the normal number of hydrogen bonds. The number of hydrogen bonds found in the FTase subunit interface may explain the unusual stability of the FTase heterodimer, which cannot be dissociated unless denatured *(6)*.

A striking feature of the FTase structure is a deep, funnel-shaped cleft formed by the central cavity of the α–α barrel. The cleft is lined with highly conserved hydrophobic residues (Fig. 2). A second cleft runs parallel to the rim of the α–α barrel and is hydrophilic in nature. These two clefts intersect at the site of a zinc ion that is required for catalysis.

1.2. Zinc Binding Site

FTase is a zinc metalloenzyme that contains one zinc atom per protein dimer *(19,20)*. Experimental evidence indicates that the zinc ion is required for catalytic activity and important for the binding of peptide, but not isoprenoid, substrates *(19)*. A direct involvement of zinc in catalysis is supported by several studies *(19,21–23)*. The most compelling is that the zinc ion coordinates the thiol of the CAAX cysteine residue in the ternary complex *(24)*. In the FTase crystal structure, there is a single zinc ion bound to the β-subunit, near the subunit interface, that marks the location of the active site. The zinc is coordinated by β-subunit residues D297β, C299β, H362β, and a well-ordered water molecule (Fig. 3). The cysteine thiol of the CAAX protein substrate coordinates the zinc ion, displacing this water molecule in a ternary enzyme complex *(3,4)*. D297β forms a bidentate ligand, resulting in a distorted penta-coordinate geometry. All three protein ligands are conserved in the β-subunits of protein prenyltransferase enzymes. C299β had previously been identified from mutational analysis and biochemical studies to affect zinc binding and abolish catalytic activity *(25)*.

Several site-directed mutagenesis studies confirm that D297β, C299β and H362β are ligands for the zinc ion *(26–28)*, and additionally suggest that D359β has a role in zinc binding *(27)*. In the crystal structure, D359β forms a hydrogen bond to H362β, possibly stabilizing a required conformation for binding zinc.

2. SUBSTRATE BINDING AND RECOGNITION

2.1. FPP Binding Site

In the X-ray co-crystal structure of rat FTase complexed with FPP, the isoprenoid moiety binds in an extended conformation along one side of the funnel-shaped hydrophobic cavity of the α–α barrel of the β-subunit *(2)*. This cavity is lined with conserved aromatic residues including W303β, Y251β, W102β, Y205β, and Y200α that make hydrophobic interactions with the isoprenoid. Strictly conserved R202β also forms a hydrophobic interaction with the isoprenoid and is stabilized by interactions with D200β and M193β. In addition, the conserved residues C254β and G250β contribute to the isoprenoid binding site (Fig. 4).

The diphosphate moiety binds in a positively charged cleft near the subunit interface at the rim of the α–α barrel and is adjacent to the catalytic zinc ion. It forms hydrogen bonds with the strictly conserved residues H248β, R291β, Y300β, K294β, and K164α and lies directly above helix 9β, whose N-terminal positive dipole contributes to the strong positive charge of this pocket (Fig. 4).

Site-directed mutagenesis of conserved residues that lie in the FTase active site support the observed location of FPP binding *(27)*. Of the 11 residues investigated, 5 increased $K_{d(FPP)}$ when mutated. Each of these residues (H248β, R291β, K294β, Y300β, and W303β) interacts with the FPP molecule in the cocrystal structure. None of the mutants investigated had an increased affinity for FPP.

A similar isoprenoid binding cavity is observed in the crystal structure of farnesyl diphosphate synthase (FPP synthase) *(29)*. This enzyme, a homodimer of 44 kDa subunits, catalyzes the synthesis of FPP using isopentenyl diphosphate and dimethylallyl diphosphate as initial substrates. This enzyme binds the diphosphate moiety of its substrate through two Mg^{2+} to conserved Asp residues *(30)*, in contrast to FTase, which binds the diphosphate moiety of FPP directly by interaction with positively charged residues. It is possible that at some stage in the reaction Mg^{2+} may interact with the diphosphate moiety because this metal ion is required for the full catalytic activity of FTase *(31)*.

2.2. Structural Basis for Isoprenoid Specificity

The co-crystal structure of FTase with bound FPP suggests how prenyltransferases discriminate between the related isoprenoid substrates, FPP and GGPP *(2)*. The depth of the hydrophobic cavity where FPP binds may function like a molecular ruler and be the primary determinant for the specificity of the 15-carbon FPP molecule over the 20-carbon GGPP molecule, the prenyl substrate of the closely related enzyme GGTase I. This hypothesis is supported by the observation that FPP and GGPP bind to FTase in a competitive manner but only FPP serves as an effective substrate *(19,21)*. The bottom of the hydrophobic cavity marks one end of the ruler, whereas the positive cleft at the top of the α–α barrel near the subunit interface marks the end where the diphosphate binds. The catalytic zinc ion is located the same distance from the bottom of the hydrophobic cavity as the C1 atom of the FPP molecule. A comparison of the observed FPP location and a model of GGPP bound to FTase made by superimposing the isoprenoid moiety of GGPP on that of FPP is shown in Fig. 5. While the C1 atom of the FPP molecule is in register with the catalytic zinc ion, the C1 atom of the GGPP molecule is out of register, preventing prenyl transfer. The depth of the isoprenoid binding cavity may be deeper in GGTase I, thereby allowing the longer GGPP molecule to bind in a catalytically competent manner. This deeper cavity would allow the strictly conserved residues that bind the diphosphate moiety in FTase to interact with the diphosphate moiety of GGPP in GGTase I.

FTase may select for the chain length of the isoprenoid in a similar mechanism as that used by FPP synthase *(30)*. A double mutation (F112A/F113S) was introduced into the hydrophobic cavity of FPP synthase thereby increasing the depth of the cavity by 5.8 Å, a distance that roughly corresponds to the difference in length between FPP and the longer GGPP *(30)*. The mutant FPP synthase was capable of producing isoprenoid products longer than FPP.

2.3. Peptide and Isoprenoid Binding Sites in a Ternary Complex

Two crystal structures of FTase with bound peptide substrates and nonreactive isoprenoid analogs have been determined. In one complex *(3)*, the peptide substrate Acetyl-Cys-Val-Ile-selenoMet-COOH was used in conjunction with the isoprenoid analog α-hydroxyfarnesylphosphonic acid (α-HFP) *(32)*. In another complex, determined to 2.1Å resolution *(4)*, the peptide substrate Thr-Lys-Cys-Val-Ile-Met-COOH was used

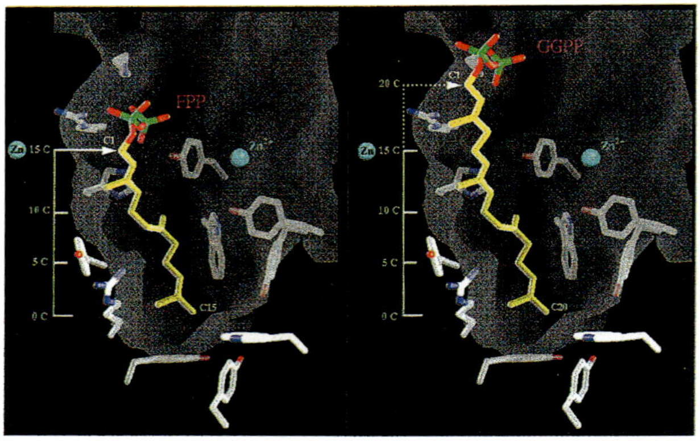

Fig. 5

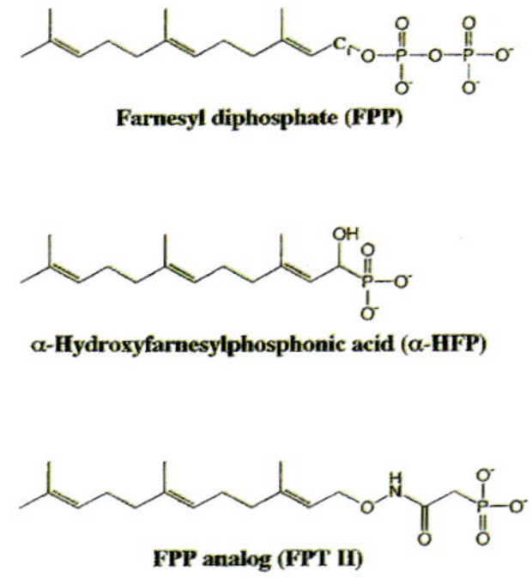

Fig. 6

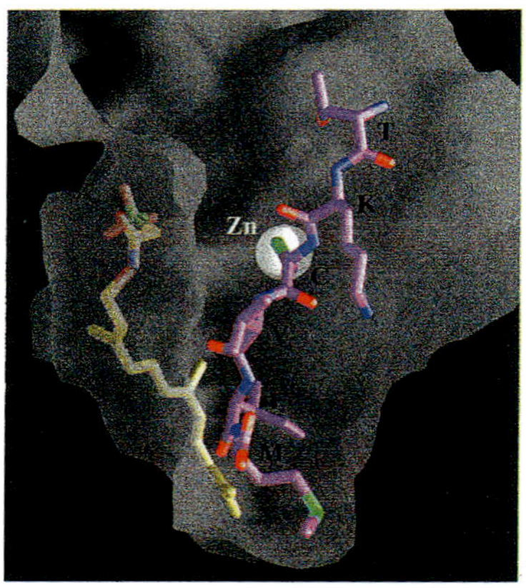

Fig. 7

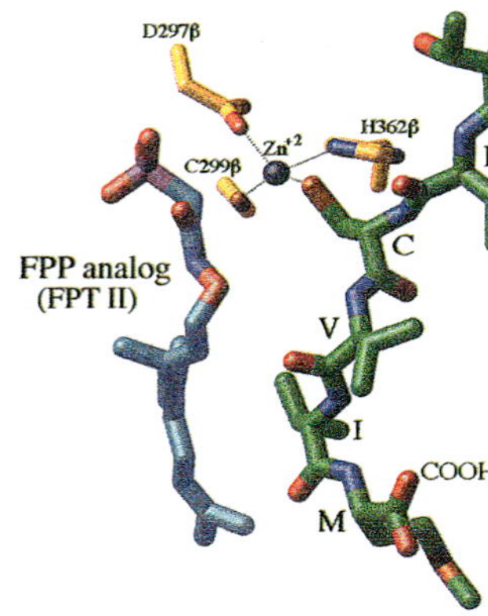

Fig. 8

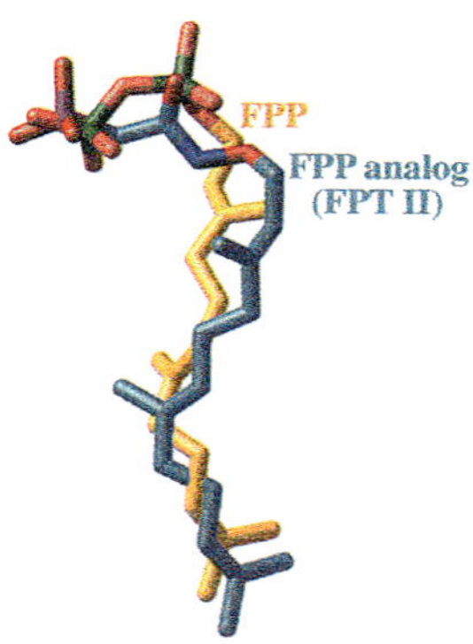

Fig. 9

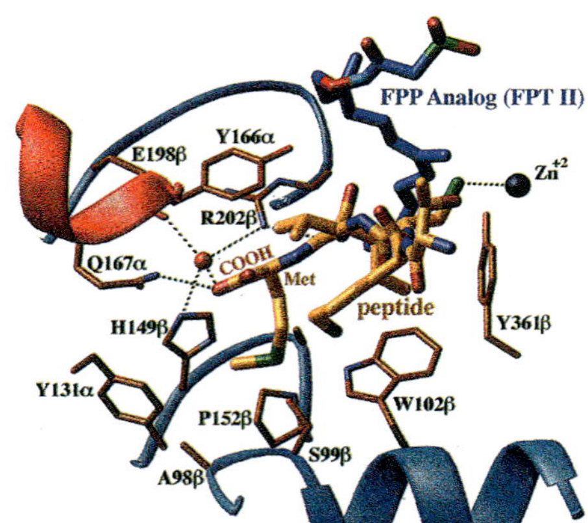

Fig. 10

Fig. 5. The molecular ruler hypothesis for isoprenoid substrate specificity. (**Left**) A portion of the solvent-accessible surface showing the observed FPP binding location in the binary complex. The catalytic zinc ion is in register with the C1 atom of the FPP molecule, the site of prenyl transfer. Trp 102β and Tyr 205β are at the bottom of the hydrophobic pocket. (**Right**) The 20-carbon long GGPP molecule, the prenyl substrate of GGTase I, is modeled into this surface based on the observed FPP location. The longer isoprenoid chain of GGPP places the C1 atom out of register with the catalytic zinc ion, accounting for the specificity of FTase for FPP over GGPP.

with the nonreactive FPP analog (E,E)-2-[2-Oxo-2-[[(3,7,11-trimethyl-2,6,10-dodecatrienyl)oxy]amino]ethyl]phosphonic acid (FPT II) *(33)*. Figure 6 shows the chemical structures of these FPP analogs. The peptides used in both studies were derived from the Ki-Ras4B protein sequence and represent the carboxyl terminal 4 or 6 residues, respectively. The two complexes are very similar. There are only minor changes in the overall structure of FTase in these ternary complexes from the crystal structure of unliganded FTase or the FTase-FPP binary complex. There are, however, a number of residues in the active-site that have different conformations from those observed in the unliganded FTase structure.

In the ternary complexes, the peptide binds in an extended conformation in the hydrophobic cavity of the β-subunit with the COOH terminus near the bottom of this cavity (Fig. 7). The Cys residue of the peptide directly coordinates the zinc ion, as predicted from spectroscopic studies *(24)*. The isoprenoid forms a large part of the peptide binding surface and makes a hydrophobic interaction with the Ile side chain of the peptide substrate, the A_2 position of the CA_1A_2X motif (Fig. 8). The side chain of the peptide Val residue (A_1) points into solvent and does not make a hydrophobic interaction with the isoprenoid. The isoprenoid is positioned between the peptide substrate and the wall of the hydrophobic cavity, consistent with the observation that the reaction is functionally ordered with isoprenoid binding preceding peptide binding *(34,35)*. The location and conformation of the isoprenoid are similar to those observed in the binary complex with FPP. Furthermore, the residues that interact with the FPP molecule (Fig. 4) were also observed to interact with the FPP analog (FPT II). Figure 9 shows a comparison of the FPP molecule and the FPP analog (FPT II) made by superimposing the Cα atoms of the binary and ternary complexes. The location of the terminal phosphate of the FPP analog matches that of the β phosphate of FPP well.

Fig. 6. Isoprenoid molecules used in the cocrystal structures.

Fig. 7. A portion of the solvent accessible surface of FTase showing the location of isoprenoid and peptide binding in the ternary complex. The isoprenoid binds along one wall of the hydrophobic cavity in the same location as the FPP molecule in the binary complex. The peptide binds in an extended conformation in the hydrophobic cavity adjacent to the isoprenoid with the COOH-terminus near the bottom of the cavity. The Cys thiol of the peptide coordinates the zinc ion. The isoprenoid forms a large part of the binding surface for the peptide and makes a hydrophobic interaction with the Ile residue of the peptide (A_2 of CA_1A_2X).

Fig. 8. The conformation of the peptide and FPP analog molecules in the ternary complex. The zinc ion coordinates the Cys residue of the peptide substrate and the β-subunit residues D297β, C299β, and H362β.

Fig. 9. Comparison of FPP in the binary complex and the FPP analog (FPT II) in the ternary complex generated by the superposition of the cα atoms of both structures.

Fig. 10. Residues interacting with the bound peptide in the ternary complex. Residues within 4Å of the peptide are shown. The peptide is within 4Å of three α-subunit residues: Y131α, Y166α, and Q167α. Only two direct hydrogen bonds are made from FTase to the peptide substrate: Q167α to the COOH terminus and R202β to the carbonyl oxygen of the Ile residue (a_2 of Ca_1a_2X). A well-ordered water molecule (shown as a red sphere) mediates the interaction of β-subunit residues H149β, E198β, and R202β with the COOH terminus. Portions of the α- and β-subunits are shown as red and blue ribbons, respectively.

Figure 10 shows the residues within 4Å of the peptide in the ternary complex. Three α-subunit residues are within 4Å of the peptide: Y131α, Y166α, and Q167α. There are only two direct FTase-peptide hydrogen bonds: from the side chain of N169α to the COOH terminus, and from R202β to the backbone carbonyl oxygen of the A_2 residue (Ile). All of the other hydrogen bonds made with the peptide are water-mediated. H149β, E198β, and R202β coordinate a well-ordered water molecule that makes a hydrogen bond with the COOH terminus. The sulfur atom of the Cys side chain is approx 7Å from the location of the C1 atom of the isoprenoid molecule. Because a covalent bond is formed between the C1 atom of FPP and the Cys thiol, movement of one or both of the substrates is required for catalysis.

3. INTERPRETATION OF MUTAGENESIS STUDIES

Site-directed mutagenesis studies of rat, human, and yeast FTase have begun to identify residues that affect substrate binding *(26,27,36–39)*, specificity *(40,41)*, and the catalytic activity of FTase *(16,26,27,36)*. Mapping of residues onto the rat FTase structure reveals that most of the mutated residues are near the site of FPP or peptide binding or stabilize residues in these regions.

3.1. Mutations Affecting Catalytic Turnover

Eleven conserved residues in the β-subunit of human FTase that lie in the active site region were mutated and their steady-state kinetic properties analyzed *(27)*. Six mutants, R202A, D297A, C299A, Y300F, D359A, and H362A (corresponding to the same residues in rat FTase), had lower k_{cat} values than the wild-type enzyme. Mutation of the corresponding residues in yeast FTase, with the exception of D359β, which was not studied, also reduced k_{cat} *(26)*.

In the crystal structures of FTase, D297β, C299β, and H362β ligate the catalytic zinc ion, and D359β stabilizes the conformation of H362β. R202β forms a hydrophobic interaction with the isoprenoid and hydrogen bonds with the peptide (Figs. 4 and 10). Y300β makes a hydrogen bond with the terminal phosphate of the FPP molecule (Fig. 3).

Five conserved residues in the α-subunit of rat FTase (K164N, Y166F, R172E, N199D, and W203H) have also been analyzed *(16)*. The K164N mutation abolished FTase activity. In the cocrystal structure of FTase with FPP, K164α lies directly above the C1 atom of the FPP molecule, the site of prenyl transfer, consistent with a direct role in catalysis *(2)*. The four other mutations reduced FTase activity from 30–75% of wild-type values. Y166α and R172α are on helix 5α. Y166α contributes to the surface that binds both substrates in the ternary complex *(4)*. Y200α also forms part of the isoprenoid binding surface, which may explain why mutation of the nearby residues N199α or W203α affected catalytic turnover.

3.2 Mutations Affecting FPP Binding Affinity

Kinetic and biochemical analyses of site-directed mutants in the β-subunit of human FTase suggest that H248, R291, K294, and W303 are involved in binding and utilization of the FPP substrate *(27)*. In addition, E256A in the β-subunit of yeast FTase (E246 in rat FTase) gave 130-fold higher K_M for the FPP substrate *(26)*. H248β, R291β, K294β, and W303β are observed to interact with the FPP molecule in the binary complex with FPP (Fig. 4). E246β stabilizes the side-chain position of R291β.

3.3. Mutations Affecting Peptide Selectivity

Random screening identified three mutants (S159N, Y362H, and Y366N) in the β-subunit of yeast FTase, with relaxed protein substrate specificity having gained the ability to farnesylate GGTase I protein substrates *(40)*. A similar effect can be observed by substitution of Y362β with a smaller side chain *(40)*. These observations suggest that these residues are in close proximity to the protein substrate binding site. In the ternary complex, Y362β (Y361β in rat) is within 4Å of the peptide substrate, and Y366β (Y365β in rat) is adjacent. Although not conserved, rat P152β (S159β in yeast) contributes to the surface that interacts with the C-terminal Met side chain of the peptide substrate in the ternary complex with isoprenoid (Fig 10). In another study on yeast FTase, residues conserved among FTase β-subunits were mutated to their conserved counterparts in GGTase I and the substrate specificities of the mutant yeast FTase enzymes were examined *(41)*. Three regions in the β-subunit of FTase were identified, which had an effect on CAAX peptide substrate selectivity: residues 74, 206–212, and 351–354 (corresponding to β-subunit residues 67, 197–203, and 350–353 in rat FTase, respectively). Residues G197β-S203β make up part of the loop connecting helices 6β and 7β and are located near the bottom of the hydrophobic cavity in the β-subunit. R202β on this loop interacts with both the isoprenoid and peptide substrates in the ternary complex, and its conformation is stabilized by hydrogen bonds made with D200β (Figs. 4 and 10). Residues L350β-K353β are on a loop connecting helices 12β and 13β and lie above the location of the catalytic zinc ion. L67β in the crystal structure is located on a loop connecting helices 1β and 2β and is distant from the active site.

Interestingly, these same mutations affected the isoprenoid selectivity of the mutant enzymes. Moreover, the isoprenoid selectivity was observed to be affected by the CAAX peptide substrate used. These observations of interdependent selectivity in yeast FTase are consistent with the observation that the substrates directly interact in a ternary complex *(3,4)*.

3.4. Mutations Affecting Protein Substrate K_M

Three mutations that increase the K_M of protein substrates (D209N, G259V, and G328S) were identified in the β-subunit of yeast FTase *(37,38)* and GGTase I *(39)*. The corresponding substitutions (D200N, G249V, and G349S) were introduced into human enzyme and the functional consequences were examined *(36)*. The D200N and G349S mutants resulted in an increase in the K_M of protein substrates without affecting the K_M of FPP substrates *(26,36)*, whereas the G249V mutation resulted in an increase in both. In a separate study, mutation of R202β in human FTase to Ala resulted in a >400-fold elevation in K_M for protein substrate *(27)*. Further analysis of this mutant using peptide derived inhibitors suggested that R202β interacts with the COOH terminus of CAAX peptide substrates *(27)*.

In the FTase-FPP structure, G249β is adjacent to H248β, which interacts with the diphosphate moiety of the FPP molecule. R202β makes a hydrophobic interaction with the isoprenoid molecule and coordinates the COOH terminus of the peptide substrate through a well-ordered water molecule. D200β stabilizes the conformation of R202β.

4. CONCLUDING REMARKS

A number of FPP analogs have been developed that are competitive inhibitors of FTase with respect to FPP *(32,33,42,43)*. The selectivity of these inhibitors for FTase over other cellular enzymes, including GGTase I, has stimulated their investigation as possible antitumor agents. The crystal structures of FTase with FPP and FPP analogs provide a structural framework from which to interpret the inhibitory potency of these inhibitor molecules. Not surprisingly, among the FPP-based inhibitor molecules, the most effective retained a hydrophobic farnesyl group and a negatively charged moiety mimicking the diphosphate *(43)*. For one of these potent and selective FTase inhibitors (designated compound 3, IC_{50} = 75 n*M*, *[43]*), a systematic structure-activity analysis was carried out that indicated that the most potent inhibitors contained a terminal phosphate group rather than other types of negatively charged groups. This is consistent with the interactions seen with the terminal phosphate in the FTase complexes, specifically with Tyr 300β that interacts with this phosphate. The length of the hydrophobic chain also had a dramatic effect on the activity in this study. Homoelongation of the farnesyl group by one carbon resulted in a decrease in IC_{50} of more than 200-fold, consistent with the FPP binding location and the molecular ruler hypothesis for prenyl substrate specificity.

In summary, the co-crystal structures of FTase with its substrates are consistent with mutagenesis data on this enzyme. However, understanding the detailed enzymatic mechanism and testing hypotheses for the molecular basis of substrate specificity will require further molecular, biochemical, and kinetic analyses in addition to further structural studies of FTase. The binding of FPP to FTase supports a hypothesis for prenyl substrate specificity whereby the length of the prenyl moiety is selected based on the depth of the hydrophobic pocket into which it binds. The binary and ternary complexes of FTase provide a framework to understand further FPP-based and peptidomimetic inhibitors of FTase, and may facilitate the rational design and optimization of cancer chemotherapeutic agents.

REFERENCES

1. Park H-W, Boduluri SR, Moomaw JF, Casey PJ, Beese LS. Crystal structure of protein farnesyltransferase at 2.25 Å resolution. *Science* 1997; 275:1800–1804.
2. Long S, Casey PJ, Beese LS. Co-crystal structure of mammalian protein farnesyltransferase with a farnesyl diphosphate substrate. *Biochemistry* 1998; 37:9612–9618.
3. Strickland CL, Windsor WT, Syto R, Wang L, Bond R, Wu Z, et al. Crystal structure of farnesyl protein transferase complexed with a CaaX peptide and farnesyl diphosphate analogue. *Biochemistry* 1998; 37: 16,601–16,611.
4. Long SB, Casey PJ, Beese LS. The basis for K-Ras4B binding specificity to protein farnesyltransferase revealed by 2 Å resolution ternary complex structures. *Structure Fold Des* 2000; 8:209–222.
5. Reiss Y, Goldstein JL, Seabra MC, Casey PJ, Brown MS. Inhibition of purified p21ras farnesyl:protein transferase by Cys-AAX tetrapeptides. *Cell* 1990; 62:81–88.
6. Chen W-J, Andres DA, Goldstein JL, Russell DW, Brown MS. cDNA cloning and expression of the peptide binding beta subunit of rat p21ras farnesyltransferase, the counterpart of yeast RAM1/DPR1. *Cell* 1991; 66:327–334.
7. Chen W-J, Andres DA, Goldstein JL, Brown MS. Cloning and expression of a cDNA encoding the alpha subunit of rat p21ras protein farnesyltransferase. *Proc Natl Acad Sci USA* 1991; 88:11,368–11,372.
8. Seabra MC, Reiss Y, Casey PJ, Brown MS, Goldstein JL. Protein farnesyltransferase and geranylgeranyltransferase share a common alpha subunit. *Cell* 1991; 65:429–434.
9. Zhang FL, Diehl RE, Kohl NE, Gibbs JB, Giros B, Casey PJ, Omer CA. cDNA cloning and expression of rat and human protein geranylgeranyltransferase Type-I. *J Biol Chem* 1994; 269:3175–3180.
10. Thunnissen AM, Dijkstra AJ, Kalk KH, Rozeboom HJ, Engel H, Keck W, Dijkstra BW. Doughnut-shaped structure of a bacterial muramidase revealed by X-ray crystallography. *Nature* 1994; 367:750–753.
11. Raag R, Appelt K, Xuong NH, Banaszak L. Structure of the lamprey yolk lipid-protein complex lipovitellin-phosvitin at 2.8 Å resolution. *J Mol Biol* 1988; 200:553–569.

12. Alzari PM, Souchon H, Dominguez R. The crystal structure of endoglucanase CelA, a family 8 glycosyl hydrolase from Clostridium thermocellum. *Structure* 1996; 4:265–275.
13. Aleshin A, Golubev A, Firsov LM, Honzatko RB. Crystal structure of glucoamylase from *Aspergillus awamori* var. X100 to 2.2-Å resolution. *J Biol Chem* 1992; 267:19,291–19,298.
14. Aleshin AE, Hoffman C, Firsov LM, Honzatko RB. Refined crystal structures of glucoamylase from *Aspergillus awamori* var. *X100*. *J Mol Biol* 1994; 238:575–591.
15. Juy M, Amit AG, Alzari PM, Poljak RJ, Claeyssens M, Beguin P, Aubert JP. Three-dimensional structure of a thermostable bacterial cellulase. *Nature* 1992; 357:89–91.
16. Andres DA, Goldstein JL, Ho YK, Brown MS. Mutational analysis of alpha-subunit of protein farnesyltransferase. *J Biol Chem* 1993; 268:1383–1390.
17. Dunten P, Kammlott U, Crowther R, Weber D, Palermo R, Birktoft J. Protein farnesyltransferase: structure and implications for substrate binding. *Biochemistry* 1998; 37:7907–7912.
18. Boguski M, Murray AW, Powers S. Novel repetitive sequence motifs in the alpha and beta subunits of prenyl-protein transferases and homology of the alpha subunit to the MAD2 gene product of yeast. *New Biol* 1992; 4:408–411.
19. Reiss Y, Brown MS, Goldstein JL. Divalent cation and prenyl pyrophosphate specificities of the protein farnesyltransferase from rat brain, a zinc metalloenzyme. *J Biol Chem* 1992; 267:6403–6408.
20. Chen W-J, Moomaw JF, Overton L, Kost TA, Casey PJ. High-level expression of mammalian protein farnesyltransferase in a baculovirus system: the purified protein contains zinc. *J Biol Chem* 1993; 268: 9675–9680.
21. Casey PJ, Seabra MC. Protein prenyltransferases. *J Biol Chem* 1996; 271:5289–5292.
22. Moomaw JF, Casey PJ. Mammalian protein geranylgeranyltransferase: subunit composition and metal requirements. *J Biol Chem* 1992; 267:17,438–17,443.
23. Zhang FL, Moomaw JF, Casey PJ. Properties and kinetic mechanism of recombinant mammalian protein geranylgeranyltransferase type I. *J Biol Chem* 1994; 269:23,465–23,470.
24. Huang C-C, Casey PJ, Fierke CA. Evidence for a catalytic role of zinc in protein farnesyltransferase: spectroscopy of Co^{2+}-FTase indicates metal coordination of the substrate thiolate. *J Biol Chem* 1997; 272:20–23.
25. Fu H-W, Moomaw JF, Moomaw CR, Casey PJ. Identification of a cysteine residue essential for activity of protein farnesyltransferase: Cys299 is exposed only upon removal of zinc from the enzyme. *J Biol Chem* 1996; 271:28,541–28,548.
26. Dolence JM, Rozema DB, Poulter CD. Yeast protein farnesyltransferase. Site-directed mutagenesis of conserved residues in the beta-subunit. *Biochemistry* 1997; 36:9246–9252.
27. Kral AM, Diehl RE, deSolms SJ, Williams TM, Kohl NE, Omer CA. Mutational analysis of conserved residues of the beta-subunit of human farnesyl:protein transferase. *J Biol Chem* 1997; 272:27,319–27,323.
28. Fu H-W, Beese LS, Casey PJ. Kinetic analysis of zinc ligand mutants of mammalian protein farnesyltransferase. *Biochemistry* 1998; 37:4465–4472.
29. Tarshis LC, Yan M, Poulter CD, Sacchettini JC. Crystal structure of recombinant farnesyl diphosphate synthase at 2.6 Å resolution. *Biochemistry* 1994; 33:10,871–10,877.
30. Tarshis LC, Proteau PJ, Kellogg BA, Sacchettini JC, Poulter CD. Regulation of product chain length by isoprenyl diphosphate synthases. *Proc Natl Acad Sci USA* 1996; 93:15,018–15,023.
31. Zhang FL, Casey PJ. Protein prenylation: molecular mechanisms and functional consequences. *Annu Rev Biochem* 1996; 65:241–269.
32. Manne V, Ricca CS, Brown JG, Tuomari AV, Yan N, Patel D, et al. Ras farnesylation as a target for novel antitumor agents: poteint and selective farnesyl diphosphate analog inhibitors of farnesyltransferase. *Drug Dev Res* 1995; 34:121–137.
33. Gibbs JB, Pompliano DL, Mosser SD, Rands E, Lingham RB, Singh SB, et al. Selective inhibition of farnesyl-protein transferase blocks ras processing *in vivo*. *J Biol Chem* 1993; 268:7617–7620.
34. Pompliano DL, Schaber MD, Mosser SD, Omer CA, Shafer JA, Gibbs JB. Isoprenoid diphosphate utilization by recombinant human farnesyl:protein transferase: interactive binding between substrates and a preferred kinetic pathway. *Biochemistry* 1993; 32:8341–8347.
35. Furfine ES, Leban JJ, Landavazo A, Moomaw JF, Casey PJ. Protein farnesyltransferase: kinetics of farnesyl pyrophosphate binding and product release. *Biochemistry* 1995; 34:6857–6862.
36. Omer CA, Kral AM, Diehl RE, Prendergast GC, Powers S, Allen CM, et al. Characterization of recombinant human farnesyl-protein transferse: cloning, expression, farnesyl diphosphate binding, and functional homology with yeast prenyl-protein transferases. *Biochemistry* 1993; 32:5167–5176.

37. Goodman LE, Judd SR, Farnsworth CC, Powers S, Gelb MH, Glomset JA, et al. Mutants of *Saccharomyces cerevisiae* defective in the farnesylation of Ras proteins. *Proc Natl Acad Sci USA* 1990; 87: 9665–9669.
38. Powers S, Michaelis S, Broek D, Santa Anna S, Field J, Herskowitz I, Wigler M. RAM, a gene of yeast required for a functional modification of RAS proteins and for production of mating pheromone a-factor. *Cell* 1986; 47:413–422.
39. Ohya Y, Goebl M, Goodman LE, Petersen-Bjorn S, Friesen JD, Tamanoi F, Anraku Y. Yeast CAL1 is a structural and functional homologue to the DPR1 (RAM) gene involved in ras processing. *J Biol Chem* 1991; 266:12,356–12,360.
40. Villar KD, Mitsuzawa H, Yang W, Sattler I, Tamanoi F. Amino acid substitutions that convert the protein substrate specificity of farnesyltransferase to that of geranylgeranyltransferase type I. *J Biol Chem* 1997; 272:680–687.
41. Caplin BE, Ohya Y, Marshall MS. Amino acid residues that define both isopreniod and CAAX preferences of the *Saccharomyces cerevisiae* protein farnesyltransferase. *J Biol Chem* 1998; 273:9472–9479.
42. Cohen LH, Valentijn ARPM, Roodenburg L, Van Leeuwen REW, Huisman RH, Lutz RJ, et al. Different analogues of farnesyl pyrophosphate inhibit squalene synthase and protein:farnesyltransferase to different extents. *Biochem Pharmacol* 1995; 49:839–845.
43. Patel DV, Schmidt RJ, Biller SA, Gordon EM, Robinson SS, Manne V. Farnesyl diphosphate-based inhibitors of ras farnesyl protein transferase. *J Med Chem* 1995; 38:2906–2921.

4 Peptidomimetic-Based Inhibitors of Farnesyltransferase

David Knowles, PHD, *Jiazhi Sun,* PHD, *Saul Rosenberg,* PHD, *Saïd M. Sebti,* PHD, *and Andrew D. Hamilton,* PHD

CONTENTS

1. INTRODUCTION

Human H-, K-, and N-ras genes encode four structurally related proteins with 188 or 189 amino acids and a molecular weight of 21 kDa *(1,2)*. Analysis of oncogenes in human tumors showed that mutated Ras existed in approx 30% of all human tumors, particularly in over 90% of human pancreatic carcinomas and 50% of human colon cancers *(1)*. The frequency of mutation is dependent on tumor type, with breast, ovary, and stomach carcinomas showing the lowest frequency Ras mutations *(3)*. Ras proteins play a crucial role as a molecular switch transducing signals from receptor tyrosine kinases to the cell nucleus *(4)*. Normal Ras exists in an equilibrium between inactive GDP- and active GTP-bound forms, with a strong preference for Ras-GDP *(5)*. When mutated at positions 12, 13, and 61, Ras proteins lose their ability to hydrolyze GTP to GDP and so deactivate the switch *(6)*. Consequently, the mutated Ras is locked in the GTP-bound form, causing uncontrolled proliferation.

Because of the role of mutated Ras in human cancers, there has been intense activity over the past decade to block or reverse its uncontrolled signaling function. The potentially most viable approach has been based on the recognition that Ras proteins must associate with the plasma membrane in order to carry out their transforming activity *(7)*. In its cytosolic state, Ras has little affinity for the membrane, however, a series of posttranslational modifications increase its hydrophobicity, particularly at the carboxyl terminus.

From: *Farnesyltransferase Inhibitors in Cancer Therapy*
Edited by: S. M. Sebti and A. D. Hamilton © Humana Press Inc., Totowa, NJ

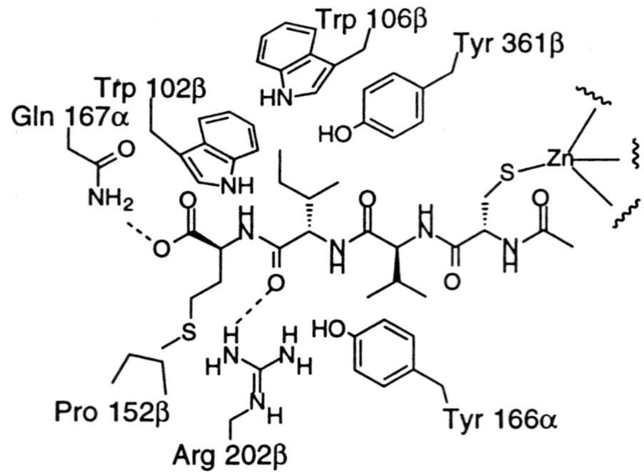

Fig. 1. Active site contacts to the CAAX substrate for FTase.

The first and obligatory step in these modifications is the attachment of a farnesyl group to the cysteine in the CAAX carboxyl terminal sequence, where C is cysteine, A is an aliphatic residue, and X is usually methionine or serine *(8)*. This reaction is catalyzed by the enzyme farnesyltransferase (FTase), a member of the family of isoprenyltransferases. Later modification steps involve peptidase cleavage of the AAX tripeptide and methylation of the resulting carboxylic acid by a methyltransferase *(9)*. The processed Ras protein translocates to the plasma membrane where it is needed for recruiting Ras effectors, such as Raf-1, which itself requires membrane association for activation. Inhibiting the farnesylation of Ras offers an attractive approach to preventing it from associating with the plasma membrane and so blocking its oncogenic signaling function. In this chapter, we describe the progress made in our groups toward the design of inhibitors of FTase and their possible applications as anticancer agents.

2. PROTEIN FARNESYLTRANSFERASE

The three isoprenyltransferases were first identified when it was discovered that tritiated mevalonic acid was incorporated into cellular polypeptides *(10)*. Protein farnesyltransferase (FTase) is a heterodimer containing α- and β-subunits with molecular weights of ~48,000 and 46,000, respectively *(11)*, which requires both Zn^{2+} and Mg^{2+} for enzyme activity. The crystal structure of mammalian FTase was determined at 2.25 angstrom Å resolution *(12)* and showed a single zinc ion at the intersection of the α- and β-subunits. A later structure of the complex between FTase, acetyl-Cys-Val-Ile-selenoMet-COOH, and α-hydroxyfarnesylphosphonic acid *(13)* showed the FPP mimic to be bound at the active site close to the CAAX peptide, which takes up an extended confirmation with the cysteine sulfur coordinating to zinc at the active site (Fig. 1). In a complex of FTase and FPP, the isoprenoid bound in an extended confirmation in a hydrophobic pocket within the β-subunit *(14)*. Although both FPP and GGPP bind to FTase with high affinity, the size of the hydrophobic pocket will only accommodate FPP for catalysis. As a consequence, very little cross-geranylgeranylation with FTase is observed.

Steady-state kinetic studies on FTase indicate that either FPP or peptide substrate may bind first, however the preferred pathway is through the FTase-FPP binary complex and is considered an ordered sequential mechanism *(15)*. The rate-limiting step is product dissociation, however, this occurs only in the presence of additional substrate *(16)*. The 3-(trifluoromethyl) analog of FPP is transferred 1000-fold more slowly than FPP itself, suggesting significant development of positive charge in the transition state *(17)*. However, analysis of the stereochemistry of the reaction using chiral deuterium-labeled derivatives showed inversion of configuration, indicating a more associative transition state *(18)*.

An important early observation was that the FTase can recognize short peptide sequences consisting of a CA_1A_2X, where C is a cysteine, A_1 and A_2 are aliphatic amino acids, and X could be any amino acid other than leucine or isoleucine *(19)*. A systematic study of the sequence requirements of the tetrapeptide inhibitors *(20)* showed that a cysteine residue was critical at the N-terminus; that highest inhibitory activity was found when A_1 and A_2 were nonpolar aliphatic or aromatic amino acids; and that X showed strong bias towards methionine or serine.

A second isoprenyltransferase, protein geranylgeranyltransferase-I (GGTase I) is responsible for the geranylgeranyl modification of proteins. GGTase I, like FTase, is a zinc metalloenzyme consisting of two subunits, the 48 kDa α-subunit, which is identical with FTase, and a 43 kDa β-subunit *(21)*. GGTase I recognizes proteins containing the C-terminal sequence CAAX with a marked preference for leucine at the terminal position *(22)*. Protein substrates for GGTase I include several heterotrimeric G-proteins, Rap1A and Rap1B. The third prenyltransferase is the type II-geranylgeranyltransferase (GGTase II), which recognizes proteins terminating in the Cys-Cys or Cys-X-Cys sequence and is particularly involved in the prenylation of the Rab proteins.

3. PEPTIDOMIMETIC INHIBITORS OF FTASE

Because CAAX tetrapeptides are farnesylated by FTase as efficiently as the corresponding full-length protein and also are potent (10–200 n*M*) competitive inhibitors of FTase *(22)*, several groups have targeted the CAAX tetrapeptide as an anticancer drug development strategy. The major efforts have involved improving stability of the tetrapeptides toward proteolytic degradation and increasing their cellular uptake *(23)*. Our strategy focused on the CA_1A_2X motif from K-Ras4B CVIM. We reasoned that the central amide bond—surrounded as it always is by aliphatic residues—would be unlikely to participate in a hydrogen bond in the enzyme-inhibitor complex. It was therefore possible that the two central aliphatic residues, A_1A_2, might be replaced by a hydrophobic spacer providing nonpeptidic and correctly spaced links between the cysteine and methionine groups in the CVIM tetrapeptide. We investigated a number of flexible and semi-rigid, aliphatic and aromatic spacers and found that the dipeptide mimetic 3-(aminomethyl) benzoic acid (3-AMBA) provided an important lead *(24)*. Significantly, incorporation of 3-AMBA in the peptide backbone reduces the number of amide bonds from four to two, neither of which are peptidic in nature. The ability of these compounds to inhibit FTase or GGTase I in vitro was assessed by measuring the incorporation of $[^3H]$-FPP into H-Ras protein using recombinant H-Ras-CVLS or $[^3H]$-GGPP using H-Ras-CVLL. These tests were performed in the presence of different concentrations of inhibitors using partially purified FTase and GGTase I from human Burkitt lymphoma (Daudi) cells and are reported as IC_{50} values. Cys-3-AMBA-Met (FTI-205) inhibited FTase with an IC_{50} of 100 n*M* *(24)*.

Because these compounds were unable to inhibit Ras farnesylation in whole cells, further modifications were made to the CA_1A_2X backbone by replacing the two central aliphatic amino acids with dipeptide mimetics 3- or 4-aminobenzoic acid and 4-aminomethylbenzoic acid (4-AMBA). By varying the distance between the C-terminal cysteine and N-terminal methionine, the conformational and structural requirements of these inhibitors were evaluated *(25)*. Peptidomimetics containing either a Cys-4-ABA-Met (IC_{50} = 50 n*M*) or Cys-3-AMBA-Met (FTI-205) maintained a similar distance between the sulfhydryl of the cysteine and the methionine carboxylate to that found for the native peptide CVIM. The regioisomer Cys-4-AMBA-Met did not maintain this distance requirement and was 17-fold less potent (IC_{50} = 2550 n*M*). The peptidomimetic Cys-3-ABA-Met (IC_{50} = 6400 n*M*) exhibited reduced activity (130-fold) relative to Cys-4-ABA-Met, suggesting a strict structural requirement. Additional derivatization and reduction of the amide bond led to Cys-4ABA-Met (FTI-249) which had an IC_{50} of 50 n*M* (Fig. 2).

Molecular modeling of inhibitor FTI-249 suggested that the optimal distance between the cysteine and methionine residues was similar to that in the extended confirmation of CVIM (10.8 Å). These results strongly suggested that FTase inhibitors (FTIs) (and indeed the peptide substrate itself) must be in an extended conformation for binding to the FTase *(26)*. An earlier study using transfer NOE measurements on a heptapeptide bound to FTase had suggested that the heptapeptide adopts a type I β-turn *(27)*. The thinking that a turn structure may be important in the active site conformation of the bound CAAX region was given further support by NOE studies on a flexible peptidomimetic inhibitor *(28)* and by the first crystal structure of FTase alone, which showed the C-terminal region of an adjacent molecule bound into the active site in a turn structure. Significantly, all of these experiments were carried out in the absence of FPP or an FPP-analog and therefore deviated from the natural setting where the CAAX terminus binds to the initially formed FTase-FPP complex. The recent X-ray structure (Fig. 1) of a ternary complex (FTase-FPP analog-CCAX analog) confirmed our suggestion for a long distance between the thiol and carboxylate groups in the CAAX peptide and showed the FTase-bound tetrapeptide to be in an extended conformation *(13)*. These results clearly suggest that in the NOE and early X-ray experiments the peptides (or inhibitors) were binding in biochemically irrelevant turn conformations within the hydrophobic pocket destined for the FPP and not the peptide substrate.

The simple inhibitor FTI-249 lacks any bulky side groups corresponding to the hydrophobic side chains (VI) of the CVIM tetrapeptide. To exploit potential hydrophobic contacts in the active site, we incorporated a phenyl group at the 2-position of the 4-aminobenzoic acid moiety to give FTI-276 (Fig. 2). This led to a dramatic improvement in potency (IC_{50} = 500 p*M*) with FTI-276, showing a 400-fold increase over FTI-249 and a 100-fold selectivity against human GGTase I (IC_{50} = 50 n*M*). The methyl ester prodrug FTI-277 potently inhibited H-Ras processing (IC_{50} = 100 n*M*) in whole cells. FTI-276 was found to inhibit the tumor growth in nude mice of H-ras transformed cells and inhibited the growth of human tumors using a human lung carcinoma expressing a K-Ras mutation and lacking the tumor suppresser gene p53 *(29)*. FTI-277 treatment of H-RasF but not H-RasGG transformed cells blocked the MAPK kinase pathway by inducing nonmembrane-bound Ras to bind with Raf in the cytoplasm where Raf becomes inactive *(30)*.

A study from Rhone-Poulenc *(31)* supported our hypothesis that peptides and peptidomimetic inhibitors of FTase take an extended confirmation rather than a β-turn. In particular, a strong correlation was found between the inhibitory activity of a flexible peptidomimetic and its propensity to adopt an extended conformation (estimated from molecular modeling) *(31)*.

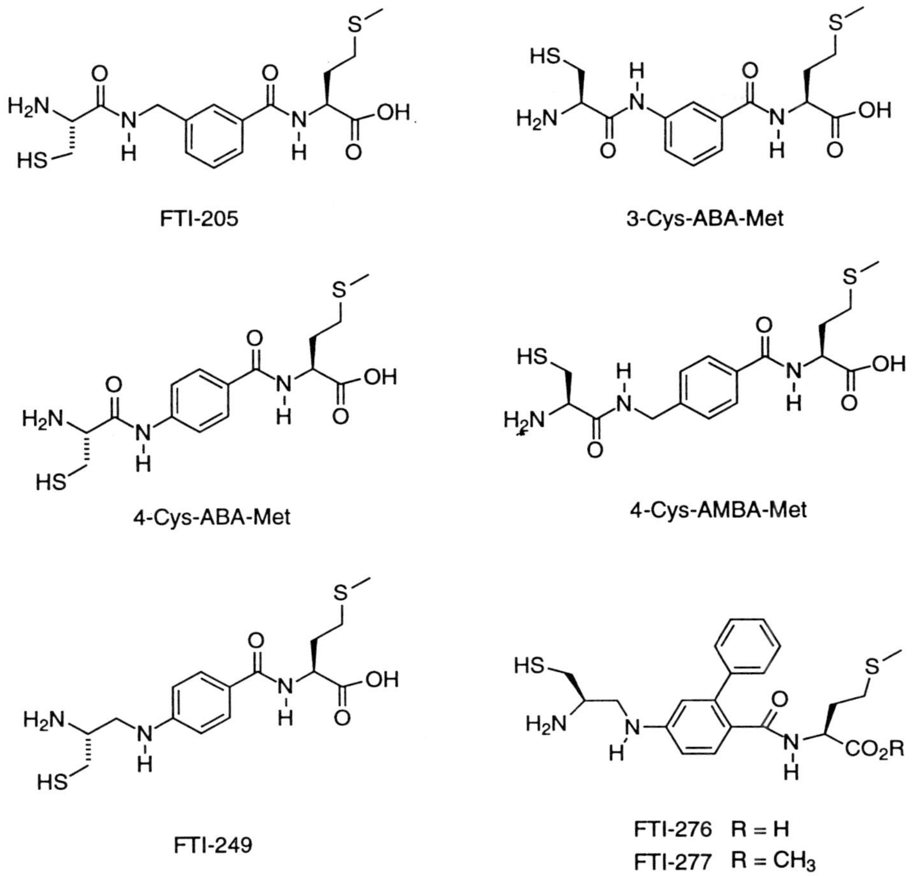

IC$_{50}$ (nM)	
Inhibitor	PFTase
4-Cys-AMBA-Met	2550
3-Cys-ABA-Met	6400
4-Cys-ABA-Met	50
FTI-205	100
FTI-249	50
FTI-276	0.50

Fig. 2. Structures and IC$_{50}$ values of FTase inhibitors.

Further investigation led to a 1,5-naphthyl scaffold that oriented the cysteine and methionine in an extended conformation. Compound **1** was a potent inhibitor (IC$_{50}$ = 48 n*M*) of human FTase and reducing the amide bond (in **2**) resulted in a 10-fold increase in potency (IC$_{50}$ = 5.6 n*M*) *(32)*. Additional modeling suggested the isomeric 1,6-naphthyl scaffold

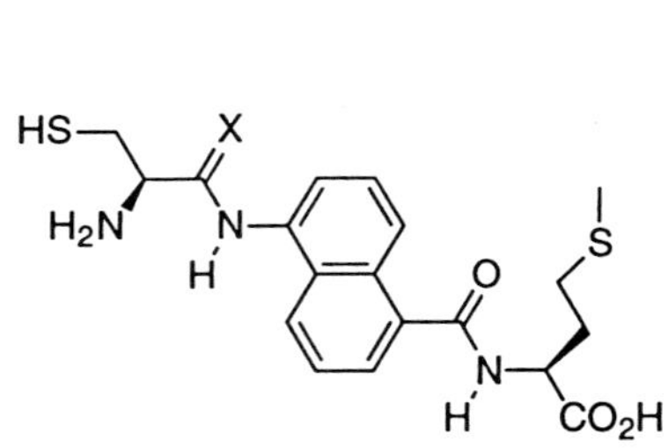

24 X = O

25 X = H, H

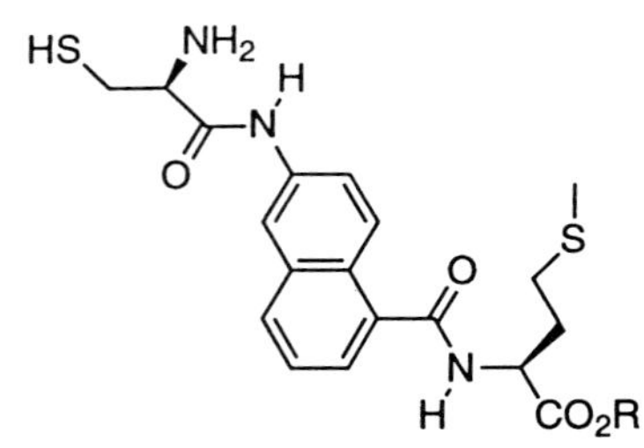

RPR 113829 R = H

RPR 114334 R = OCH_3

Inhibitor	IC_{50} (nM) PFTase
24	48
25	5.6
RPR 113829	1.8
RPR 114334	80

Fig. 3. Structures and IC_{50} values of peptidomimetics.

RPR 113829 (IC_{50} = 1.8 n*M*) and its methyl ester prodrug RPR 114334 (IC_{50} = 80 nM) as potent inhibitors of FTase (Fig. 3). In addition RPR 114334 was found to have activity against several cell lines including a Ki-ras transformed cell line (IC_{50} = 17.5 n*M*). This was an important result because activated Ki-Ras tumors are more prevalent in human cancers than Ha-Ras tumors. RPR 114334 also inhibited the anchorage-independent growth of c-src-transformed NIH 3T3 cells.

Until this time, very little work had been done with GGTase I inhibitors (GGTIs). However, our success with the peptidomimetic approach to FTIs suggested that a similar strategy might target GGTIs based on a CAAL tetrapeptide motif corresponding to the C-terminus of many geranylgeranylated proteins. GGTI-279 contains a C-terminus leucine linked to a reduced cysteine by a 4-aminobenzoate spacer and inhibited GGTase I with an IC_{50} of 100 n*M* (Fig. 4). Incorporation of a hydrophobic 2-phenyl group onto the spacer led to GGTI-287, a potent inhibitor of GGTase I (IC_{50} = 5 n*M*) with good selectivity against FTase (IC_{50} = 25 n*M*) *(33)*. The methyl ester prodrug GGTI-286 inhibited the geranylgeranylation of Rap1A (IC_{50} = 2 μ*M*) with good selectivity over H-Ras processing (IC_{50} = 30 μ*M*). Further modifications to the hydrophobic spacer in GGTI-286 by incorporation of a 2-naphthyl group for the 2-phenyl group led to GGTI-297 and GGTI-298, its methyl ester. GGTI-297 was less potent towards inhibiting GGTase-I (IC_{50} = 50 n*M*) and FTase (IC_{50} = 250 n*M*), however, GGTI-298 completely blocked Rap1A processing in whole cells without affecting H-Ras processing *(34)*.

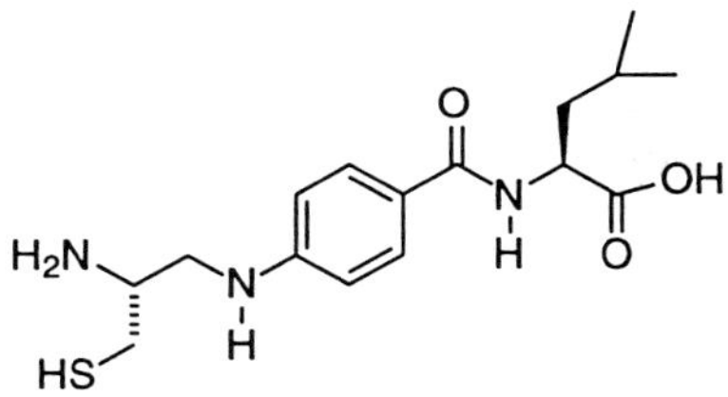

GGTI-279

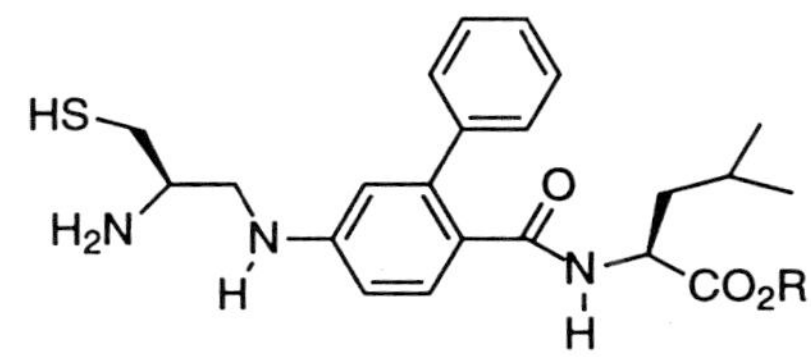

GGTI-287 R = H

GGTI-286 R = CH_3

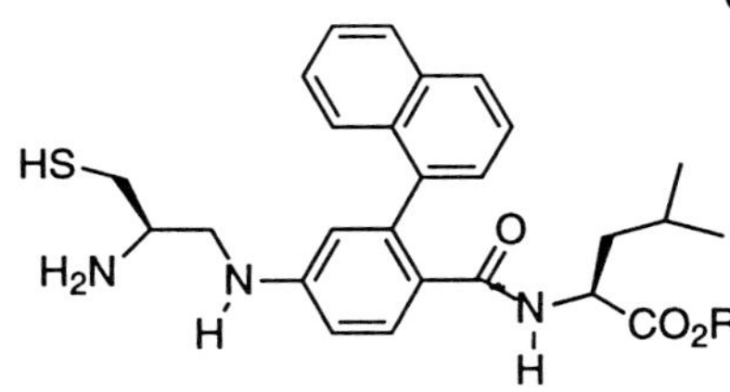

GGTI-297 R = H

GGTI-298 R = CH_3

	IC_{50} (nM)	
Inhibitor	PFTase	PGGTase-I
GGTI-279	460	100
GGTI-287	25	5
GGTI-297	250	50

Fig. 4. Structures and IC_{50} values for GGTase I inhibitors.

3.1. Non-Thiol-Containing Peptidomimetics of CA_1A_2X

A serious problem in these early inhibitors is the presence of a thiol group that is rapidly oxidized to the corresponding disulfide derivative. Although cysteine appeared to be an important structural feature of FTIs, we began to explore other nonsulfhydryl-based inhibitors to determine if replacement of the cysteine could provide potent inhibitors of FTase. A key intermediate for screening different N-terminal groups was the common 2-phenyl-4-aminobenzoylmethionine derivative **1** whose synthesis is shown in Scheme 1 *(35)*.

A series of derivatives with different potential zinc binding groups **(2–5)** were prepared from the corresponding carboxylic acids using EDCI as the coupling reagent (Table 1). Phenol **2** and pyrazine **3** were only modest inhibitors of FTase ($IC_{50} > 1\ \mu M$), presumably because of weak coordination to the zinc ion. Improvements in inhibition potency were observed with benzimidazole **4** and imidazole **5**, which have pKa values (5.68 and 7.0, respectively) close to that of thiol of cysteine. Imidazole **5** showed significant potency particularly when compared not to FTI-276 but the corresponding cysteine amide derivative (**6**). The only fivefold difference in FTase inhibition suggested that both were functioning in a similar manner by binding to the active site zinc. An important breakthrough however was the much higher selectivity of **5** for FTase over GGTase I (1000-fold) compared to only 60-fold for **6**.

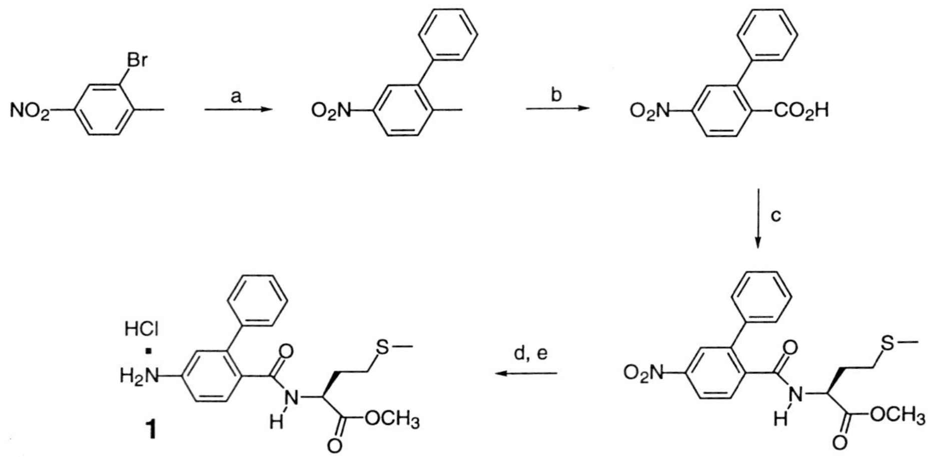

Scheme 1. Synthesis of a key intermediate for FTase inhibitors. Reagents: (**A**) phenylboronic acid, $Pd(OAc)_2$, aqueous acetone; (**B**) $KMnO_4$, aqueous pyridine; (**C**) (L)-methionine methyl ester, EDCI, HOBT; (**D**) $SnCl_2$ hydrate; (**E**) 3 *N* HCl in diethyl ether.

Table 1
Non-Thiol Replacements of Cysteine

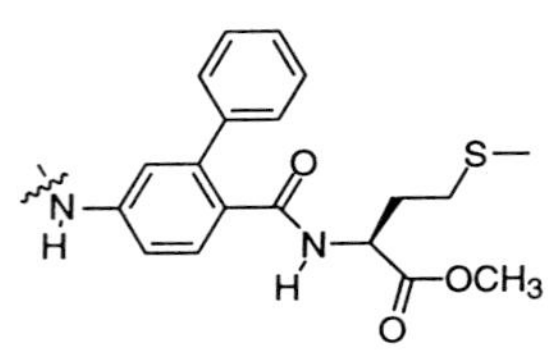

Compound	Thio replacement	IC50 (nM) PFTase	PGGTase-I
2	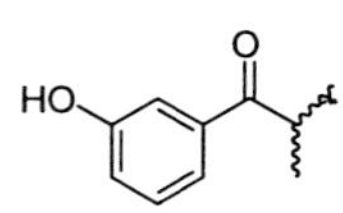	> 10,000	>100,000
3	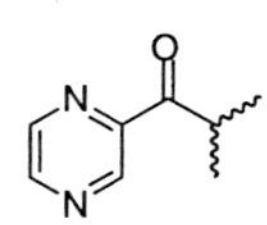	7,000	nd
4	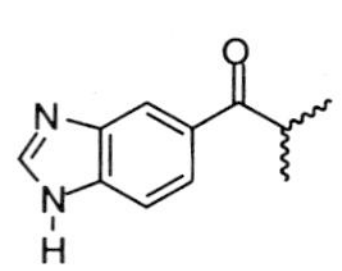	340	nd
5	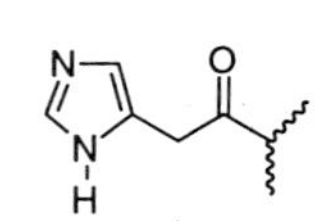	25	24,000
6	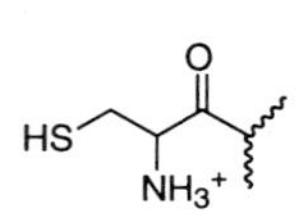	4.5	267

Table 2
Effects of Distance and Methylene Isosteres on the Imidazole-Containing Inhibitors

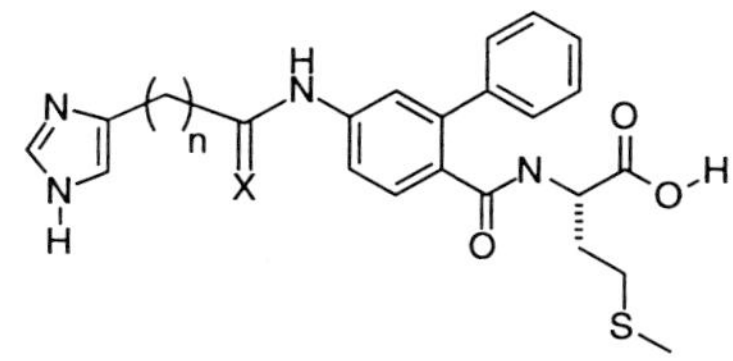

Inhibitor	n	X	in vitro IC_{50} (nM) PFTase	in vitro IC_{50} (nM) PGGTase-I	Processing IC_{50} (uM) H-Ras	Processing IC_{50} (uM) Rap1A
5	1	O	25	24,000	>10	>10
7	2	O	600	80,000	>100	>100
8	0	H,H	32	1,300	75	>100
9	1	H,H	10	160	10	>10

In order to determine the optimum distance between the imidazole and the hydrophobic spacer we prepared compounds **7–9**. In comparing compounds **5**, **8**, and **9**, little difference in FTase potency was observed, suggesting that the enzyme is tolerant of one and two carbon spacers. However, lengthening the linker to three (**7**) resulted in a significant drop (24-fold) in FTase potency relative to **5**. These results confirm the strict distance and structural requirements of FTase when binding CAAX peptidomimetics (*see* Table 2). When the amide bond-reduced isostere was incorporated into the peptididomimetic (**9**) only a 2.5-fold decrease in inhibition of FTase over **5** was observed, however, there was a significant loss in selectivity over GGTase-1 (16- vs 1000-fold). This would suggest that the increased rigidity present in the amide derivative (**5**) allows the imidazole subunit to bind to zinc in FTase but prevents it from taking up the necessary active conformation in GGTase-1. In contrast, the more flexible methylene isosteres (**8** and **9**) show relatively good activity in both enzymes. The advantage of the reduced cysteine, however, is seen in the whole cell activities of the compounds. The amide **5** shows no effect on the processing of H-Ras in NIH-3T3 cells whereas the corresponding methylene isostere **9** blocks H-Ras farnesylation with an IC_{50} of 10 μ*M*.

3.2. Probing the Effect of Substitution on the Imidazole Ring

In order to probe both the hydrophobic binding pocket at the N-terminus and potentially to influence zinc binding properties of the peptidomimetics, we prepared a series of different substituted derivatives of the imidazole. A series of imidazole aldehydes were prepared and reacted using a reductive amination procedure with the core spacer **1**. One hope was that subtle modifications to the pKa of imidazole might lead to an enhancement of its ability to bind zinc.

Compound **10** incorporating 2-methylimidazole (pKa = 7.5 vs 6.9) resulted in a threefold loss in FTase potency (IC_{50} = 110 n*M*) when compared to **8**, but exhibited better selectivity for FTase over GGTase I (IC_{50} = 10,000; *see* Table 3). Compound **11** with a 2-phenyl substituent has a slightly lower pKa on the imidazole ring (pKa = 6.4) but a

Table 3
Effects of Substitution on the Imidazole-Containing Inhibitors

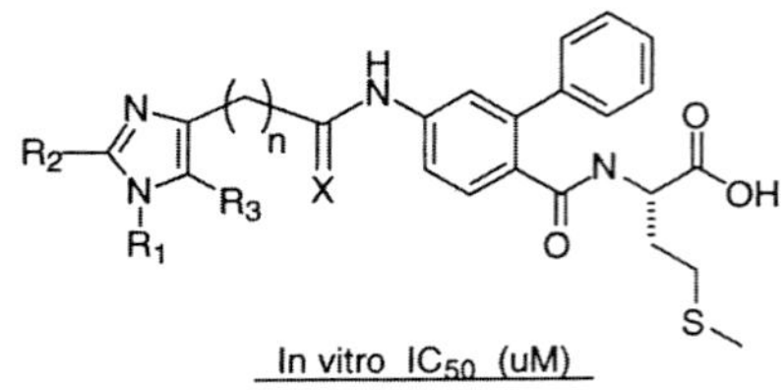

Inhibitor	R_1	R_2	R_3	n	X	In vitro IC_{50} (uM) PFTase	PGGTase-I
8	H	H	H	0	H,H	0.032	1.1
10	H	CH3	H	0	H,H	0.11	10
11	H	Ph	H	0	H,H	5.2	85
12	H	2-Pyr	H	0	H,H	2.0	nd
13	H	H	CH3	0	H,H	2.0	10
14	CH3 (4-yl)	H	H	1	O	0.046	nd
15	CH2C6H5 (4-yl)	H	H	0	H,H	0.6	3.9
16	CH2C6H5 (5-yl)	H	H	0	H,H	0.046	4.8

significant loss in potency, presumably owing to an unfavorable steric interaction with the FTase active site. The N-methyl derivative **14** was comparable in potency to the unsubstituted analog compound **8** and encouraged us to explore other hydrophobic groups on the nitrogen of the imidazole ring. Two compounds (**15** and **16**) were prepared incorporating a benzyl group via reductive amination of 1-benzyl-4- and 1-benzyl-5-imidazole carboxaldehydes with **1** followed by deesterification of the methionine methyl ester. Table 3 shows a significant difference in FTase potency between the isomers. Compound **15** was only a moderate inhibitor of FTase (IC_{50} = 600 n*M*), whereas **16** was significantly better (IC_{50} = 46 n*M*), implying a favorable interaction with FTase. Selectivity for FTase over GGTase I was the same for both isomers.

3.3. Probing the Hydrophobic Pocket of the Enzyme FTase

In addition to modifications on the imidazole nucleus, modifications to the 2-phenyl spacer to probe the hydrophobic pocket of the protein farnesyltransferase were also investigated. An important breakthrough in this area came from the demonstration that positioning a 2-methyl substituent on the phenyl group in our 2-phenyl-4-aminobenzoate spacer could lead to significant improvements in inhibition activity *(36)*. In order to test this effect within the context of our imidazole-based inhibitors, we prepared several ortho-substituted analogs (*see* Table 4).

Compounds **17** and **18** incorporating an ortho substituent on the phenyl ring exhibited a significant improvement in potency against FTase in vitro, compared to the unsubstituted phenyl **8**. The 10-fold increase in potency for **17** could only be attributed to the ortho methyl substitution on the phenyl group. The ^{1}H-NMR spectrum of **17** showed that two diastereomers are present in solution at room temperature. This mixture presumably arises from rotational isomerism owing to restricted bond rotation around the biphenyl

Table 4
Effects of Hydrophobic Substituents on the Imidazole-Containing Inhibitors

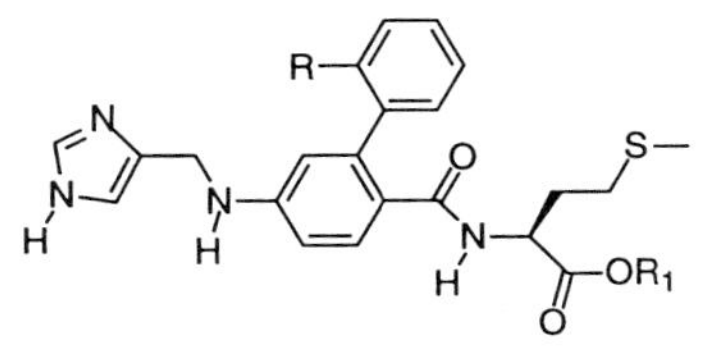

17	R = CH_3	R_1 = H
18	R = OCH_3	R_1 = H
19	R = CH_3	R_1 = CH_3

	IC50 (nM)	
Inhibitor	PFTase	PGGTase
8	32	1300
17	2.8	257
18	10	420

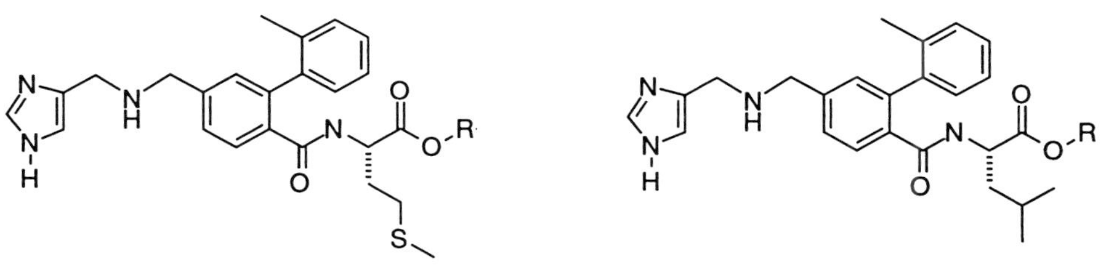

20 R = H **21** R = Me **22** R = H **23** R = Me

Fig. 5. Structures of aminomethylbenzoate-based inhibitors.

C-C bond. The increased bulk of the Me-group (relative to H- in **8**) leads to an increase in the barrier to rotation about the biphenyl group, which was calculated by variable temperature nuclear magnetic resonance (NMR) be approx 16.9 kcal/mol *(36)*.

3.4. Probing the Hydrophobic Spacer in the Peptidomimetics

The potency of the two carbon linker to the imidazole in **9** and the 2-tolyl substituted derivative in **17** prompted us to combine these features in another family of inhibitors. We also wanted to investigate the position of the N-atom in the two carbon linker, with particular focus on the effect of an aniline-N or alkylamine-N in the linker between the benzoate spacer and the imidazole unit. We prepared a series of compounds based on the 4-aminomethyl-2-(2-methylphenyl)-benzoate spacer (Fig. 5). These were synthesized by the reduction of 4-cyano-2-(2-methylphenyl)-benzoylmethionine methyl ester, followed by reductive amination with *N*-trityl-4-formyl imidazole.

The imidazole-aminomethylene-methionine derivative **20** (FTI-2148) is an extremely potent and selective inhibitor of FTase. Although threefold potency is lost relative to FTI-276, a significant increase in selectivity over GGTase (>1000-fold) is gained (*see* Table 5). Most importantly, peptidomimetic **21** (FTI-2153), the methyl ester derivative of **20**, retains this high potency and selectivity in whole cells. Compound **21** was able to

Table 5
Inhibition Potency of Aminomethylbenzoate-Based Inhibitors

	In vitro IC_{50} *(nM)*		*Processing* IC_{50} *(μM)*	
Inhibitor	*FTase*	*GGTase I*	*H-Ras*	*Rap1A*
9	10	160	10	>10
20	1.4	1,700	—	—
21	—	—	0.01	30
22	5,600	21	—	—
23	—	—	>30	0.3

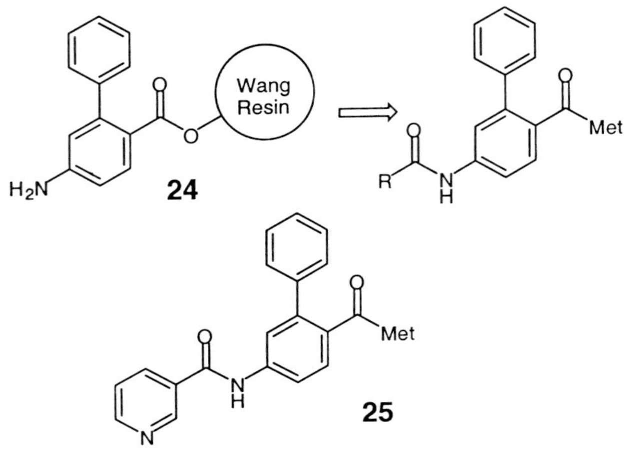

Fig. 6. Antitumor activity of FTI-276 and FTI-2153.

block the processing of H-Ras in NIH-3T3 cells with an IC_{50} of 10 n*M* but had no effect on the geranylgeranylation of Rap1A at concentrations over 3 μ*M*. The corresponding imidazole-aminomethylene-leucine derivative **22** was also prepared and showed potent inhibition of GGTase I with little effect on FTase *(37)*. Again, the selective inhibition was carried over into whole cell experiments with the methyl ester derivative **23** blocking the geranylgeranylation of Rap1A at n*M* level concentrations with little effect on the farnesylation of H-Ras at much higher concentrations.

3.5. Focused Combinatorial Libraries for the Discovery of Cysteine Replacements

A second strategy for the identification of noncysteine N-termini for the biphenyl-derived peptidomimetics involved the use of a focused combinatorial library *(36)*. Amide coupling to Wang resin-linked amine **24** provided a series of nonthiol structures from which nicotinamide **25** rapidly emerged as a promising lead (Fig. 6; Table 6). Reduction of the amide bond and retro-inversion provided 3-aminopyridine derivative **29** and the corresponding ether **30,** which were 10-fold and 20-fold more active than the original amide **25**. The pyridine nitrogen appears to participate in a critical interaction, perhaps with the active-site zinc, because positional isomers **26** and **27** and phenyl-derivative **31** *(38)* were significantly less active. Interestingly, the pyridine-derived inhibitors were essentially devoid of activity against GGTase I (IC_{50} > 10 μ*M* for **25–37**).

Table 6
Farnesyltransferase Inhibitors Incorporating an N-Terminal Pyridine

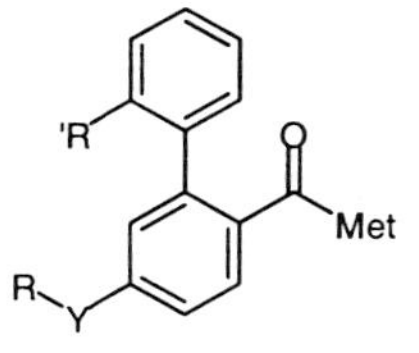

no.	R	Y	R'	*in vitro* IC_{50} (n**M**)	cellular EC_{50} (μ**M**)
25	pyridin-3-yl	CONH	H	70	50[a]
26	pyridin-2-yl	CONH	H	> 1000	-
27	pyridin-4-yl	CONH	H	> 1000	-
28	pyridin-3-yl	CH_2NH	H	20	40[a]
29	pyridin-3-yl	$NHCH_2$	H	6.8	3
30	pyridin-3-yl	OCH_2	H	4.0	5
31	phenyl	OCH_2	H	> 1000	
32	Pyridin-3-yl	CONH	CH_3	1.6	-
33	Pyridin-3-yl	CH_2NH	CH_3	1.5	-
34	Pyridin-3-yl	$NHCH_2$	CH_3	0.42	-
35	Pyridin-3-yl	OCH_2	CH_3	0.40	0.35
36	Pyridin-3-yl	$NBnCH_2$	CH_3	0.10	0.013
37	Phenyl	$NBnCH_2$	CH_3	1.4	0.4

[a] Methyl ester

A structure-activity study of the biphenyl core revealed that incorporation of a methyl group at the *ortho*-position of the 2-phenyl ring uniformly enhanced activity 10–44 fold (*36*). The most potent compound from this series, **35**, was not only equipotent with FTI-276, but was active in intact cells as the free carboxylate at submicromolar concentrations (Ras processing EC_{50} = 350 n*M*). Further optimization led to the discovery of *N*-benzyl derivative **36** (A-192630; *39*). In vitro potency increased fourfold compared to **35**, while cellular activity improved by >25-fold (EC_{50} = 13 n*M*), suggesting that increased lipophilicity enhanced cellular penetration. That the phenyl analog **37** retains significant activity indicates that the pyridine contributes less to overall binding in the presence of the *N*-benzyl substituent.

4. BIOLOGICAL EVALUATION OF THE PEPTIDOMIMETIC INHIBITORS

The full biological evaluation of these selective FTase and GGTase inhibitors is discussed in detail in Chapter 13. However, an important test for our approach was the effect

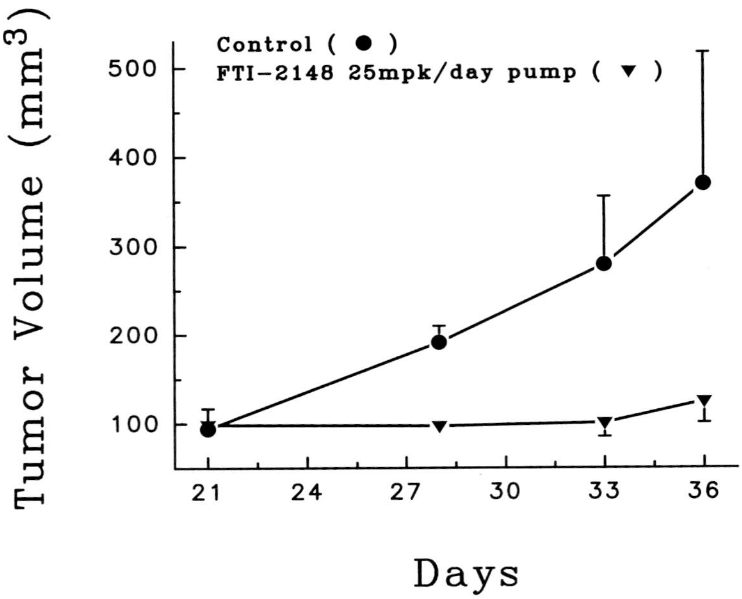

Fig. 7. Antitumor activity of FTI-276 and FTI-2153.

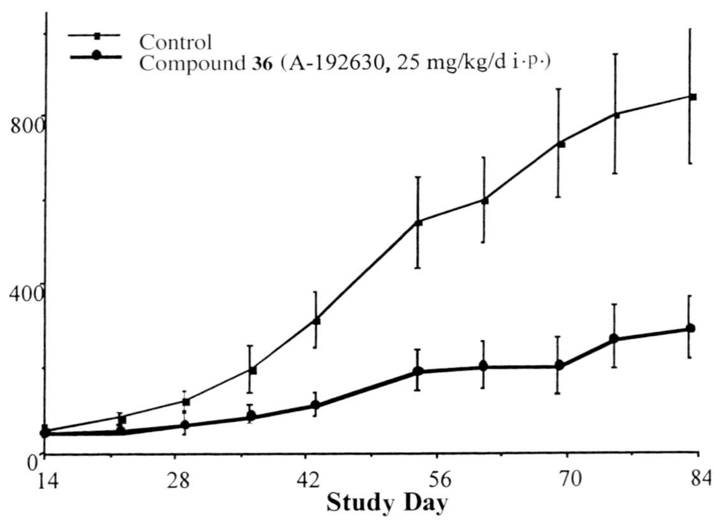

Fig. 8. Effects of compound **36** against A549 human lung cancer xenografts in nude mice.

of the inhibitors on tumor growth in nude mouse models. An investigation of the growth of A-549 human lung carcinoma cells in nude mice showed that peptidomimetic **20** (FTI-2148) is highly effective at slowing tumor growth when administered intraperitoneally (i.p.) in a daily schedule over a 10-wk period. Figure 7 shows that FTI-2148 inhibited growth by 91% after an 80-d period compared to 82% for a corresponding treatment with FTI-276. These results confirm the potential of FTase as a novel and highly attractive target for anti-cancer therapy. Compound **36** exhibited similar antitumor activity in this model (Fig. 8).

ACKNOWLEDGMENT

We thank the National Institutes of Health (CA-67771) for support of this work.

REFERENCES

1. Barbacid M. Ras genes. *Annu Rev Biochem* 1987; 56:779–827.
2. Owen D. The biochemistry of Ras p21. *Biochem J* 1991; 279:609–631.
3. Bos JL. Ras oncogenes in human cancer: a review. *Cancer Res* 1989; 49:4682–4689.
4. McCormick F. Signal transduction. How receptors turn ras on. *Nature* 1993; 363:15,16.
5. McCormick F. Ras GTPase activating protein:signal transmitter and signal terminator. *Cell* 1989; 56:5-8.
6. Trahey M, McCormick F. A cytoplasmic protein stimulates normal N-Ras p21 GTPase, but does not affect oncogenic mutants. *Science* 1987; 238:542–545.
7. Willingham MC, Pastan I, Shih TY, Scolnick EM. Localization of the src gene product of the Harvey strain of msr to plasma membrane of transformed cells by electron-microscopic immunocytochemistry. *Cell* 1980; 19:1005–1014.
8. Casey PJ, Solski PA, Der CJ, Buss JE. Ras is modified by a farnesylisoprenoid. *Proc Natl Acad Sci USA* 1989; 86:8323–8327.
9. Gutierrez L, Magee AI, Marshall CJ, Hancock JF. Posttranslational processing of Ras is two step and involves carboxyl methylation and carboxyl terminal proteolysis. *EMBO J* 1989; 8:1093–1098.
10. Goldstein JL, Brown MS. Regulation of the mevalonate pathway. *Nature* 1990; 343:425–430.
11. Reiss Y, Seabra MC, Armstrong SA, Slaughter CA, Goldstein JL, Brown MS. Non-identical subunits of p21H-Ras farnesyltransferase. *J Biol Chem* 1991; 266:10,672-10,677.
12. Park H-W, Boduluri SR, Moomaw JF, Casey PJ, Beese LS. Crystal structure of protein farnesyltransferase at 2.25 Angstrom resolution. *Science* 1997; 275:1800–1804.
13. Strickland CL, Windsor WT, Syto R, Wang L, Bond R, Wu Z, et al. Crystal structure of farnesyl protein transferase complexed with a CaaX peptide and farnesyl diphosphate analogue. *Biochemistry* 1998; 37: 16,601–16,611.
14. Long SB, Casey PJ, Beese LS. Cocrystal structure of protein farnesyltransferase complexed with farnesyl diphosphate substrate. *Biochemistry* 1998; 37:9612–9618.
15. Pompliano DL, Rands E, Schaber MD, Mosser SD, Anthony NJ, Gibbs JB. Steady state kinetic mechanism of Ras protein farnesyltransferase. *Biochemistry* 1992; 31:3800–3807.
16. Tschantz WR, Furfine ES, Casey PJ. Substrate binding is required for product release from protein farnesyltransferase. *J Biol Chem* 1997; 272:9989–9993.
17. Dolence JM, Poulter CD. A mechanism for posttranslational modifications of proteins by yeast protein farnesyltransferase. *Proc Natl Acad Sci USA* 1995; 92:5008–5011.
18. Mu Y, Omer CA, Gibbs RA. On the stereochemical course of human protein farnesyltransferase. *J Am Chem Soc* 1996; 188:1817–1823.
19. Hancock JF, Magee AI, Childs JE, Marshall CJ. All Ras proteins are polyisoprenylated but only some are palmitoylated. *Cell* 1989; 57:1167–1177.
20. Reiss Y, Goldstein JL, Seabra MC, Casey PA, Brown MS. Inhibition of purified p21Ras protein farnesyltransferase by Cys-AAX tetrapeptides. *Cell* 1990; 62:81–88.
21. Seabra MC, Goldstein JL, Sudhof TC, Brown MS. Rab geranylgeranyltransferase: a multisubunit enzyme that prenylates GTP-binding proteins terminating in Cys-X-Cys or Cys-Cys. *J Biol Chem* 1992; 267:14,497–14,503.
22. Casey PJ, Thissen JA, Moomaw JF. Enzymatic modification of proteins with a geranylgeranyl isoprenoid. *Proc Natl Acad Sci USA* 1991; 88:8631–8635.
23. Sebti SM, Hamilton AD. Inhibition of Ras prenylation: a novel approach to cancer chemotherapy. *Pharmacol Therapeut* 1997; 71:1–12.
24. Nigam M, Seong CM, Qian Y, Blaskovich MA, Hamilton AD, Sebti SM. Potent inhibition of human tumor p21ras farnesyltransferase by A_1A_2-lacking CA_1A_2X peptidomimetics. *J Biol Chem* 1993; 268: 20,695–20,698.
25. Qian Y, Blaskovitch M, Saleem M, Seong CM, Wathen S, Hamilton AD, Sebti SM. Design and structural requirements of potent peptidomimetic inhibitors of p21RAS farnesyltransferase. *J Biol Chem* 1994; 269:12,410–12,413.
26. Hamilton AD, Sebti SM. Inhibitors of Ras farnesyltransferase as novel antitumor agents. *Drug News Perspectives* 1995; 3:138–145.

27. Stradley SH, Rizo J, Gierasch LM. Conformation of a heptapeptide substrate bound to protein farnesyltransferase. *Biochemistry* 1993; 32:12,586–12,590.
28. Koblan KS, Culberson JC, deSolms SJ, Giuliani EA, Mosser SD, Omer CA, et al. NMR studies of novel inhibitors bound to farnesy-protein transferase. *Protein Sci* 1995; 4:681–688.
29. Sun S, Qian Y, Hamilton AD, Sebti SM. Ras CAAX peptidomimetic FTI-276 selectively blocks in nude mice the growth of a human lung carcinoma with a K-Ras mutation and a p53 deletion. *Cancer Res* 1995; 55:4243–4247.
30. Lerner EC, Qian Y, Blaskovitch M, Fossum R, Vogt A, Sun J, et al. Ras CAAX peptidomimetic FTI277 selectively blocks oncogenic Ras signaling by inducing cytoplasmic accumulation of inactive Ras/Raf complexes. *J Biol Chem* 1995; 270:26,802–26,806.
31. Clerc FF, Guitton J-D, Fromage N, Lelievre Y, Duchesne M, Tocque B, et al. Constrained analogs of KCVFM with improved inhibitory properties against farnesyltransferase. *Bioorg Med Chem Lett* 1995; 5:1779–1784.
32. Burns CJ, Guitton JD, Baudoin B, Lelievre Y, Duchesne M, Parker F, et al. Novel conformationally extended naphthalene based inhibitors of farnesyl transferase. *J Med Chem* 1997; 40:1763–1767.
33. Lerner EC, Qian Y, McGuire TF, Hamilton AD, Sebti SM. Disruption of oncogenic k-ras4B processing and signaling by a potent geranylgeranyltransferase inhibitor. *J Biol Chem* 1995; 270, 26,770–26,773.
34. McGuire TF, Qian Y, Vogt A, Hamilton AD, Sebti SM. Platelet-derived growth factor receptor tyrosine phosphorylation requires protein geranylgeranylation but not farnesylation. *J Biol Chem* 1996; 271: 27,402–27,407.
35. Qian Y, Vogt A, Sebti SM, Hamilton AD. Design and synthesis of non-peptide ras CAAX mimetics as potent farnesyltransferases inhibitors. *J Med Chem* 1996; 39:217–223.
36. Augeri DJ, O'Connor SJ, Janowick D, Szczepankiewicz B, Sullivan G, Larsen J, et al. Potent and selective non-cysteine containing inhibitors of protein farnesyltransferase. *J Med Chem* 1998; 41: 4288–4300.
37. Vasudevan A, Qian Y, Vogt A, Blaskovich M, Ohkanda J, Sebti SM, Hamilton AD. Potent and highly selective inhibitors of geranylgeranyltransferase-I. *J Med Chem* 1999; 42:1333–1340.
38. Augeri DJ, Janowick D, Kalvin D, Sullivan G, Larson J, Dickman D, et al. Potent and orally bioavailable noncysteine-containing inhibitors of protein farnesyltransferase. *Biorg Med Chem Lett* 1999; 9: 1069–1074.
39. O'Connor SJ, Barr KJ, Wang L, Sorensen BK, Tasker AS, Sham H, et al. Second-generation peptidomimetic inhibitors of protein farnesyltransferase demonstrating improved cellular potency and significant in vivo efficacy. *J Med Chem* 1999; 42:3701–3710.

5 Antitumor Efficacy of a Farnesyltransferase Inhibitor in Transgenic Mice

Jackson B. Gibbs, PHD,
Samuel L. Graham, PHD,
George D. Hartman, PHD,
Kenneth S. Koblan, PHD,
Nancy E. Kohl, PHD,
Charles Omer, PHD,
Angel Pellicier, MD, PHD,
Jolene Windle, PHD,
and Allen Oliff, MD

CONTENTS

1. INTRODUCTION

The *ras* genes, Harvey (H-), Kirsten (K-), and N-*ras* were among the first oncogenes found to be mutated in human cancers *(1)*. In particular, a splice variant of Ki-*ras*, termed "Ki4B-*ras*," is the most commonly mutated form of *ras* found to be altered in the majority of colon and pancreatic carcinomas. Because solid tumors are often the most difficult to treat clinically, in academic and pharmaceutical settings, great effort has been placed on developing inhibitors to the function of the products of *ras* genes, the Ras proteins. Many

From: *Farnesyltransferase Inhibitors in Cancer Therapy*
Edited by: S. M. Sebti and A. D. Hamilton © Humana Press Inc., Totowa, NJ

of the early efforts to modify Ras GTP binding, GTPase activity, or Ras interactions with putative effector molecules such as GAP, neurofibromin, or Raf were not successful *(2)*. In 1989, however, new insights were realized when the chemical modifications were identified that convert Ras from a biologically inactive precursor protein in the cytoplasm into a mature and biologically active protein in the plasma membrane *(3,4)*. The key modification, farnesylation of Cys on the Ras C-terminal CAAX sequence (C, Cys; a, aliphatic amino acid; X, another amino acid) is catalyzed by protein farnesyltransferase (FTase). FTase has afforded a target for drug discovery that has yielded compounds with properties appropriate for clinical testing *(5–7)*.

A number of FTase inhibitors have been described *(8)*. These compounds have been discovered either by random screening of chemical and natural product sources or by rational medicinal chemistry using as templates the substrates of FTase, farnesyl diphosphate (FPP), and tetrapeptides having the appropriate CAAX sequence. The compounds were subsequently optimized based on primary biochemical activity against FTase followed by characterization against transformed cells in culture. Compounds that met the appropriate criteria were then tested for antitumor efficacy in animal models. The purpose of this chapter is to summarize our efforts to use transgenic mice as a tumor model to evaluate the biology and efficacy of FTase inhibitors (FTIs) *(9–11)*.

2. ANIMAL MODELS

The most common animal model used in evaluating antitumor efficacy of potential cancer therapeutics utilizes tumor xenografts in immune-compromised nude mice. Indeed, several FTase inhibitors have been reported to be active in this model using *ras*-transformed fibroblast cells or human tumor-derived cell lines *(12–18)*. Mouse xenografts are useful because the tumors often arise quickly, making for ease of experimental design as new inhibitors are discovered. Also, this animal model affords the opportunity to study tumor cells previously characterized for their sensitivity to the test compound in a cell culture assay. One possible drawback of this assay is that tumors arise from a homogeneous population of cells in a biologically artificial compartment, typically subcutaneously in the flank of the mouse. The pathophysiology of these tumors does not mimic the properties of human cancer, in which a genetically heterogeneous population of cells arise in a specific organ type and subsequently develop invasive properties. The differences in the pathology between human tumors and mouse xenograft tumors may account in part for the poor clinical predictive value of antitumor efficacy observed in mouse xenografts *(19)*.

In the mid-1980s, an alternative model was created by Leder and colleagues *(20,21)*. These investigators genetically placed oncogenes such as v-Ha-*ras*, c-*myc*, or c-*neu* under the control of a mouse mammary tumor virus (MMTV) promoter into mice. The transgenic animals that resulted developed mammary and salivary tumors in a stochastic manner, implying that additional genetic events were necessary for the tumors to arise. Furthermore, the pathology of the resultant tumors had similarities with human tumors. The application of this model to the characterization of therapeutic agents was first reported by Dexter et al. *(22)*. Much of our efforts to characterize the antitumor efficacy of FTase inhibitors have focused on transgenic mice. Here we describe studies with the FTase inhibitor L-744,832 (Fig. 1). This compound is a CAAX-mimetic with properties (described elsewhere *[10]*), suitable for preclinical proof-of-concept experiments.

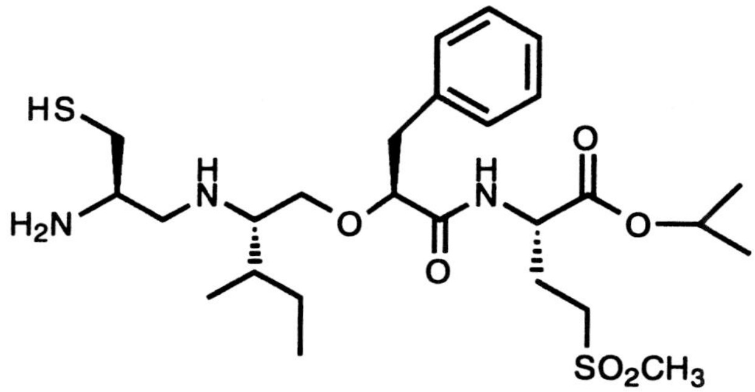

Fig. 1. Structure of L-744,832.

Table 1
Effect of L-744,832 on Tumors in Transgenic Mice

	Tumor mean growth rate (mm^3/d)	
Relevant genotype[a]	*Control*	*40 mg/kg L-744,832*
v-Ha-*ras*	11.8	−7.7
N-*ras*	28.2	−0.7
c-*neu*	25.6	15.5
v-Ha-*ras*/c-*myc*	43.6	−10.2
v-Ha-*ras*/p53$^{-/-}$	26.3	−12.3

[a] Tumor-bearing mice having the indicated relevant genotype were treated with daily subcutaneous injections of L-744,832, and the changes in tumor size were monitored by caliper over a 14-d period. (The primary data and details can be found in refs. *9–11*.)

3. TUMORS IN TRANSGENIC MICE

We have used transgenic mice bearing different oncogenes for the evaluation of FTase inhibitors (Table 1). The tumors are monitored twice weekly by caliper to yield a mean tumor growth rate. Tumor growth is scored as a positive mean growth rate, tumor stasis is indicated by a mean growth rate near zero, and tumor regression is evident by a negative mean growth rate.

Our first studies focused on v-Ha-*ras* mice *(10)*. These mice contain Ha-*ras* derived from the Harvey sarcoma virus. The mice developed mammary and salivary carcinomas that were highly sensitive to the action of L-744,832, as seen by the negative mean growth rate (Table 1). This regression was dependent on the dose of L-744,832, and the action of L-744,832 was superior to that observed with doxorubicin (mean growth rate of 7.5 mm^3/d; *see* ref. *10*). As striking as the results were with the v-Ha-*ras* mice, this model is likely the most sensitive to inhibition with L-744,832. Of the three Ras proteins, Ha-Ras protein has the poorest affinity for FTase, implying that it would be the easiest to inhibit. Furthermore, Ha-Ras is not a substrate for geranylgeranyl-protein transferase type I (GGTase I), whereas both N-Ras and Ki4B-Ras are substrates if FTase is inhibited *(23–26)*. Geranylgeranylation of N-Ras and Ki4B-Ras restores function to these proteins and could potentially confer resistance to an FTase inhibitor. We are currently preparing a transgenic mouse model of the clinically relevant Ki4B-*ras* gene using the MMTV promoter in order to test the activity of L-744,832 and other FTase inhibitors; however, we turned our attention to an existing N-*ras* model.

It was previously demonstrated that transgenic mice develop mammary tumors in response to overexpression of wild-type N-*ras* *(27)*. Although L-744,832 did not induce tumor regression in N-*ras* mice *(11)*, this compound still exhibited a significant antitumor effect relative to vehicle control-treated animals (Table 1). Biochemical analysis of these tumors showed no evidence that L-744,832 was inhibiting the processing of N-Ras protein, most likely because N-Ras can become geranylgeranylated and fully processed in cells treated with an FTI. The ability of an FTI to exhibit antitumor activity in cells having functionally processed Ras has been interpreted to mean that other farnesylated proteins may also be the biological targets of FTIs *(5,6)*. Interestingly, mammary tumors arising from overexpression of c-*neu* were poorly responsive to L-744,832 (Table 1). Although Ras is among the mediators of Neu signaling, it is possible that parallel pathways not involving Ras confer resistance to the action of L-744,832 *(9)*.

A hallmark of human tumor malignancy is genomic instability that leads to multiple genetic alterations. Although the stochastic manner in which *ras*-mediated tumors arise in transgenic mice infers that additional alterations are present in the tumors, the genetics of this event have not been defined. We therefore generated mice that had two defined genetic alterations: v-Ha-*ras*/c-*myc* and v-Ha-*ras*/p53$^{-/-}$ *(9)*. Alterations in the function of *myc* and p53 are commonly found in solid carcinomas. In particular, loss of wild-type p53 function is often associated with resistance to therapy. As summarized in Table 1, L-744,832 induced tumor regression in both v-Ha-*ras*/c-*myc* and v-Ha-*ras*/p53$^{-/-}$ mice.

4. MECHANISMS OF TUMOR REGRESSION

There are several mechanisms by which the Ha-*ras* tumors could have regressed. Histologically, there was no evidence for an infiltrate, arguing against a T-cell-mediated response, nor any evidence for necrosis *(10)*. Thus, the most reasonable explanation for tumor regression was apoptosis. This idea proved to be true based on the studies of Barrington et al. *(9)* and Mangues et al. *(11)*. Interestingly, the extent to which L-744,832 induced apoptosis varied with the genetic background that gave rise to the tumor (Table 2). The tumors most prone to undergo apoptosis in response to L-744,832 were those having the v-Ha-*ras* genotype. Importantly, the induction of apoptosis did not require wild-type p53 function, because significant apoptosis was seen in the v-Ha-*ras*/p53$^{-/-}$ mice.

Only a slight induction of apoptosis was evident in the tumors of N-*ras* and v-Ha-*ras*/c-*myc* mice (Table 2). The small magnitude of the response in N-*ras* mice may not be surprising given the observation that L-744,832 appeared to inhibit further tumor progression rather than induce regression (Table 1). In contrast, v-Ha-*ras*/c-*myc* mice treated with L-744,832 exhibited a large tumor regression. To determine the mechanisms that might have led to this regression, Barrington et al. *(9)* characterized cells in tumors for cell cycle distributions. They found that L-744,832 appeared to restore a G1/S checkpoint in tumors of v-Ha-*ras*/c-*myc* mice, as evidenced by a decrease in the percentage of tumor cells in S phase. L-744,832 also restored a G1/S checkpoint in tumor cells of v-Ha-*ras*/p53$^{-/-}$ but not in the tumor cells of v-Ha-*ras* having wild-type p53 function. Thus, alterations in the cell cycle distribution of tumor cells varied with the genotype of the tumor, as was seen for the induction of apoptosis in tumors in response to L-744,832.

Table 2
Induction of Tumor Apoptosis in Mice Treated with L-744,832

Relevant genotype[a]	*Fold increase over control*
v-Ha-*ras*	16
N-*ras*	2.6
c-*neu*	ND[b]
v-Ha-*ras*/c-*myc*	1.4
v-Ha-*ras*/p53$^{-/-}$	7.8

[a] Tumor-bearing mice of the relevant genotype were injected subcutaneously with either vehicle or 40 mg/kg L-744,832 for 48 h. The animals were sacrificed 24 h after the second dose and tumors were analyzed for apoptosis. The amount of tumor apoptosis in animals treated with L-744,832 relative to that seen in vehicle-treated animals is expressed as "fold increase over control." (Primary data and the details of the experimental methods can be found in refs. *9–11*.)

[b] ND, not determined.

5. CONCLUSIONS

Tumor-bearing transgenic mice offer a reasonable alternative to the more traditional nude mouse tumor xenograft model for the evaluation of potential therapeutic agents. Studies of L-744,832 in *ras* transgenic mice led to the key observation that tumor regression could occur by mechanisms involving the induction of apoptosis. This has not been reported for FTIs tested in xenograft tumor models. The critical issue to be tested now is how predictive these data will be to the current clinical trials of FTIs.

REFERENCES

1. Lowy DR, Willumsen BM. Function and regulation of *ras*. *Annu Rev Biochem* 1993; 62:851–891.
2. Gibbs JB. Pharmacological probes of Ras function. *Seminars Cancer Biol* 1992; 3:383–390.
3. Hancock JF, Magee AI, Childs JE, Marshall CJ. All *ras* proteins are polyisoprenylated but only some are palmitoylated. *Cell* 1989; 57:1167–1177.
4. Casey PJ, Solski PA, Der CJ, Buss JE. p21*ras* is modified by a farnesyl isoprenoid. *Proc Natl Acad Sci USA* 1989; 86:8323–8327.
5. Cox AD, Der CJ. Farnesyltransferase inhibitors and cancer treatment: targeting simply Ras? *Biochim Biophys Acta* 1997; 1333:F51–F71.
6. Gibbs JB, Oliff A. The potential of farnesyltransferase inhibitors as cancer chemotherapeutics. *Ann Rev Pharmacol Toxicol* 1997; 37:143–166.
7. Sebti SM, Hamilton AD. Inhibition of Ras prenylation: a novel approach to cancer chemotherapy. *Pharmacol Ther* 1997; 74:103–114.
8. Williams TM. Inhibitors of protein farnesylation. *Exp Opin Therapeut Patents* 1998; 8:553–569.
9. Barrington RE, Subler MA, Rands E, Omer CA, Miller PJ, Hundley JE, et al. A farnesyltransferase inhibitor induces tumor regression in transgenic mice harboring multiple oncogenic mutations by mediating alterations in both cell cycle control and apoptosis. *Mol Cell Biol* 1998; 18:85–92.
10. Kohl NE, Omer CA, Conner MW, Anthony NJ, Davide JP, deSolms SJ, et al. Inhibition of farnesyltransferase induces regression of mammary and salivary carcinomas in ras transgenic mice. *Nature Med* 1995; 1:792–797.
11. Mangues R, Corral T, Kohl NE, Symmans WF, Lu S, Malumbres M, et al. Antitumor effect of a farnesyl transferase inhibitor in mammary and lymphoid tumors overexpressing N-*ras* in transgenic mice. *Cancer Res* 1998; 58:1253–1259.

12. Hara M, Akasaka K, Akinaga S, Okabe M, Nakano H, Gomez R, et al. Identification of Ras farnesyltransferase inhibitors by microbial screening. *Proc Natl Acad Sci USA* 1993; 90:2281–2285.
13. Ito T, Kawata S, Tamura S, Igura T, Nagase T, Miyagawa J-I, et al. Suppression of human pancreatic cancer growth in BALB/c nude mice by manumycin, a farnesyl:protein transferase inhibitor. *Jpn J Cancer* 1996; 87:113–116.
14. Kohl NE, Wilson FR, Mosser SD, Giuliani EA, deSolms SJ, Conner MW, et al. Farnesyltransferase inhibitors block the growth of *ras*-dependent tumors in nude mice. *Proc Natl Acad Sci USA* 1994; 91: 9141–9145.
15. Leftheris K, Kline T, Vite GD, Cho YH, Bhide RS, Patel DV, et al. Development of highly potent inhibitors of Ras farnesyltransferase possessing cellular and in vivo activity. *J Med Chem* 1996; 39:224–236.
16. Nagasu T, Yoshimatsu K, Rowell C, Lewis MD, Garcia AM. Inhibition of human tumor xenograft growth by treatment with the farnesyl transferase inhibitor B956. *Cancer Res* 1995; 55:5310–5314.
17. Sun J, Qian Y, Hamilton AD, Sebti SM. Ras CAAX peptidomimetic FTI 276 selectively blocks tumor growth in nude mice of a human lung carcinoma with K-*ras* mutation and p53 deletion. *Cancer Res* 1995; 55:4243–4247.
18. Williams TM, Ciccarone TM, MacTough SC, Bock RL, Conner MW, Davide JP, et al. 2-Substituted piperazines as constrained amino acids. Application to the synthesis of potent, non carboxylic acid inhibitors of farnesyltransferase. *J Med Chem* 1996; 39:1345–1348.
19. Plowman J, Dykes DJ, Hollingshead M, Simpson-Herren L, Alley MC. Human tumor xenograft models in NCI drug development, in *Anticancer Drug Development Guide: Preclinical Screening, Clinical Trials, and Approval* (Teicher B, ed.). Humana Press, Totowa, NJ, 1997, pp. 101–125.
20. Sinn E, Muller W, Pattengale P, Tepler I, Wallace R, Leder P. Coexpression of MMTV/v-Ha-*ras* and MMTV/c-*myc* genes in transgenic mice: synergistic action of oncogenes in vivo. *Cell* 1987; 49:465–475.
21. Muller WJ, Sinn E, Pattengale PK, Wallace R, Leder P. Single-step induction of mammary adenocarcinoma in transgenic mice bearing the activated c-*neu* oncogene. *Cell* 1988; 54:105–115.
22. Dexter DL, Diamond M, Creveling J, Chem S-F. Chemotherapy of mammary carcinomas arising in *ras* transgenic mice. *Investig New Drugs* 1993; 11:161–168.
23. Zhang FL, Kirschmeier P, Carr D, James L, Bond RW, Wang L, et al. Characterization of Ha-Ras, N-Ras, Ki-Ras4A, and Ki-Ras4B as in vitro substrates for farnesyl protein transferase and geranylgeranyl protein transferase type I. *J Biol Chem* 1997; 272:10,232–10,239.
24. James GL, Goldstein JL, Brown MS. Polylysine and CVIM sequences of K-RasB dictate specificity of prenylation and confer resistance to benzodiazepine peptidomimetic in vitro. *J Biol Chem* 1995; 270: 6221–6226.
25. Whyte DB, Kirschmeier P, Hockenberry TN, Nunez-Oliva I, James L, Catino JJ, Bishoip WR, Pai J-K. K- and N-Ras are geranylgeranylated in cells treated with farnesyl protein transferase inhibitors. *J Biol Chem* 1997; 272:14,459–14,464.
26. Rowell CA, Kowalczyk JJ, Lewis MD, Garcia AM. Direct demonstration of geranylgeranylation and farnesylation of Ki-Ras in vivo. *J Biol Chem* 1997; 272:14,093–14,097.
27. Mangues R, Seidman I, Gordon JW, Pellicer A. Overexpression of the N-*ras* proto-oncogene, not somatic mutational activation, associated with malignant tumors in transgenic mice. *Oncogene* 1992; 7: 2073–2076.

6 Development of Farnesyltransferase Inhibitors as Potential Antitumor Agents

Veeraswamy Manne, PHD, *Frank Lee*, PHD,
Ning Yan, PHD, *Craig Fairchild*, PHD,
and William C. Rose, PHD

CONTENTS

1. INTRODUCTION

Activating mutations of the ras genes are among the more common genetic aberrations known in human cancers, particularly in pancreatic and colon carcinomas *(1)*. Three highly homologous members of the ras protooncogene family have been identified in higher mammals; H-ras, K-ras, and N-ras *(2)*. All three ras genes code for proteins that end in a sequence called the "CAAX box," which is the recognition sequence for post-translational farnesylation *(3)*. Farnesylation is catalyzed by farnesyltransferase (FTase), a cytosolic enzyme that utilizes farnesyl pyrophosphate (FPP) as a farnesyl donor to modify the cysteine residue of the ras CAAX terminus *(4)*. Anchoring of the Ras proteins to the inner surface of the plasma membrane is required for normal functions in signal transduction as well as for transforming activities and signaling would not occur when farnesylation is blocked *(5)*. Although farnesylation is not entirely specific for Ras proteins, only a few other cellular proteins undergo this posttranslational modification *(6,7)*. Our goal has been to develop potent and selective inhibitors of FTase in an attempt to inhibit and/or reverse Ras-mediated malignant transformation in rodent and human cancers, in cell culture as well as in animal models and ultimately in humans.

From: *Farnesyltransferase Inhibitors in Cancer Therapy*
Edited by: S. M. Sebti and A. D. Hamilton © Humana Press Inc., Totowa, NJ

2. MATERIALS AND METHODS

2.1. *In Vitro and Cell-Based Assays*

The in vitro and cell-based assays used for characterization of Bristol-Myers Squibb (BMS) FTase inhibitors (FTIs) have been described previously *(8–10)*.

2.2. *In Vivo Evaluations*

2.2.1. Mice and Tumors

Athymic Balb/c-background female mice, 18–22 gm, were purchased from Harlan-Sprague Dawley (Indianapolis, IN). RASK, a K-ras transformed mouse 3T3 cell line, Rat-1 tumor, an H-Ras transformed rat fibroblast line, and MCF-7, a human breast carcinoma, were used as in vivo tumor models in athymic mice. The Rat-1 tumors were sourced from in vitro propagated cells. Experiments were initiated by intraperitoneal (i.p.), intraveneous (i.v.), or subcutaneous (s.c.) implantation of 1×10^6 cells, except in control groups given titrated cell inocula. The RASK tumors were maintained in vivo via s.c. tumor passage at approx 2 wk intervals. The experiment was begun by implanting 0.5 mL of a 2% (w/v) tumor brei, i.p., except those control groups given titrated inocula. The MCF-7 breast tumor was propagated in vivo via s.c. passage of tumor fragments at approx 3-wk intervals. Initiation of the MCF-7 experiment involved s.c. implantation of tumor fragments. In vivo growth of the estrogen-dependent MCF-7 tumor was supported by the s.c. insertion of an estradiol pellet into each host mouse 1 d prior to (contra lateral) tumor implantation; the release of adequate levels of estradiol (to support tumor growth) had previously been determined to occur for 60 d, sufficient for the planned duration of the experiment. All experiments began on day 1 posttumor implant except for the MCF-7 study which was begun within 1 h of tumor implant.

2.2.2. Compounds

BMS-188222, BMS-191563, and BMS-192331 were dissolved in sterile water. They were injected i.p., s.c., or i.v. within 1 h of dissolution. BMS-191563 was also administered s.c. via Alzet pump (Alza Co., Palo Alto, CA) at concentrations of 105, 35, 11.5, and 4 mg/mL. The Alzet pumps were inserted in a manner that allowed the effluent to flow toward the adjacent tumor fragments. With regard to parenteral treatment regimens in which twice-a-day injections were indicated, they were given 6–8 h apart on weekdays; on weekends, a single injection was given at twice the indicated dose (i.e., the total intended daily amount of compound was given once, *not* as a split dosage, as was done during the rest of the week).

The stability of BMS-191563 was evaluated in water (100 mg/mL) at 22°C and only 5–10% decomposition was noted after 12 h. The stability of BMS-191563 contained in Alzet pumps held at 37°C for 4 d also was evaluated and about 50% decomposition was found (the disulfide was assumed to be the major decomposition product).

2.2.3. Assessment of Antitumor Activity

Detailed description of the basic assay and evaluation methods used for the experiments conducted have been reported *(11)*. Briefly, therapeutic results are presented in terms of: (a) increases in lifespan reflected by the relative median survival time (MST) of treated (T) vs control (C) groups (i.e., % T/C values) and any long-term survivors; and (b) primary tumor growth inhibition determined by calculating the relative median times for T and C mice to grow tumors of a predetermined size (250 mg for MCF-7 tumors, 1.0 g for Rat-1 tumors) (i.e., T/C values). Tumor weights were interchangeable with tumor size on the basis of 1 mm^3 = 1 mg. Statistical evaluations of data were performed using the Gehan's generalized Wilcoxon test *(12)*.

The activity criterion for increased lifespan was a T/C of ≥125% and was applicable for both s.c. and i.p. tumor implant experiments. The activity criterion for tumor inhibition was a delay in tumor growth consistent with one gross $\log_{10}$ cell kill (LCK). The absolute T/C value needed to attain this level of efficacy varied from experiment to experiment and depended on the tumor volume doubling time of the control mice in each study.

Group sizes typically consisted of six mice in i.p.-implanted tumor treatment groups, eight mice in s.c.-implanted tumor treatment groups, and eight mice in all control groups. The various drug treatments used are described in conjunction with the experimental data. Drug-treated mice dying before their tumors reached target size, or before the first death in parallel control mice implanted with the same tumor inoculum, were considered to have died from drug toxicity. Groups of mice with more than one death owing to drug toxicity were not used in the evaluation of antitumor efficacy, and for all compounds, the highest dose tested that did not cause such lethality was termed the maximum tolerated dose (MTD).

3. TESTING STRATEGY FOR FTIs

All compounds were screened in vitro against purified porcine FTase using recombinant H-Ras as the acceptor of radiolabeled FPP. As the substrate specificity of geranylgeranyltransferase I (GGTase I) is similar to that of FTase, selected inhibitors were tested for inhibition of GGTase I in order to characterize their selectivity for FTase over GGTase I. Compounds with FTase IC_{50} values ≤ 1 μ*M* were tested in cell assays. The Ras transformation inhibition (RTI) assay is based on transformation of NIH 3T3 cells by oncogenic ras DNA transfection. In the absence of an inhibitor, transfected cells grow aggressively and form foci. In the presence of a cell permeable inhibitor, the number of foci are either reduced or transformation is completely inhibited. Gross cytotoxicity of compounds is also readily observed with this assay. The soft agar growth (SAG) assay is based on the ability of Ras-transformed NIH 3T3 cells to grow in soft agar and form discrete colonies in 14–21 d. Initially, inhibitors were incorporated into the soft agar layer and fresh inhibitors were layered on top of the soft agar every 48 h for 8 d. In the presence of potent and cell permeable inhibitors, Ras transformed cells fail to form colonies. The relative potency of inhibitors (EC_{50}) reflects the concentration of inhibitor that reduces the number of colonies by 50%. Inhibitors that show promising activity (EC_{50} ≤ 0.3 μ*M*) in the SAG assay were evaluated for in vivo activity using Rat-1 tumors in nude mice. Selected compounds with potent activity in either the NIH 3T3 SAG assay or in in vivo Rat-1 tumors were subjected to the more stringent task of inhibiting the growth of human tumor cells, which usually contain multiple genetic aberrations. For testing of human cell lines, both the SAG and the colony growth on plastic (CGP) assays were used. The overall strategy for evaluating and advancing compounds through the testing protocols is shown in Fig. 1.

4. BISUBSTRATE INHIBITORS

Our extensive studies of the structure-activity relationships for the phosphonic acid series of bisubstrate inhibitors is represented by BMS-184467 *(13)*. Although it is a potent inhibitor of FTase (IC_{50} = 4 n*M*), the doubly charged nature of BMS-184467 made it an unlikely candidate for cell penetration. The phosphinic acid bisubstrate, BMS-185878,

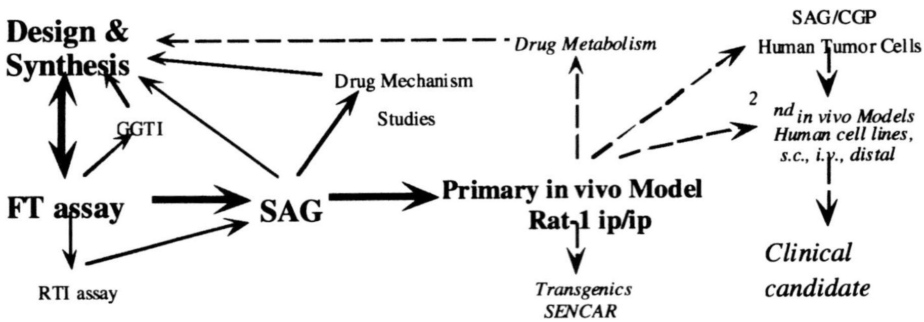

Fig. 1. Compound evaluation strategy. Abbreviations: FTase, farnesyl transferase; RTI, Ras transformation inhibition assay; SAG, soft agar growth assay; GGTase I; geranylgeranyltransferase I assay; CGP; colony growth on plastic.

was identified as equipotent to BMS-184467 and because the phosphinic acid moiety of BMS-185878 is less acidic than the corresponding phosphonic acid of BMS-184467, we speculated that carboxylic ester "prodrugs" of BMS-185878 might be better able to penetrate cells (*see* Scheme 1).

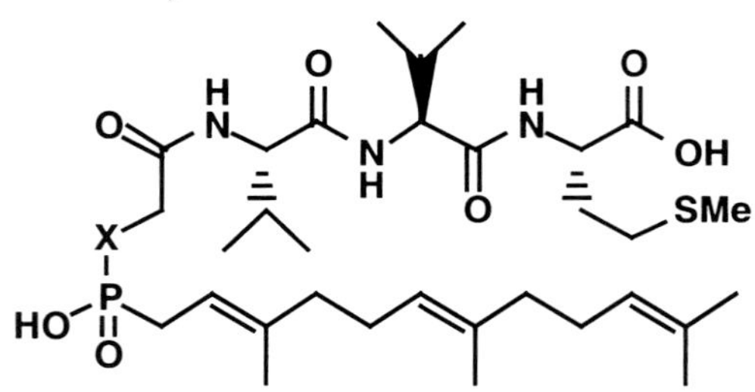

BMS-184467 X=O
BMS-185878 X=CH_2

In order to evaluate the effect of FTIs on Ras transformation in cells, an RTI assay was developed: In the RTI assay, compounds are added 1 d after oncogenic H-*ras* DNA transfection of NIH 3T3 cells and replenished every 48 h until the end of the experiment (typically 7–9 d). In the absence of an inhibitor, transfected cells grow aggressively and initially give the appearance of abnormal "foci" surrounded by normal cells. If the experiment is allowed to progress, the transformed cells dominate and completely cover the surface of the plate. In the presence of an effective inhibitor, the number of transformed cells would be reduced or transformation completely prevented. The gross cytotoxicity of inhibitors can be evaluated at the same time by visual inspection.

Several carboxylic and carboxylic/phosphinic acid mono- and diesters were prepared and tested. The most effective at preventing transformation were the methyl and pivaloyloxymethyl carboxylic acid esters, BMS-186511 and BMS-188031, respectively (*see* Scheme 2). Cell transformation proceeded to only 10% of the control value at 100 μ*M* and 50–70% of the control at 10 μ*M*. There were no signs of cytotoxicity at 10 or 100 μ*M*. Although BMS-186511 was very effective at preventing transformation at 100 μ*M*, BMS-187993, the methyl ester prodrug of the diastereomeric inhibitor BMS-187203 (FTase IC_{50} = 820 n*M*) was not effective at this concentration (*see* Scheme 3). This observation is consistent with the mode of action of BMS-186511 and BMS-188031 involving inhibition of FPT by the parent diacid, BMS-185878 [IC_{50} = 6 n*M*].

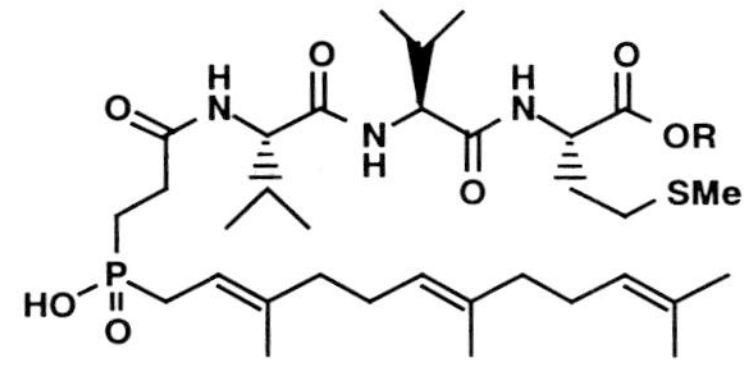

BMS-186511 R = CH_3

BMS-188031 R = 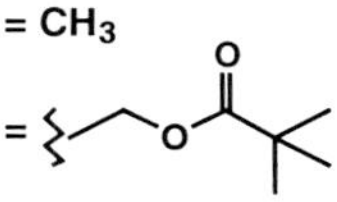

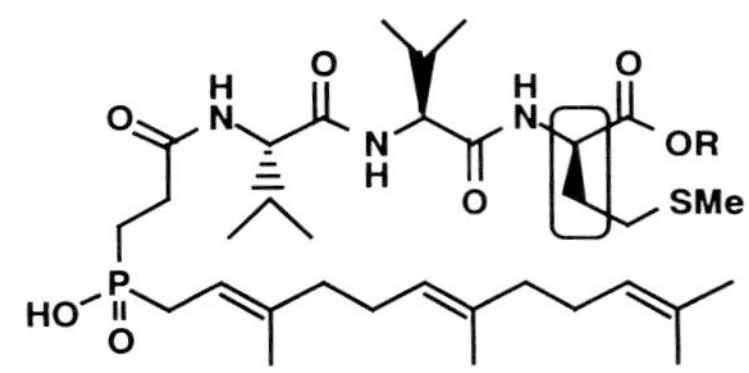

BMS-187203 R=H
BMS-187993 R=CH_3

To address the question of specificity, the ability of BMS-186511 to prevent a *trk* oncogene-mediated transformation was tested. BMS-186511 had no effect on transformation by this tyrosine kinase oncogene. This result indicates that inhibition of FTase may selectively interfere with *Ras*-mediated transformation.

Next, BMS-186511 and BMS-188031 were subjected to the more difficult task of reversing *Ras* transformation. NIH 3T3 cells were allowed to transform for 7–8 d following *ras* DNA transfection, by which time the transformation process was well under way. Compounds were then added every 48 h for 8 d. BMS-186511 was effective at "freezing" transformation to that observed at d 7–8. BMS-188031 partially reversed transformation, with clear morphological changes (including flattened appearance) apparent upon examination by light microscopy.

We previously reported that 100 μ*M* BMS-186511 or BMS-188031 had no effect on the growth of NIH 3T3 cells. In contrast, the doubling times of the *Ras*-transformed cell lines were increased by a factor of 1.5–1.9 by BMS-186511 and BMS-188031 *(8)*. These observations, coupled with the change in morphology of *Ras* transformed cells in the presence of BMS-188031 indicate that *Ras*-transformed cells are regaining normal growth control mechanisms in the presence of FTIs as a consequence of suppression of Ras function.

In summary, several prodrug esters of the bisubstrate, farnesyl protein transferase inhibitor BMS-185878 were shown to inhibit and reverse oncogenic *Ras*-induced transformation of NIH 3T3 cells without affecting the growth of normal cells. These prodrugs also selectively inhibited protein farnesylation without affecting protein geranylgeranylation in normal and *Ras*-transformed NIH 3T3 cells and inhibited oncogenic Ras-induced oocyte maturation *(8)*. However, we considered the level of whole cell activity of these compounds to be not potent enough to warrant in vivo evaluation of antitumor activity.

5. CAAX BOX ANALOGS

Our program was also focused on inhibitors structurally related to the CAAX box tetrapeptide *(14)*. In this series, BMS-188222 was the first compound demonstrating activity in the whole cell assay systems although there was a large discrepancy between the micromolar concentrations needed to inhibit cellular transformation and growth in soft agar and its nanomolar potency against the enzyme. This disparity suggested that BMS-188222 possessed poor membrane permeability and/or poor chemical and metabolic stability. Because BMS-188222 displayed excellent intrinsic inhibitory potency vs FTase, we focused our synthetic studies on delineating the factors potentially important to potency in the whole cell assay systems and, presumably, to potency in vivo.

A number of modifications to BMS-188222 led to improved biological activity. In particular, BMS-191563 and BMS-192331 displayed significantly improved cell activity (*see* Scheme 4).

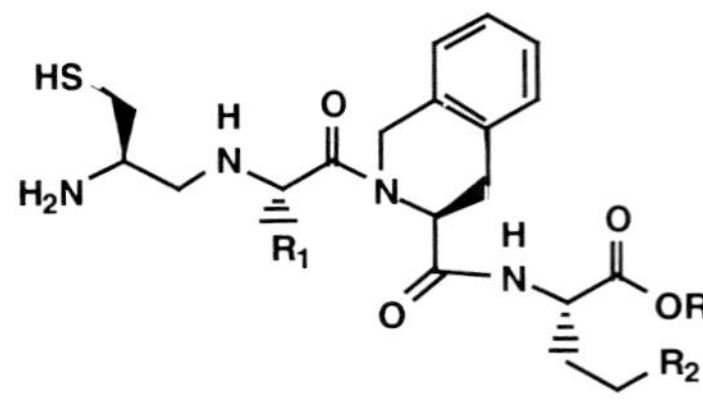

BMS #	R_1	R_2	R_3	FT IC_{50} (nM)	GGTI IC_{50} (nM)	SAG EC_{50} (μM)
188222	$CH(CH_3)_2$	SCH_3	H	0.6	110	5
190622	$C(CH_3)_3$	SCH_3	H	0.57	4.8	0.5
191563	$C(CH_3)_3$	SCH_3	CH_3	85	201	0.25
192331	$C(CH_3)_3$	$CONH_2$	H	2.8	1400	0.19

When 100 μ*M* BMS-188222 was added to NIH 3T3 cells 1 d after transfection with oncogenic H-*Ras*, it was completely effective at preventing transformation. Even at 10 μ*M*, BMS-188222 was ≥80% effective in this assay. More significantly, when BMS-188222 was added 8 d post-transfection (when transformation was already well established), it was completely effective at reversing transformation at 100 μ*M*, and partially effective at 10 μ*M*. H-*Ras*-transformed cells treated with 100 μ*M* BMS-188222 showed a perfectly normal, flat phenotype. However, these cultures had more cells than in the case of non-transfected cells, suggesting that at the concentrations tested, BMS-188222 can revert transformation but not kill the *Ras*-transformed cells.

BMS-188222 was also tested at 100 μ*M* in an agar assay against K-NIH and 44-9-1-1 cell lines; two NIH 3T3 cell lines transformed by v-K-*ras* and human H-*ras* oncogenes, respectively. BMS-188222 completely inhibited the agar growth of 44-9-1-1 cells and was 70–90% effective at inhibiting the growth of K-NIH cells (colonies were scored after 10 d). The K-NIH cell line was directly transformed by the Kirsten-MSV virus. Therefore, the v-K-*ras* oncogene present in these cells is driven by the retroviral LTR, which is a promoter at least one order of magnitude more powerful than the endogenous H-*ras* promoter used to drive the H-*ras* oncogene in the 44-9-1-1 cell line. BMS-188222 was also 70 and 95% effective at preventing oocyte GVBD at intracellular concentrations of 2.5 and 25 μ*M*, respectively.

5.1. Membrane Localization

Because farnesylation is required for the association of Ras with the cell membrane, unprocessed Ras protein should localize to the cytosolic fraction when cells are treated

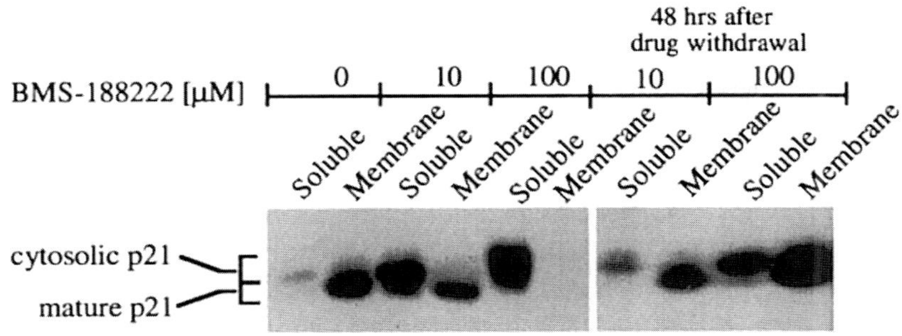

Fig. 2. Effect of BMS-188222 on processing and membrane localization.

with FTIs. To test whether this process occurs, H-Ras transformed NIH 3T3 cells were treated for 6 d with BMS-188222. Total cellular extracts were prepared, fractionated into soluble and membrane fractions and analyzed by sodium dodecyl sulfate-polyacrylamide gel electrophoresis (SDS-PAGE) and Western blots using an H-Ras specific antibody. The results presented in Fig. 2 show that following treatment of cells with BMS-188222, Ras protein processing is blocked and unprocessed Ras, with slower mobility, accumulates in the cytosolic fraction. This effect is reversible as removal of BMS-188222 and culturing the cells for an additional 48 h in the absence of inhibitor led to resumption of Ras protein processing and association of Ras with the membrane (Fig. 2).

5.2. Specificity

The specificity of our FTIs was examined in cell assays using Rat-1 cell lines transformed by ras genes encoding RasH (61L,CVLS), RasH (61L,SVLS,N-myristoyl), or RasH (61L, CVLL). All three Ras mutants are transforming but depend on different mechanisms for membrane attachment. RasH (61L,CVLS) undergoes farnesylation leading to membrane association. RasH (61L,SVLS,N-myristoyl) is not farnesylated because Cys^{186}, the site of farnesylation, is replaced with Ser; however, a myristoylation signal sequence attached to the N-terminus allows myristoylation and subsequent membrane association. RasH (61L, CVLL), owing to the changes made in the "CAAX box" that alter prenylation specificity, is geranylgeranylated, which leads to membrane association. All three cell lines were tested in the SAG assay to determine the specificity of BMS-188222. BMS-188222 inhibited SAG of Rat-1 RasH (61L,CVLS) cells in a dose-dependent manner, and at 100 μ*M* the inhibition is complete. In contrast, BMS-188222 had no effect on SAG of myristoylated Ras or geranylgeranylated Ras transformed cells (Fig. 3A). These results suggest that BMS-188222 reverses or inhibits transformation by specifically inhibiting FTase.

In order to provide biochemical evidence supporting the high specificity of the FTIs, Ras protein analysis was carried out in cells treated with BMS-190622 (free-acid version of BMS-191563). Total cellular extracts rather than subcellular fractions were analyzed in this experiment because unprocessed Ras could easily be recognized as a slower migrating species. Ras processing and thus its membrane association was blocked in cells carrying the farnesylated 61L oncogenic Ras and the slower migrating cytosolic form accumulated (Fig. 3B). Ras processing was not significantly affected, as seen by lack of the slower migrating form, in cells containing geranylgeranylated or myristoylated Ras

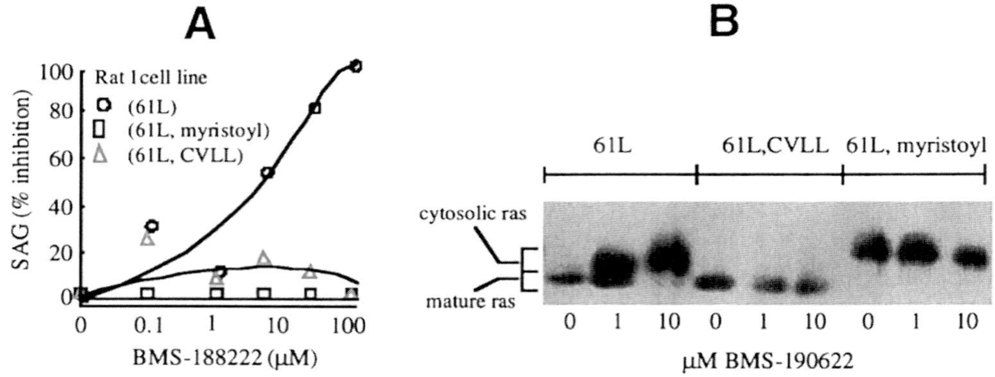

Fig. 3. (**A**) Specificity of BMS-188222. (**B**) Effect of BMS-190622 on Ras processing.

(myristoyl Ras migrates slower in all the lanes owing to 11 extra amino acids). These biochemical results support and confirm the specificity of our FTIs described previously in the SAG assays utilizing these various Ras forms.

5.3. Reversal of Ras-Transformed Morphology and Actin Cable Restoration

Ras-transformed rodent cells are highly refractile, rounded, and pile up owing to the loss of contact inhibition. These gross morphological changes, not found in their untransformed counterparts, are also manifested in a general disruption in the actin cytoskeletal network. We investigated reversal of the Ras-transformed phenotype by FTIs as a complementary system (in addition to RTI and SAG) to study oncogenic Ras function. Treatment of Ras transformed cells with 100 μ*M* BMS-188222 continuously for 4 d resulted in complete reversal of the transformed morphology to that of normal as reflected by the nonrefractile, flattened appearance of cells in a monolayer (not shown). In addition, the integrity and appearance of the actin cytoskeleton, assessed using the fluorescent reagent rhodamine phalloidin (which binds specifically to F-actin filaments), was restored to normal (not shown). FTI-induced morphological changes and the actin cable restoration are specifically observed with farnesylated Ras but not with myristoylated Ras or geranylgeranylated Ras mutants. These results are in complete agreement with specificity studies presented previously.

5.4. Effects on Human Cell Lines with Ras Mutations

Initially, our studies with FTIs focused on mouse and rat fibroblasts manipulated genetically by introduction of specific ras oncogenes as model systems. However, human tumors are the result of multiple genetic aberrations and hence it is important to test FTIs on human cancer cells. Certain established human tumor cell lines contain mutated ras genes that presumably contribute significantly to the malignant properties of these cells. As a first step to study the effects of FTIs on human cancers, several cell lines with reported ras mutations were screened for sensitivity to FTIs using the SAG assay. BMS-188222 and BMS-192331 induced inhibition of growth of MiaPaCa-2, a pancreatic carcinoma with a reported G to C mutation in K-ras codon 12, and HCT-116, a colon carcinoma with a G to N mutation in K-ras codon 13. However, compared to the Ras transformed rodent cells, growth inhibition of these two human tumor cell lines carrying ras mutations

Table 1
Sensitivity of Human Tumor Cell Lines to BMS-188222

Cell line	*Tumor type*	*ras mutation status*	*CGP EC_{50} (μM)*
MCF-7	Human breast	wt	10
A549	Human lung	K-12 gly to ser	100
L2987	Human lung	ND	>100
HCT116	Human colon	K-13 gly to asn	60
RCA	Human colon	ND	60
HCT15	Human colon	ND	100
A2780/1A9	Human ovarian	ND	15

required higher concentrations of the FTIs (data not shown), which may be owing to the presence of additional genetic alterations, or to differences in permeability between human and rodent cells.

5.5. Effects on Other Human Cell Lines

As the SAG assay is time consuming and labor intensive, we developed a rapid assay to quickly screen and identify human tumor cell lines sensitive to FTIs. This assay, termed colony growth on plastic (CGP), is based on the ability of some human tumor cell lines to form discrete colonies on plastic when plated at low densities. Cells are plated and cultured for 7–8 d with the inhibitor. Fresh media and inhibitor are added every 48 h. Colonies are fixed, stained with crystal violet, and the EC_{50} of FTIs determined. Using the CGP assay, a comparative study of cell lines with reported ras mutations, with wild-type ras and with unknown ras status was carried out using BMS-188222 (Table 1). Cell lines with wild-type ras were included because they may have tumorigenic events upstream of the Ras signal transduction pathway and hence could respond to Ras FTIs. MCF-7 cells and A2780/1A9 cells are more sensitive than the human colon or lung cell lines tested. Whether this is owing to variable compound stability, permeability differences, or different relative importance of the Ras pathway for malignant growth properties in the cell lines tested remains to be determined.

5.6. Susceptibility of Other Breast Cancer Cells

In addition to MCF-7, two other breast cancer cell lines (ZR-75 and H3396) were found to be sensitive to BMS-191563. Though preliminary, these studies suggest that FTIs may be useful for a large number of breast carcinomas, including breast carcinomas without a genetic defect in ras. Growth factors and their receptors of the epidermal growth factor family play an important role in the regulation of the growth of human breast cancer and many of these growth factor receptors apparently use the Ras signaling pathway. A substantial number of breast carcinomas are associated with overexpression and amplification of growth factor receptors, possibly resulting in Ras activation (higher Ras-GTP levels). Therefore, it is not surprising that some breast cancer cell lines are highly sensitive to FTIs. If so, FTIs could be therapeutically useful against a wider spectrum of human cancers than previously envisioned. Thus, our focus is on identifying the genetic aberrations upstream of Ras and the status of Ras-GTP levels in the highly sensitive human breast cancer cell lines. We are also extending the studies to a larger panel of breast cancer cell lines.

Table 2
Selected Effects of Ras FTIs on Mice Bearing Rat-1 Tumors: Experiment No. 2

Compound (BMS no.)	*Optimum dose[a] (mg/kg/inj)*	*Schedule, route*	*MST (d)*	*%T/C*	*T-C (d)[b]*
		I.P. Tumors[c]			
191563	45	2qd 1_11, i.p.	18.5	154	—
192331	45	2qd 1_11, i.p.	17.0	142	—
Control	—	—	12.0	100	—
1/10th Control[d]	—	—	16.0	133	—
		S.C. Tumors[c]			
191563	45	2qd 1 _11, i.p.	31.0	124	2.5
	6.9	qd1_14, scap[e]	26.0	104	3.3
Control	—	—	25.0	100	—
1/10th Control[d]	—	—	34.0	136	4.5

[a] Or maximum tolerated dose, if inactive.
[b] Difference in median times for treated (T) and control (C) mice to reach 1 g tumors.
[c] 10^6 cells were implanted on d 0; indicated treatments were begun on d 1.
[d] 10^5 cells were implanted on d 0.
[e] Subcutaneously inserted Alzet pump.

6. EFFECT OF FTIs ON TUMORS IN ANIMALS

6.1. In Vivo Testing of BMS-188222 in K-Ras Tumor Model

Soon after the discovery of its whole cell activity, BMS-188222 was tested i.p. against i.p. tumor in nude mice. Administered once daily at doses up to 90 mg/kg/inj for 9 d, this compound showed no significant activity. However, no toxicity was observed. At this time, we reasoned that demonstration of in vivo antitumor activity would require a more potent cell active compound (10–50-fold more potent than BMS-188222). In addition, because H-Ras transformed cells are more sensitive to FTIs than K-Ras transformed cells in the SAG assay, we decided to develop an H-Ras based tumor model (Rat-1), which should be less stringent than the K-Ras model.

6.2. Rat-1 Tumor Experiments

BMS-191563 and BMS-192331 were selected for testing in vivo, based on their potency as well as differences in chemical structure (ester vs acid at C-terminus; Met vs Gln at X position). An initial study was performed in which BMS-191563 was evaluated i.p. in mice implanted i.p. with Rat-1 tumor cells. The compound was administered at a single dose level of 45 mg/kg/inj, twice a day, for 11 d (i.e., 2qdx11). Control mice (n = 10) had a MST of 14 d and the 8 mice treated with BMS-191563 had a 9 d greater MST. Thus, this therapy resulted in a %T/C of 164%. Weight loss associated with the therapy reached a maximum of 3.2 g on d 7 of the treatment but then stabilized and recovered. One mouse survived to d 35 post-implant, at which time the experiment was terminated; no overt sign of tumor was found at necropsy.

In the next Rat-1 tumor experiment (No. 2) conducted, cells were implanted either i.p. or s.c. into different host mice (Table 2). In the i.p.-implanted portion of the study, both BMS-191563 and -192331 were administered i.p., 2qdx11, at varying dose levels.

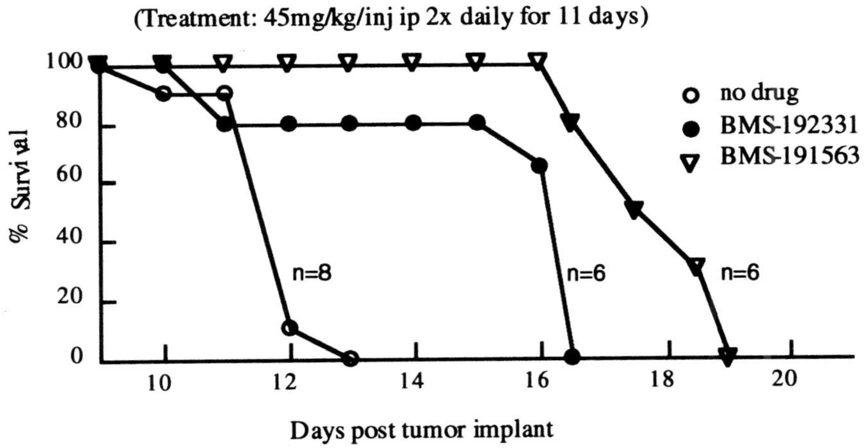

Fig. 4. Anti-tumor activity of BMS-191563 and BMS-192331 vs i.p. Rat-1 tumors. Both compounds were administered twice daily, 45 mg/kg/inj, for nine consequtive daily treatments, beginning on d 1 post-tumor implant.

At the highest dose level tested, 45 mg/kg/inj, both compounds produced an active result; BMS-191563 achieved a T/C of 154% and BMS-192331 caused a T/C of 142%. These results are depicted graphically in Fig. 4. Lower doses of both compounds were without activity. A single dose level of 90 mg of BMS-191563/kg/inj, given once daily for 9 d (qdx9), also was evaluated but it proved too toxic, causing the deaths of the majority of treated mice by the end of the scheduled therapy.

Against s.c. Rat-1 tumors, i.p. treatment with BMS-191563, 2qdx11, resulted in a maximum T/C of 124% at the highest dose tested, 45 mg/kg/inj. This increase in lifespan just missed our criterion for activity, and it was also less than the increase in lifespan (136%) of the titration control mice receiving only 10^5 Rat-1 cells s.c. The tumor growth delay associated with this therapy was 2.5 d, also an inactive result (considering the tumor volume doubling time of 1.8 d). Included in this experiment were groups of mice that received BMS-191563 delivered via Alzet pumps implanted s.c. (scap) adjacent to the growing tumors. The MTD of BMS-191563 administered in this manner was 6.9 mg/kg/d (the higher doses evaluated resulted in nearly total lethality) and it produced an inactive delay in tumor growth of 3.3 d. The effects of these treatments against s.c. Rat-1 tumors is shown in Fig. 5.

In the Rat-1 tumor experiment (No. 3), BMS-191563 and BMS-192331 were evaluated in various configurations of tumor implant site and route of injection (Table 3). Against i.p. Rat-1 tumors, BMS-191563 was evaluated using several i.p. regimens beginning on d 1 post-implant. A dose of 45 mg/kg/inj administered once a day for 9 d proved to be ineffective. We knew from the previous experiment that twice this dose level proved too toxic, so we can conclude that once daily injection of a maximum tolerated dose of BMS-191563 is ineffective against i.p. Rat-1 tumors. Two regimens involving twice daily i.p. injections were found, however, to be efficacious. When 45 mg/kg/inj of BMS-191563 was given twice daily for 7 or 14 d, the %T/C values obtained were 140 and 180%, respectively, once again confirming the activity of the compound in this setting.

BMS-191563 was also evaluated i.v. against i.v.-implanted Rat-1 cells beginning on d 1 post-implant. At the highest dose tolerated, 11.3 mg/kg/inj, given twice daily for a total of nine injections (i.e., 4 1/2 d), and without any double dose administered (e.g., as

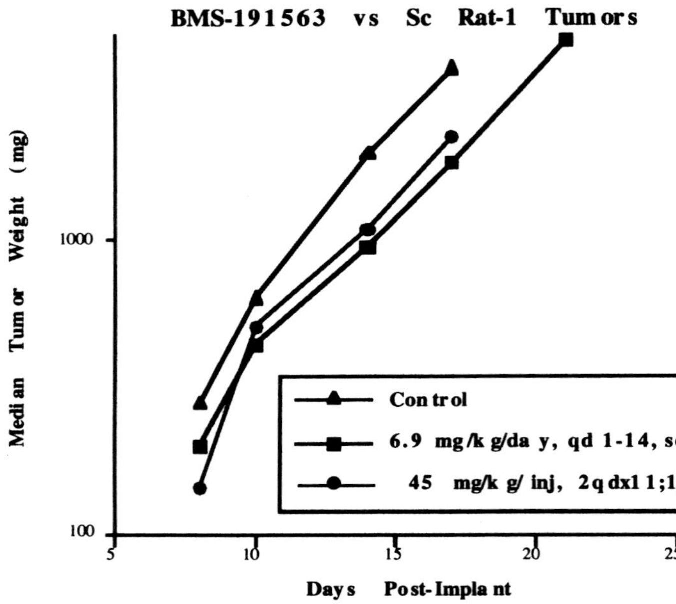

Fig. 5. Effect of BMS-191563 on s.c.-implanted Rat-1 tumors (Experiment No. 2). Tumor-bearing mice received 45 mg/kg/inj, 2qdx11, i.p., beginning on d 1 post-implant; or 6.9 mg/kg/d for 14 d subcutaneously via Alzet pump beginning on d 1 post-implant; or untreated tumor-bearing controls.

Table 3
Effect of Ras FTIs on Mice Bearing Rat-1 Tumors: Experiment No. 3

Compound (BMS no.)	*Optimum dose[a] (mg/kg/inj)*	*Schedule, route*	*MST (d)*	*%T/C*	*T-C (d)[b]*
		I.P. Tumors[c]			
191563	45	qd 1_9, i.p.	10.0	100	—
	45	2qd 1_14, i.p.	18.0	180	—
	45	2qd 1_7, i.p.	14.0	140	—
Control	—	—	10.0	—	—
1/10th Control[d]	—	—	11.5	115	—
		I.V. Tumors[c]			
191563	11.3	2qd 1_5, i.v.	14.5	116	—
Control	—	—	12.5	100	—
1/10th Control[d]	—	—	17.0	136	—
		S.C. Tumors[c]			
191563	45	2qd 1 _7, scpt[e]	ND[g]		7.8
	45	2qd _7, sccl[f]	ND		4.0
	11.3	2qd 1 _5, i.v.	ND		1.5
192331	45	2qd 1_7, scpt[e]	ND		5.5
	45	2qd 1_7, sccl[f]	ND		4.0
Control	—	—	ND		—
1/10th Control[d]	—	—	ND		1.3

[a–d]See footnotes in Table 1.
[e]Subcutaneous peritumoral injections.
[f]Subcutaneous contralateral injections.
[g]ND, not determined.

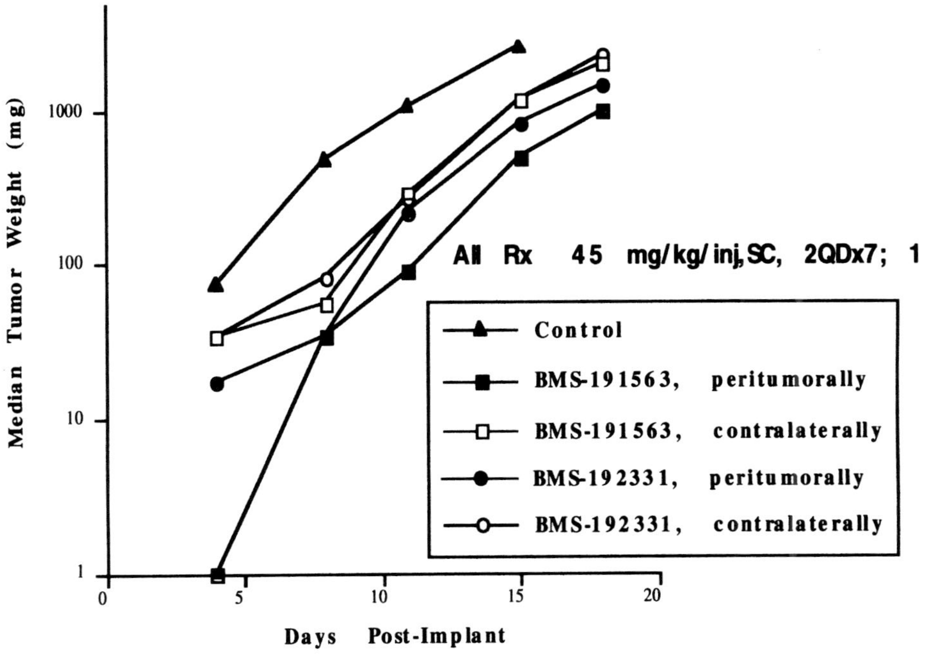

Fig. 6. Effect of FTIs on s.c.-implanted Rat-1 tumors (Experiment No. 3). All treatments were 45 mg/kg/inj, 2qdx7, s.c., beginning on d 1 post-implant given peritumorally or contralaterally.

typically applied on a weekend), no activity was observed as reflected by a T/C of 116% (Table 3). Twice this dose level caused excessive deaths.

Also included in Rat-1 Experiment No. 3 were the evaluations of both BMS-191563 and BMS-192331 vs s.c. Rat-1 tumors (Table 3). Intravenous administration of BMS-191563, 11.3 mg/kg/inj twice daily for nine injections, failed to inhibit the growth of s.c. Rat-1 tumors compared to untreated tumor-bearing control mice. All other treatment regimens evaluated consisted of the compounds administered s.c., either peritumorally or contralaterally from the tumor site, at a dose level of 45 mg/kg/inj, twice daily, for 7 d beginning d 1 post-implant. The tumor volume doubling time of s.c. Rat-1 tumors in this experiment was 2.2 d; thus, a delay of 7.3 d was the criterion for activity. Contralateral s.c. injections of both compounds resulted in a delay in tumor growth of 4 d, an inactive result. Peritumoral s.c. injections of BMS-192331 yielded a 5.5-d delay in tumor growth, and only peritumoral s.c. injections of BMS-191563 achieved a borderline active result of 7.8 d (Fig. 6).

6.3. RASK Tumor Experiment

BMS-191563 was evaluated i.p. vs i.p.-implanted RASK tumors. At the only dose level evaluated, 45 mg/kg/inj, given twice a day for 11 d beginning on d 1 post-implant, BMS-191563 was inactive.

6.4. MCF-7 Tumor Experiment

One experiment was performed in which both BMS-191563 and BMS-192331 were evaluated in mice bearing s.c. MCF-7 tumors. The effect of various treatments on delaying tumor growth is shown in Table 4. Control mice grew 250-mg tumors in a median time

Table 4
Optimal Effects of Ras FTIs on Sc MCF-7 Breast Carcinoma

Compound (BMS no.)	*MTD[a] (mg/kg/inj)*	*Schedule, route*	*T-C (d)[b]*
191563	45	2qd 0_21, i.p.	9.8[d]
	19.5	qd 0_14, scap[c]	4.8
192331	45	2qd 0_21, i.p.	9.5
Control	—	—	—

[a]Maximum tolerated dose.
[b]Difference in median time for treated (T) and control (C) mice to growth tumors of 250 mg. The values shown are the maximum delays obtained with the compound and treatment regimen shown, but not necessarily at the MTD.
[c]Subcutaneously inserted Alzet pump.
[d]$p < 0.05$ compared to control group.

of 19.8 d. Mice administered BMS-191563 twice a day, i.p., for 21 d, beginning on the day of implant (d 0), had a median delay in tumor growth of between 7.8 and 9.8 d, with the best effect achieved at a dose of 5 mg/kg/inj (although nine times as much compound was tolerated in another treatment group). BMS-191563 was also administered s.c. via Alzet pump for 14 d beginning on d 0. Doses of 2.6–19.5 mg/kg/d were within the acceptable limits of toxicity. The maximum T/C value achieved using tolerated doses of BMS-191563 was 4.8 d following administration of the lowest dose evaluated, 2.6 mg/kg/d. BMS-192331 was assayed using the same 21 d, twice a day, i.p. injection regimen described previously for BMS-191563. At the highest dose tested, 45 mg/kg/inj, BMS-192331 caused a T/C of 9.5 d. Considering the 5.8-d tumor volume doubling time in this experiment, none of these maximum delays in tumor growth was consistent with activity (i.e., 1 LCK), although the 21-d i.p. injection regimens involving both compounds did produce statistically significant ($p < 0.05$) perturbations in tumor growth (Fig. 7), determined either in terms of T/C values and/or relative median tumor weights on certain days of measurement (data not shown).

7. CONCLUSIONS

BMS has synthesized several different classes of FTIs *(13–14)*. Bisubstrate mimics and CAAX box analogs afforded us the most potent compounds. Selected on the basis of their performance in several in vitro assays for inhibition of Ras farnesyltransferase, inhibition of Ras-induced cell transformation, reversal of Ras-induced cell transformation, and inhibition of tumor cell growth on plastic or in agarose, BMS-191563 and BMS-192331 were subjected to in vivo antitumor testing.

Against an i.p.-implanted H-Ras transformed rat cell line, Rat-1, both compounds were active when administered i.p. The levels of activity observed, T/C of 140–180%, depending on the compound and the particular treatment regimen used, were mild to moderate. Nonetheless, these data represent the clear demonstration of in vivo antitumor activity for FTIs. Of particular note are the findings that lower doses, below the 45 mg/kg/inj (2qdx11) level, were devoid of activity and that the same daily exposure to BMS-191563, but given in a single injection, was excessively toxic (BMS-192331 was not similarly evaluated). Furthermore, 45 mg/kg/inj of BMS-191563 given only once a day

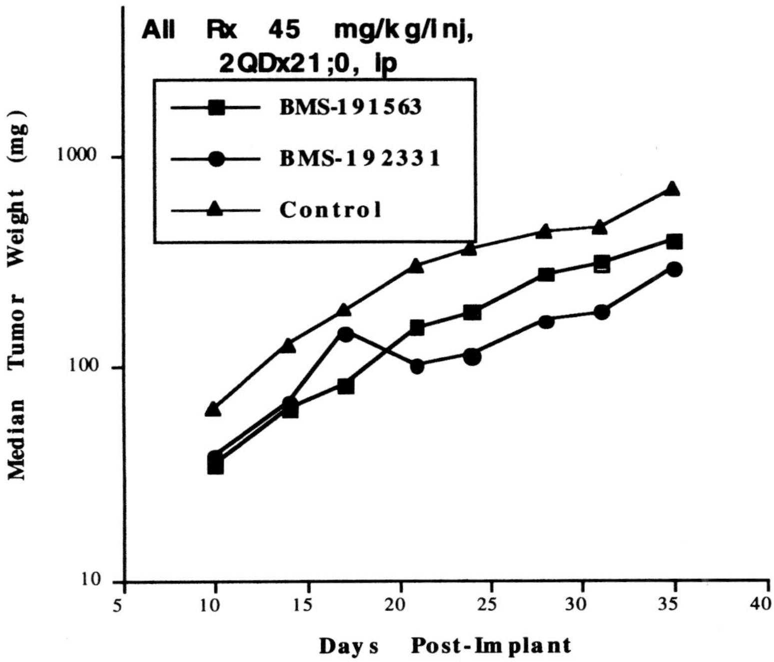

Fig. 7. Effect of BMS-191563 and BMS-192331 on s.c.-implanted MCF-7 tumors. All treatments were 45 mg/kg/inj, 2qdx21, i.p., beginning on d 0 (i.e., 1 h post-tumor implant).

was also ineffective. Thus, although we have established the in vivo activity of two Ras farnesyltransferase inhibitors, the therapeutic window is extremely narrow and it was fortuitous that we selected for evaluation the particular dose and regimen we did.

Additional antitumor assays were performed using i.v. and s.c. Rat-1 tumor implants. BMS-191563 was not active when given i.v. to mice implanted i.v. with Rat-1 tumor cells. BMS-191563 was not active when given i.p. or i.v. to mice bearing s.c. Rat-1 tumors, nor was it active when given s.c. via Alzet pump inserted adjacent to the s.c. growing tumor. Neither BMS-191563 nor BMS-192331 was active when administered s.c. contralateral to the site of the s.c. growing tumor. Peritumoral s.c. injections of the former compound, but not the latter, did result in a borderline active delay in tumor growth. Despite the activity of BMS-191563 in the i.p. H-Ras Rat-1 tumor model, identical treatment was not effective against i.p.- implanted K-Ras RASK tumor cells.

The human breast cancer cells, MCF-7, were found in vitro to be quite sensitive to BMS-191563 relative to all other tumor cells assayed. Accordingly, mice bearing s.c. MCF-7 tumors were treated with both BMS-191563 and BMS-192331. Statistically significant perturbations in tumor growth were obtained following i.p. treatments with these compounds, but the extent of tumor growth delays were insufficient to qualify as *active* results according to our standard criterion (i.e., 1 LCK).

Of unusual character was the dose-response/schedule dependent toxicity profile of BMS-191563. Once daily i.p. administrations of 90 mg/kg/inj of BMS-191563 resulted in lethality as early as d 7 post-initiation of treatment, but 45 mg/kg/inj, twice a day, was tolerated for as many as 21 d. Furthermore, such twice a day injections were the only means found to achieve an antitumor effect, suggesting rapid clearance or biotransformation of the active principle. Yet, when it appeared that peak plasma levels of BMS-191563 (or a metabolite) were responsible for toxicity, slow s.c. infusion of as little as 24 mg/kg/d for 14 d proved to be lethal (and not effective at tolerated doses vs adjacent s.c. tumors). The cause of deaths owing to BMS-191563 is not known, and although the compound showed good selectivity in vitro with regard to the relative concentrations associated with inhibition of growth in normal and tumor cells, in vivo the therapeutic window is extremely narrow.

In conclusion, the in vivo antitumor activity of Ras FTIs was reproducibly demonstrated in an i.p. H-Ras transformed tumor model (Rat-1). No meaningful antitumor effects were obtained in disseminated or distal site models of this tumor, nor in an i.p. K-Ras tumor model (RasK), nor in a s.c. human tumor model (MCF-7) whose cells were particularly sensitive in vitro to the inhibitors.

Future efforts should include providing compounds with improved pharmacokinetic properties in an attempt to establish distal site antitumor activity and to eliminate the need for twice-daily injections.

REFERENCES

1. Bos JL. ras oncogenes in human cancer: a review. *Cancer Res* 1989; 49:4682–4689.
2. Barbacid M. ras genes. *Annu Rev Biochem* 1987; 56:779–827.
3. Khosravi-Far R, Cox AD, Kato K, Der CJ. Protein prenylation: key to ras function and cancer intervention? *Cell Growth Differ* 1992; 3:461–469.
4. Zhang FL, Casey PJ. Protein prenylation: molecular mechanisms and functional consequences. *Annu Rev Biochem* 1996; 65:241–269.
5. Hancock JF, Magee AI, Childs JE, Marshall CJ. All ras proteins are polyisoprenylated but only some are palmitoylated. *Cell* 1989; 57:1167–1177.
6. Farnsworth CC, Gelb MH, Glomset JA. Identification of geranylgeranyl-modified proteins in HeLa cells. *Science* 1990; 247:320–322.
7. Rilling HC, Bruenger E, Leining LM, Buss JE, Epstein WW. Differential prenylation of proteins as a function of mevalonate concentration in CHO cells. *Arch Biochem Biophys* 1993; 301:210–215.
8. Manne V, Yan N, Carboni JM, Tuomari AV, Ricca CS, Brown JG, et. al. Bisubstrate inhibitors of farnesyltransferase: a novel class of specific inhibitors of ras transformed cells. *Oncogene* 1995; 10:1763–1779.
9. Yan N, Ricca C, Fletcher J, Glover T, Seizinger BR, Manne V. Farnesyltransferase inhibitors block the neurofibromatosis type I (NF1) malignant phenotype. *Cancer Res* 1995; 55:3569-3575.
10. Carboni JM, Yan N, Cox AD, Bustelo X, Graham SM, Lynch MJ, et. al. Farnesyltransferase inhibitors are inhibitors of Ras but not R-Ras2/TC21, transformation. *Oncogene* 1995; 10:1905–1913.
11. Geran RI, Greenberg NH, MacDonald MM, Abbott BJ. Modified protocol for the testing of new synthetics in the L1210 lymphoid leukemia murine model in the DR&G program, DCT, NCI. *Natl Cancer Inst Monogr* 1977; 45:151–153.
12. Gehan GA. A generalized Wilcoxon test for comparing arbitrarily singly-censored samples. *Biometrika* 1985; 52:203–233.
13. Patel DV, Gordon EM, Schmidt RJ, Weller HN, Young MG, Zahler R, et al. Phosphinyl acid based bisubstrate analog inhibitors of farnesyl protein transferase. *J Med Chem* 1995; 38:435–442.
14. Leftheris K, Kline T, Natarajan S, DeVirgilio MK, Cho YH, Pluscec J, et al. Peptide based p21 ras farnesyl transferase inhibitors: systematic modification of the tetrapeptide CA_1A_2X motif. *Bioorg Med Chem Lett* 1994; 4:887–892.

7 Tricyclic Farnesyl Protein Transferase Inhibitors

Antitumor Activity and Effects on Protein Prenylation

W. Robert Bishop, PHD, *James J.-K. Pai,* PHD, *Lydia Armstrong,* PHD, *Marguerite B. Dalton,* PHD, *Ronald J. Doll,* PHD, *Arthur Taveras,* PHD, *George Njoroge,* PHD, *Michael Sinensky,* PHD, *Fang Zhang,* PHD, *Ming Liu,* PHD, *and Paul Kirschmeier,* PHD

Contents

1. INTRODUCTION

Over the last decade, the posttranslational processing pathway of the Ras oncoprotein has been elucidated and the enzyme that catalyzes the initial and critical step in this pathway, protein farnesyltransferase (FTase), has been characterized *(1–6)*. These advances led to a great deal of effort, both in the pharmaceutical industry and in academia, focused on the discovery of selective FTase inhibitors (FTIs). This chapter discusses two aspects of the FTI program at Schering-Plough Research Institute: 1) the discovery and development of the tricyclic series of inhibitors, and 2) studies on the substrate specificity of the prenyl transferases. The latter studies are relevant to our understanding of the in vivo effects of FTIs on prenylation of Ras and other proteins.

From: *Farnesyltransferase Inhibitors in Cancer Therapy*
Edited by: S. M. Sebti and A. D. Hamilton © Humana Press Inc., Totowa, NJ

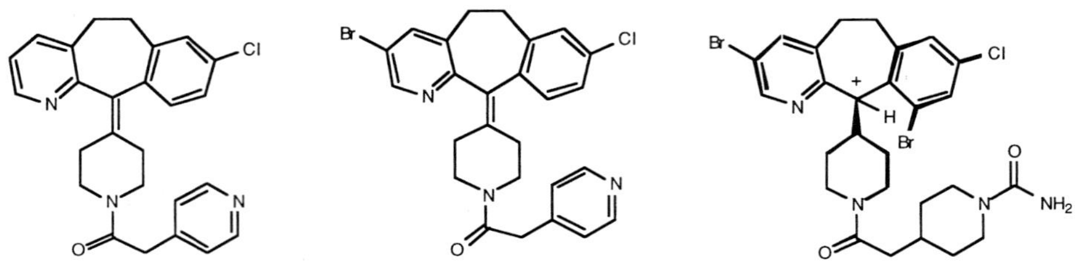

SCH 44342 **SCH 56582** **SCH 66336**

Fig. 1. Structures of key tricyclic FTIs. (**Left**) SCH 44342 is an early-generation tricyclic FTI that inhibits FTase with an IC_{50} value of 250 n*M*. Further modification—in particular, halogenation of the tricyclic ring system—enhanced potency. (**Middle**) SCH 56582 inhibits FTase with an IC_{50} value of 60 n*M*. (**Right**) SCH 66336 inhibits FTase with an IC_{50} value of 1.9 n*M*. This compound is currently being evaluated in Phase I clinical trials.

2. DISCOVERY AND INITIAL CHARACTERIZATION OF TRICYCLIC FTIs

A number of inhibitors of FTase have been reported (reviewed in refs. *7,8*). Many of these are derived from substrate-mimetic approaches. The design of CAAX peptidomimetics *(9–14)* has resulted in potent and selective FTIs capable of blocking H-Ras processing in cells, inhibiting cellular transformation induced by oncogenic Ras proteins, and slowing the growth of Ras-dependent tumors in nude mice *(15–17)*.

SCH 44342 (Fig. 1) was one of the initial lead compounds in the tricyclic FTI series *(18)*. This class of FTIs was discovered in 1991 by random screening of the Schering-Plough library of compounds. Unlike many other FTIs, the tricyclic class is entirely nonpeptidic and lacks a sulfhydryl function. SCH 44342 inhibits recombinant human FTase with an IC_{50} value of approx 0.25 μ*M*. This inhibition is competitive with respect to the protein or peptide CAAX substrate. Tricyclic FTIs are structurally related to compounds possessing histamine H1 and platelet-activating factor antagonist activity *(19)*, however, the FTase inhibitory activity is separable from these other activities. Most of the tricyclic FTIs are inactive or weakly active as inhibitors of the related protein prenyltransferase, protein geranylgeranyltransferase-1 (GGTase I). For example, SCH 44342 has no inhibitory activity against GGTase I at concentrations up to 100 μ*M*.

When the biological effects of SCH 44342 and related tricyclics were examined in cellular or animal systems driven by an activated H-Ras oncogene, the results observed were straightforward. H-Ras driven systems allow the use of a powerful negative control *(20,21)*. Although H-Ras with its native CAAX sequence of CVLS is only a substrate for FTase, H-Ras in which the CAAX sequence is changed to CVLL is only a GGTase I substrate. Thus, selective effects on CVLS vs CVLL H-Ras transformed cells is a clear indication that compounds are exerting their biological effects by selective inhibition of FTase.

SCH 44342 inhibits the post-translational processing of H-Ras-CVLS in Cos-7 monkey kidney cells with an IC_{50} value of approx 3.0 μ*M (18)*. In contrast, it does not inhibit H-Ras-CVLL processing. Additionally, SCH 44342 prevents the phenotypic change that occurs in Cos cells after transient expression of an activated [Val^{12}] form of H-Ras-CVLS. SCH 44342 does not prevent this phenotypic change in cells overexpressing [Val^{12}]-H-Ras-CVLL. It is important to note that SCH 44342 and its analogs have no apparent cytotoxic effects on Cos cells at the concentrations that exert this morphological effect.

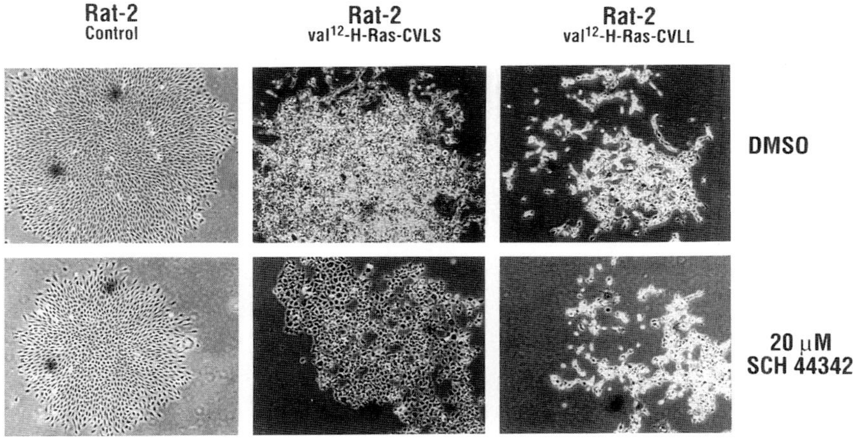

Fig. 2. Effects of SCH 44342 on morphology of normal and Ras-transformed Rat2 fibroblasts. Control Rat 2 fibroblasts and fibroblasts transformed with Val[12]-activated forms of H-Ras-CVLS or H-Ras-CVLL were plated at 250 cells/100 mm per dish in the presence or absence of 20 μ*M* SCH 44342 in Dulbecco's modified Eagle's medium containing 10% fetal calf serum. Plates were re-fed after 5 d and photographed on d 14. Drug treatment had no effect on plating efficiency of any of the cell lines.

The effect of SCH 44342 on the growth properties of Rat 2 fibroblasts stably transformed with activated H-Ras was also examined. SCH 44342 does not effect the plating efficiency of parental Rat 2 cells or Rat 2 cells transformed by H-Ras-CVLS or H-Ras-CVLL, confirming its lack of cytotoxicity at concentrations where H-ras processing is fully inhibited. Figure 2 shows the morphology of clones grown in the presence of either DMSO or 20 μ*M* SCH 44342. The morphology of the normal Rat-2 cells is unaffected by the presence of drug, although the clones are somewhat smaller on average compared to those grown in its absence. Cells transformed by H-Ras-CVLS undergo a pronounced morphological change in response to SCH 44342, displaying a flatter, less refractile, more contact inhibited phenotype than vehicle control-treated cells. Morphological reversion is not observed in cells transformed with geranylgeranylated H-Ras-CVLL. Consistent with these selective effects on the transformed phenotype, treatment of H-Ras-CVLS (but not H-Ras-CVLL) transformed cells with SCH 44342 results in the reformation of stress fibers and a decrease in membrane extensions (Fig. 3). In addition, SCH 44342 dose-dependently reduces anchorage-independent soft agar growth of the H-Ras-CVLS transformants, but not of Rat-2 cells transformed with the geranylgeranylated protein.

During the characterization of early generation FTIs, we and others *(see* refs. *21,22)* found that cellular functions that were predicted to be altered by blocking farnesylation were spared. These included signaling pathways thought to involve Ras activation and assembly of the nuclear envelope, despite the fact that both lamin B and prelamin A are FTase substrates (*see below*). These findings are consistent with the lack of general cytotoxicity observed with FTIs. There are several possible explanations for these paradoxical observations. One possibility is that redundant signaling pathways exist that bypass Ras inactivation. A second possibility is that inhibition of Ras farnesylation is incomplete owing either to incomplete inhibition of FTase or to alternative prenylation of Ras by another cellular prenyl transferase. The discussion to follow focuses on the second possibility; however, the contribution of potential redundant pathways has not been fully explored and needs to be addressed further experimentally.

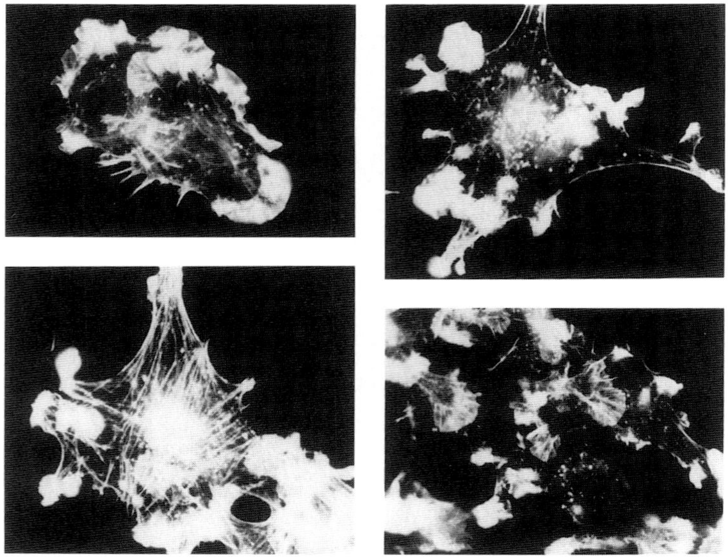

Fig. 3. Effects of SCH 44342 on actin filaments in normal and Ras-transformed Rat2 fibroblasts. Rat 2 fibroblasts transformed with Val^{12}-activated forms of H-Ras-CVLS (**A**) or H-Ras-CVLL (**B**) show many membrane ruffles and filopodia (arrows). Organized actin cables are seen only in the cell periphery. After treatment with 10 μM SCH 44342, Ras-CVLS-transformed cells (**C**) show well-organized stress fibers extending across the entire cell (arrow heads). Ras-CVLL-transformed cells (**D**) show little change in the actin cytoskeleton.

3. ALTERNATIVE PRENYLATION OF RAS PROTEINS IN THE PRESENCE OF FTIs

A number of cellular proteins undergo C-terminal prenylation; the majority of these are modified with the 20-carbon isoprene, geranylgeranyl *(23–25)*. Two distinct protein geranylgeranyltransferases (GGTase I and II) have been identified *(26–29)*. GGTase II utilizes protein substrates terminating in Cys-Cys or Cys-X-Cys whereas GGTase I, like FTase, recognizes substrates with C-terminal CAAX motifs. The primary determinant of specificity of FTase and GGTase I is the protein substrate's carboxy terminal amino acid *(30–32)*. Proteins ending in Ser or Met are preferred FTase substrates, whereas proteins terminating in Leu are preferred substrates for GGTase I. As indicated previously, substitution of leucine for serine at the C-terminus of the H-Ras CAAX box makes this protein a substrate for geranylgeranylation (rather than farnesylation) both in vitro and in cells *(30)*.

Despite this known preference for distinct C-terminal residues, the rules that govern protein utilization by prenyl transferases are not fully defined. The specificity of FTase can be modified by site-directed mutagenesis or by replacement of the catalytic zinc. The zinc atom is an integral component of FTase, and its removal by prolonged dialysis against chelating agents results in complete inactivation of the enzyme *(33,34)*. Metal-depleted FTase retains high-affinity binding of the isoprenoid substrate, but is no longer able to bind the protein substrate. Restoring activity to apo-FTase requires simultaneous addition of both zinc and magnesium ions. Zhang et al. *(35)* demonstrated that Cd^{2+} can

Table 1
Ras Isoform Carboxy-Terminal Sequences

H-ras:	D	E	S	G	P	G	C	M	S	C	K	C	V	L	S
N-ras:	D	D	G	T	Q	G	C	M	G	L	P	C	V	V	M
K-ras4A:	E	K	T	P	G	C	V	K	I	K	K	C	I	I	M
K-ras4B:	G	K	K	K	K	K	K	S	K	T	K	C	V	I	M

The carboxy-terminal amino acid sequences from the 4 Ras isoforms are shown. H-Ras terminates in serine, while the other isoforms all terminate in methionine. Note also the basic character of the carboxy-terminus of K-ras4B. All Ras isoforms, with the exception of H-Ras serve as substrates for GGTase I both in vitro and in FTI-treated cells.

substitute for Zn^{2+} to reconstitute this activity. Using standard substrates, H-Ras-CVLS and FPP, the specific activity of Cd^{2+}-reconstituted FTase (Cd-FTase) is about 50% that of the Zn^{2+}-containing enzyme. In addition, Cd-FTase has gained the ability to prenylate efficiently leucine-terminated substrates.

Even without structural modification, the prenyl transferases are not absolute in their specificity. This has been elucidated in part by examining the four isoforms of Ras (H-Ras, N-Ras, K-Ras4A, and K-Ras4B) as prenyl transferase substrates. These proteins are highly homologous to each other, with most of the differences residing in the last 24 residues *(36)*. These studies have potential implications for the clinical effects of FTIs because oncogenic mutations of the different isoforms predominate in different tumors *(37)*. For example, H-ras mutations are found in carcinomas of the bladder, kidney, and thyroid; N-ras mutations are found in myeloid and lymphoid disorders, liver carcinoma, and melanoma; K-ras mutations predominate in lung, colon and pancreatic carcinoma. The C-terminal sequence of each isoform is shown in Table 1. All Ras proteins are normally modified by FTase intracellularly *(1,38,39)*.

James et al. *(40)* reported that the K-Ras4B protein could serve as a substrate for both FTase and GGTase I in vitro. To follow up on this observation, we produced all four Ras isoforms as His-tagged proteins in *Escherichia coli* and tested them as in vitro substrates for human FTase and GGTase I *(41)*. As FTase substrates, the K_m for H-Ras, N-Ras, and K-Ras4A are 0.6, 0.4, and 0.4 μ*M*, respectively. The K_m for K-Ras4B is 30 n*M*. The higher affinity for K-Ras4B accounts for the fact that higher concentrations of CAAX-competitive FTIs are needed to inhibit K-Ras farnesylation.

We also found that N-Ras and K-Ras4A, in addition to K-Ras4B, are substrates for GGTase I. All of these proteins terminate in methionine. Overall, the catalytic efficiencies for the GGTase I reactions with N-Ras or either K-Ras isoform are similar and only about twofold lower than that for a leucine-terminated substrate (Table 2). It is also clear from this data that farnesylation of these substrates is the preferred reaction. In the case of K-Ras4B, the catalytic efficiency of its farnesylation is about 140-fold higher than that of its reaction with GGTase I.

To further explore recognition of the Ras proteins by FTase and GGTase I, we utilized peptides comprising their 15 C-terminal residues (Table 1). Consistent with whole protein studies, the K-Ras4B peptide is the highest affinity FTase substrate. Also as anticipated, the K-Ras4A and K-Ras4B peptides are substrates for GGTase I. Surprisingly, the N-Ras peptide was not a GGTase I substrate, suggesting that upstream sequences present

Table 2
Kinetic Constants for the Ras Isoforms as FTase and GGTase I Substrates

	FTase			*GGTase I*		
Substrate	K_m (μM)	k_{cat} min^{-1}	k_{cat}/K_m $\mu M.min^{-1}$	K_m (μM)	k_{cat} min^{-1}	k_{cat}/K_m $\mu M.min^{-1}$
H-Ras	0.6	1.2	2.0	–	inactive	–
N-Ras	0.4	1.9	4.7	2.1	0.68	0.32
K-Ras4A	0.4	5.0	12.4	8.8	4.0	0.50
K-Ras4B	0.03	1.6	53.3	12.0	4.6	0.38

All 4 Ras isoforms were produced as recombinant His-tagged proteins in E. coli, purified by metal chelate chromatography and evaluated as in vitro substrates for human FTase and GGTase I. The kinetic parameters are shown. H-Ras is not a substrate for GGTase I. All other Ras isoforms are GGTase I substrates, however, the farnesylation of these proteins occurs with a much higher catalytic efficiency.

in this protein may play a role in its recognition by GGTase I. Previous studies have suggested a role for upstream structural elements in the recognition of protein substrates by prenyl transferases *(42,43)*.

Focusing on the C-terminus of K-Ras4B, the peptide GKKKKKKSKTKCVIM is a good GGTase I substrate. Removing four of the six contiguous lysine residues greatly decreases the affinity of GGTase I for this peptide. Furthermore, changing the C-terminal methionine to serine completely abolishes its ability to serve as a GGTase I substrate. Therefore, consistent with results reported using a chimeric protein approach *(40)*, both the lysine residues and the C-terminal methionine contribute to the utilization of K-Ras4B by GGTase I.

It was critical to extend these in vitro observations to the intact cell. James et al. *(44)* reported that a chimeric Ras protein consisting of the first 164 amino acids of H-Ras followed by the carboxy terminal 24 amino acids of K-Ras4B continues to incorporate [^{3}H]mevalonate in Rat-1 cells grown in the presence of the peptidomimetic FTI BZA-5B. The prenyl group attached to this protein was reported to be farnesyl, even in the presence of the FTI. We performed similar [^{3}H]mevalonate-labeling studies to explore the effects of tricyclic FTIs on Ras prenylation in DLD1 human colon carcinoma cells *(45)*. DLD1 cells express predominantly N-Ras and K-Ras isoforms by Western-blot analysis. Mevalonate labeling efficiency was enhanced by introduction of a cDNA encoding a mevalonate transport protein and the use of the HMG-CoA reductase inhibitor, mevastatin. Ras proteins were immunoprecipitated from the cell lysate with the pan-ras antibody Y13-259. Mevalonate incorporation into Ras was not substantially inhibited by increasing concentrations of SCH 56582. The nature of the prenyl group attached to Ras was then evaluated using methyl iodide cleavage followed by high-pressure liquid chromatography (HPLC) analysis. Ras-associated prenyl groups derived from untreated DLD1 cells comigrated with a farnesol standard. Following treatment with SCH 44342, the majority of Ras-derived prenyl groups migrated with the geranylgeraniol standard. The shift from farnesol to geranylgeraniol incorporation is dose-dependent. Rowell et al. *(46)* reached a similar conclusion using a distinct peptidomimetic FTI.

To examine the effect of FTI treatment on prenylation of individual Ras isoforms, we cotransfected Cos cells with the mevalonate transporter construct and the various human ras genes (H-Ras, N-Ras, K-Ras4A, K-Ras4B, or the GGTase I substrate, H-Ras-CVLL). In these cells, K-Ras4A, K-Ras4B, N-Ras, and H-Ras-CVLL continued to incorporate

[^{3}H]mevalonate in the presence of SCH 56582. In contrast, H-Ras labeling was completely inhibited. Isoprene analysis was then performed on the Ras immunoprecipitates (Fig. 4). In the absence of FTI, N-Ras, K-Ras 4A, and H-Ras contained farnesol exclusively. In contrast, in the absence of FTI, K-Ras4B contained a mixture of approx 80% farnesol and 20% geranylgeraniol. This result was unexpected because previous studies failed to detect incorporation of geranylgeranyl groups into K-Ras proteins in untreated cells *(1)*. Our result may be a consequence of the high level of expression of K-Ras4B in Cos cells. Following treatment with SCH 56582, immunoprecipitates of both N-Ras and K-Ras proteins contained only geranylgeraniol.

The effect of SCH 56582 on membrane-association of the Ras proteins was also analyzed. Transfected Cos cells were fractionated into a particulate (membrane) and a soluble fraction. In control cells, each of the Ras proteins was associated exclusively with the particulate fraction. Treatment with mevastatin releases all forms of Ras from this fraction. In cells treated with SCH 56582, H-Ras is released from the particulate fraction. In contrast, both isoforms of K-Ras remain entirely particulate. Most of the N-Ras protein also remains associated with the particulate fraction in the presence of SCH 56582. Some (~10%) of the N-Ras protein, however, consistently appears as unprocessed precursor in the soluble fraction. The unique sensitivity of H-Ras processing to inhibition suggests that cells whose transformed phenotype depends on H-Ras activity may be particularly responsive to FTI treatment. In agreement with this, Rat-2 cells transformed with mutant H-Ras are approx 5–10-fold more sensitive to tricyclic FTIs in soft agar cloning experiments than are cells transformed with mutant K-Ras.

In light of alternative prenylation, the mechanism by which FTIs inhibit anchorage-independent growth of K-Ras transformed fibroblasts or human tumor cell lines expressing an activated K-Ras oncogene is unclear. The incorporation of geranylgeranyl rather than farnesyl into N- and K-Ras proteins may cause subtle changes in Ras signaling, perhaps altering subcellular localization or protein–protein interaction, leading to reduced ability to grow in an anchorage-independent fashion. Alternatively, the mechanism of action of FTIs may involve inhibiting the farnesylation of a target protein distinct from Ras, which lies downstream in the Ras signaling pathway (*see below*). These mechanistic questions were recently reviewed by Der and Cox *(47)*.

The mechanism of action of FTIs is clearly more complex than originally envisioned. The overall result of FTI treatment must reflect the combined effects of inhibiting the prenylation of some proteins, such as H-Ras, and altering the pattern of prenylation of others, such as N- and K-Ras. The overall changes in protein prenylation that occur after FTI treatment must largely determine not only their antitumor activity but also their side-effect profile.

4. OPTIMIZATION OF TRICYCLIC INHIBITORS: IDENTIFICATION OF SCH 66336 AS A CLINICAL CANDIDATE

Since the discovery of SCH 44342, a large medicinal chemistry effort was undertaken to improve both the intrinsic potency and the pharmacokinetics in this series *(48–52)*. Introduction of a halogen substituent at the 3-position of the tricyclic ring (e.g., SCH 56582) enhanced the potency in this series on the order of 10-fold. In vivo antitumor activity of these dihalogenated compounds was recently reported *(53)*. Substitution of the tricyclic ring with a third halogen gave a further potency increase of approx 10-fold.

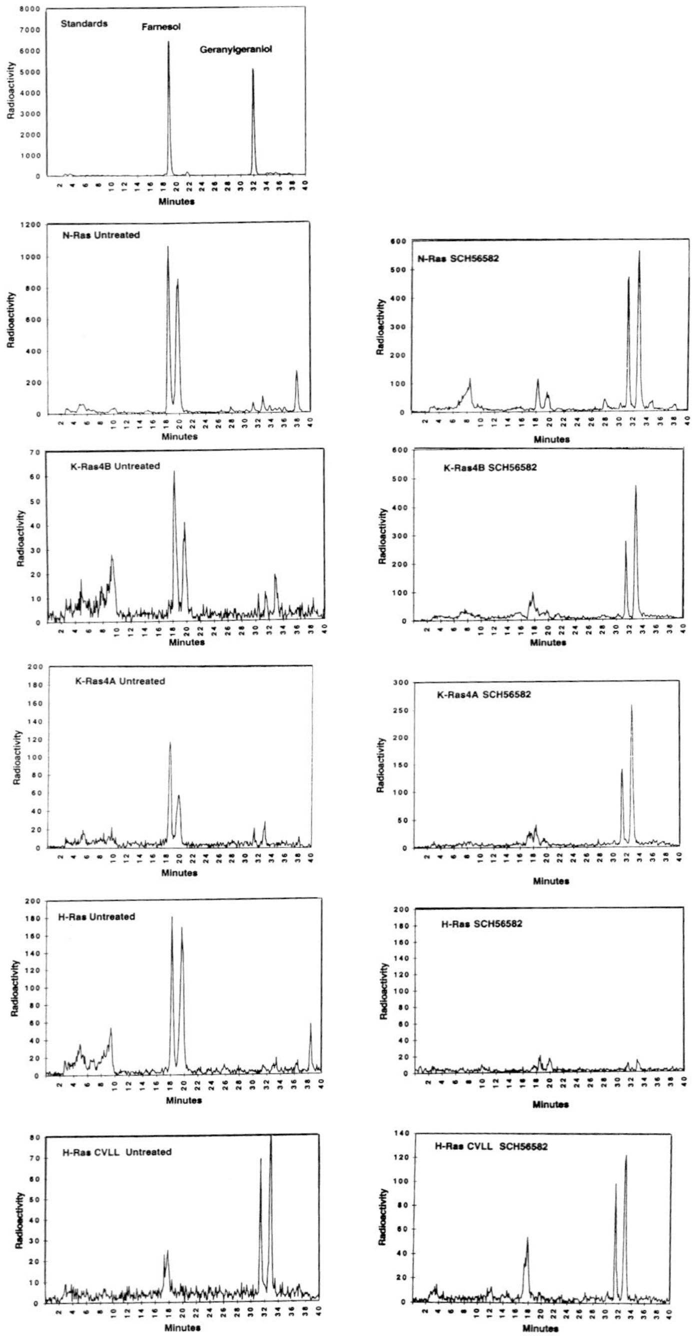

Fig. 4. Prenylation of Ras proteins in COS Cells. Various Ras proteins were expressed in Cos cells and labeled with [^{3}H]mevalonate as described in the text. Cells were treated with 5 μ*M* SCH 56582 (+) or dimethyl sulfoxide (DMSO) (−) for 48 h after transfection. Ras proteins were immunoprecipitated, and isoprene groups were released by methyl iodide cleavage and analyzed by high-pressure liquid chromatography (HPLC) analysis with radiochemical detection. H-Ras prenylation is blocked by SCH 56582. After FTI treatment, N-Ras, K-Ras4B, and K-Ras4A are alternatively prenylated by GGTase I. H-Ras-CVLL is geranylgeranylated in the absence and presence of FTI.

A prototypic compound in the trihalogenated tricyclic series is SCH 66336. This compound possesses potent anti-tumor activity against a wide variety of tumors following oral administration in mice *(54)* and is currently undergoing Phase I human clinical trials as an anticancer chemotherapeutic agent.

SCH 66336 blocks farnesylation of Ha-Ras in vitro by human FTase with an IC_{50} value of 1.9 n*M*. SCH 66336 also inhibits farnesylation of N-Ras (2.8 n*M*) and K-Ras-4B in vitro (5.2 n*M*), although it does not inhibit GGTase I at concentrations up to 50 μ*M*. SCH 66336 inhibits farnesylation of H-Ras in Cos cells with an IC_{50} value of 10 n*M*, indicating that it penetrates the cell membrane readily.

SCH 66336 also blocks anchorage-independent, soft agar growth of rodent fibroblasts expressing an activated form of H-Ras-CVLS with an IC_{50} of 75 n*M*. Anchorage-independent growth of fibroblasts expressing activated K-Ras is also inhibited by SCH 66336 with an IC_{50} value of 400 n*M*. The anchorage-independent, soft agar growth of a variety of human tumor cell lines containing activated H-Ras and K-Ras is also inhibited by SCH 66336 (Kirschmeier et al., in preparation). This includes human tumor cell lines derived from a variety of tissues of origin. As observed with other FTIs *(55)*, some tumor cell lines lacking Ras mutations are also growth inhibited in soft agar by SCH 66336. Sensitive cells with a wild-type *ras* genotype nevertheless may be dependent on Ras-mediated signaling for maintenance of the transformed phenotype. Ras-dependence may be owing to mutational activation of an oncogene(s) lying upstream of Ras in a signaling pathway (e.g., *her-2/neu*) or reliance on autocrine growth factors that activate Ras signal transduction pathways. Overall, the concentration of SCH 66336 needed to achieve 50% reduction of soft agar growth was ≤500 n*M* for approx 60% of the human tumor cell lines tested.

SCH 66336 also displayed significantly improved pharmacokinetics in mice, rats, and cynomolgus monkeys compared to earlier compounds in the tricyclic series *(56,57)*. The oral bioavailability of SCH 66336 in the mouse is 76% and the half-life after intravenous administration is 1.4 h. In the cynomolgus monkey, the half-life is 3 h and oral bioavailability is approx 50%. Following chronic dosing in tumor-bearing mice, significant steady-state levels of SCH 66336 are recovered in the xenograft tumor tissue, indicating that it readily reaches the target tissue.

SCH 66336 displays in vivo antitumor activity in a number of mouse models when dosed orally on either a four-times-a-day or twice-a-day schedule *(54)*. Activity was observed vs subcutaneous (s.c.) implants of NIH 3T3 fibroblasts transformed with activated H-Ras-CVLS and in a variety of human tumor xenografts, including carcinomas of the lung, pancreas, colon, bladder and prostate. In some models, 100% tumor growth inhibition or tumor regression were observed at the high dose (80 mg/kg, twice a day).

We have initiated studies to explore combinations of SCH 66336 with various cytotoxic chemotherapeutic agents using xenograft models. There were concerns that signal transduction inhibitors, such as FTIs, would be cytostatic agents, inducing G1 cell cycle arrest and potentially antagonizing the action of classical cytotoxic agents. Our data indicate that this is not the case. In the studies performed to date, greater antitumor activity has been seen when combinations of SCH 66336 with cyclophosphamide, 5-fluorouracil (5-FU), or vincristine are employed compared to single agent treatment. The greater efficacy observed with the combinations indicates that there is no antagonism upon combining SCH 66336 with the cytotoxics. Each of these agents has a distinct mechanism of action and acts at different points within the cell division cycle. Recently, Moasser et al. *(58)* examined in vitro effects of combining FTI treatment with various cytotoxic agents, including doxorubicin, cisplatin, 5-FU, and vinblastine. They reported no antagonism in the in vitro setting and, in fact, a synergistic interaction was noted when the FTIs were combined with microtubule stabilizing agents, such as taxol. Synergy between SCH 66336 and taxanes has also been observed in numerous preclinical models (Nielsen et al., submitted). Results such as these indicate that the use of such combination approaches may enhance the clinical efficacy of FTIs.

SCH 66336 was also evaluated in a transgenic model in which an activated [Val^{12}] H-Ras oncogene is expressed from the whey acidic protein promoter *(59–61)*. When dosed prophylactically, SCH 66336 delayed tumor onset and reduced both the average number of tumors and the average tumor weight per mouse especially when dosed four times a day at 40 mg/kg. At this dose, animals remained tumor-free throughout the dosing period and for a minimum of 20 d after treatment was terminated.

When SCH 66336 was dosed to transgenic animals already bearing palpable tumors, significant dose-dependent tumor regressions were observed. Tumor regression was essentially complete by d 14. Interestingly, cyclophosphamide or a suboptimal dose of SCH 66336 —which did not result in tumor regression when used as single agents—yielded significant regressions when used in combination. This was accompanied by a marked increase in apoptotic cells and a decrease in proliferation. Interestingly, induction of apoptosis and tumor regression has also been observed in the EJ human bladder carcinoma model. FTIs have previously been reported to induce an apoptotic response in vitro in H-Ras-transformed Rat1 fibroblasts denied substratum attachment *(62)* and in vivo in MMTV-H-*ras* mice *(63)*. The importance of the presence on an H-Ras mutation to these apoptotic responses remains to be established. Efforts to demonstrate a similar response and/or tumor regression in various K-Ras-driven models are ongoing.

5. EFFECTS OF FTIs ON OTHER SUBSTRATES

A growing number of potential cellular FTase substrates are being identified based on labeling studies or on cloning of novel proteins containing an appropriate CAAX motif. This list will continue to grow as the pace of sequencing of the human genome accelerates. Some of the FTase substrates, in addition to Ras, include prelamin A and lamin B, the γ-subunit of the retinal trimeric G protein transducin, rhodopsin kinase, the α-subunit of retinal cGMP phosphodiesterase and PxF, a peroxisomal protein of unknown function *(23,64)*. We have been examining the effects of tricyclic FTIs on the prenylation and function of several of these proteins.

5.1. Nuclear Lamins

Farnesylation of the nuclear lamina proteins, prelamin A and lamin B, is required for assembly of these proteins into the nuclear envelope *(65–70)*. Prelamin A is converted to mature lamin A by a proteolytic event resulting in the loss of the isoprenylated C-terminus. Isoprenylation is a prerequisite for this maturational processing and the farnesylation-dependent endoprotease responsible for this cleavage has been characterized *(71)*. The effect of SCH 44342 on prenylation of nuclear lamins was assessed by sequential immunoprecipitation of Ras, prelamin A, and lamin B from [^{3}H]mevalonate-labeled CHO cell lysates. Prenylation of both prelamin A and lamin B was inhibited by SCH 44342; however, the incorporation of [^{3}H]mevalonate into Ras was more sensitive. SCH 44342 blocked [^{3}H]mevalonate-labeling of Ras with an IC_{50} between 2.5 and 5 μM, whereas at

10 μ*M* lamin prenylation was only inhibited about 40%. Other analogs in the tricyclic series also displayed selective inhibition of Ras labeling; however, specificity towards Ras relative to the lamin proteins was not observed using two peptidomimetic FTIs *(22)*. The results with the tricyclic FTIs may reflect poor accessibility into the nuclear compartment or differential sensitivity of a nuclear form of FTase to this compound. Both cytoplasmic and nuclear pools of FTase appear to exist *(70,72)*, but only a single isoform of FTase has been purified and cloned.

Following a 24-h exposure to SCH 44342, HeLa cells were examined by indirect immunofluorescence to detect any alteration in nuclear lamina structure. Immunostaining with an antibody specific for prelamin A *(73)* indicates that prelamin A accumulates in the nucleus of cells treated with SCH 44342. Despite prelamin A accumulation and inhibition of lamin B farnesylation, no derangement of the nuclear lamina structure was observed when cells were examined by indirect immunofluorescence staining with antilamin B or antilamin A/C. Similar results were obtained with the peptidomimetic FTI, BZA-5B *(72)*.

What is the mechanism by which nuclear lamina structure is spared in the presence of an FTI? Is there alternative processing of the lamins to support lamin assembly in the absence of farnesylation? In vitro studies indicate that lamin B is not subject to alternative prenylation by GGTase I (Cavalchire and Bishop, unpublished data), despite the fact that it terminates in methionine. Therefore, a distinct escape mechanism must be postulated.

5.2. Rho Family Members

Effects of FTIs on other cellular proteins may also be important in eliciting the observed biological responses. Included among these are other members of the small GTPase family. One protein suggested to play a role in cellular responses to FTIs is RhoB, which has an unusual C-terminal sequence of CCKVL *(74)*. Consistent with the presence of a terminal leucine, this protein is a preferred substrate for GGTase I vs FTase. We have found that in cells, RhoB is predominantly (≥80%) geranylgeranylated, the remainder being farnesylated. After FTI treatment, this small farnesol peak is lost. These results are qualitatively similar to those reported by Lebowitz et al. *(75)*. These observations suggest that FTI treatment only perturbs the prenylation pattern of a small subset of cellular RhoB, making RhoB less likely to be a candidate protein for conferring FTI-sensitivity.

Other members of the Rho family are intriguing as potential targets of FTI action. Although many Rho family members terminate in leucine, some of these proteins are predicted to be primarily FTase substrates based on the presence of a C-terminal Met residue (Rho6, Rho7, RhoE). The effects of FTIs on prenylation of these proteins as well as the distribution of their expression in tumor cell lines is being explored.

5.3. CAAX Phosphatase

Another protein family of interest is the CAAX motif-containing protein tyrosine phosphatases. These proteins were cloned as FTase substrates from a human breast carcinoma cell line and termed PTP CAAX1 and PTP CAAX2 *(76)*. Their function(s) are unknown although they are reported to be transforming when overexpressed. Our results to date suggest that PTP CAAX1 is a substrate for alternative prenylation but that this reaction occurs inefficiently in Cos cells. The majority of mevalonate-labeling of this protein is blocked by FTI treatment. We are continuing to explore the biological effects of this protein and its modulation by FTI treatment.

6. FUTURE PERSPECTIVES AND CLINICAL EVALUATION OF FTIs

Inhibitors of the function of oncogenic Ras proteins may have utility in the treatment of human cancers. The elucidation of the pathway by which Ras is post-translationally modified and the isolation, cloning, and recent structural studies *(77)* of the enzyme responsible for this modification have opened up a new and promising approach for the development of anti-Ras therapeutics.

The observations that some Ras isoforms are subject to alternative processing and studies of the effects of FTIs on other cellular prenylated proteins clearly suggest that the mechanism of this class of inhibitors is more complex than initially thought. The biological data to date with various FTIs provide further evidence of this complexity, especially in light of the wide differences in FTI sensitivity observed with various human tumor lines and the lack of correlation of this sensitivity to Ras mutational status.

SCH 66336 is being evaluated in early phase clinical trials to establish proof-of-principle for farnesyl transferase inhibition in man. Current Phase I trials with SCH 66336 are being carried out with twice-a-day oral dosing in cancer patients suffering from a variety of refractory solid tumors. Preclinical studies to date suggest that SCH 66336 is efficacious against a wide array of human cancers and may, at least in some cases, lead to tumor regression. How these preclinical studies will translate into clinical benefit is an open question. It is critical that the outstanding mechanistic questions be addressed in order to exploit fully the potential of this novel class of agents in cancer chemotherapy.

REFERENCES

1. Casey PJ, Solski, PA, Der CJ, Buss JE. p21ras is modified by the isoprenoid farnesol. *Proc Natl Acad Sci USA* 1989; 86:8323–8327.
2. Reiss Y, Goldstein JL, Seabra MC, Casey PJ, Brown MS. Inhibition of purified p21^{ras} farnesyl:protein transferase by Cys-AAX tetrapeptides. *Cell* 1990; 62:81–88.
3. Manne V, Roberts D, Tobin A, O'Rourke E, DeVirgilio M, Meyers C, et al. Identification and preliminary characterization of proteine-cysteine farnesyltransferase. *Proc Natl Acad Sci USA* 1990; 87: 7541–7545.
4. Chen W-J, Andres DA, Goldstein JL, Russell DW, Brown MS. cDNA cloning and expression of the peptide-binding β subunit of rat p21^{ras} farnesyltransferase, the counterpart of yeast DPR1. *Cell* 1991; 66:327–334.
5. Chen W-J, Andres DA, Goldstein JL, Brown MS. Cloning and expression of a cDNA encoding the α subunit of rat p21^{ras} protein farnesyltransferase. *Proc Natl Acad Sci USA* 1991; 88:11,368–11,372.
6. Omer CA, Kral AM, Diehl RE, Prendergast GC, Powers S, Allen CM, et al. Characterization of recombinant human farnesyl-protein transferase: cloning, expression, farnesyl diphosphate binding, and functional homology with yeast prenyl-protein transferases. *Biochemistry* 1993; 32:5167–5176.
7. Tamanoi F. Inhibitors of Ras farnesyl transferase. *Trends Biol Sci* 1993; 18:349–353.
8. Singh SB, Lingham RB. Farnesyl-protein transferase inhibitors in early development. *Exp Opin Invest Drugs* 1996; 5:1589–1599.
9. James GL, Goldstein JL, Brown MS, Rawson TE, Somers TC, McDowell RS, et al. Benzodiazepine peptidomimetics: potent inhibitors of Ras farnesylation in animal cells. *Science* 1993; 260:1937–1942.
10. Kohl NE, Mosser SD, DeSolms J, Giuliani EA, Pompliano DL, Graham SL, et al. Selective inhibition of ras-dependent transformation by a farnesyltransferase inhibitor. *Science* 1993; 260:1934–1937.
11. Garcia AM, Rowell C, Ackermann K, Kowalczyk JJ, Lewis MD. Peptidomimetic inhibitors of Ras farnesylation and function in whole cells. *J Biol Chem* 1993; 268:18,415–18,418.
12. Vogt A, Qian Y, Blaskovich MA, Fossum RD, Hamilton AD, Sebti SM. A non-peptide mimetic of Ras-CAAX: selective inhibition of farnesyltransferase and Ras processing. *J Biol Chem* 1995; 270:660–664.
13. Hunt JT, Lee VG, Leftheris K, Seizinger B, Carboni J, Mabus J, et al. Potent cell active non-thiol tetrapeptide inhibitors of farnesyl-transferase. *J Med Chem* 1996; 39:353–358.

14. Graham SL, deSolms SJ, Giuliani EA, Kohl NE, Mosser SD, Oliff AI, et al. Pseudopeptide inhibitors of Ras farnesyl-protein transferase. *J Med Chem* 1994; 37:725–732.
15. Kohl NE, Wilson FR, Mosser SD, Giuliani E, DeSolms SJ, Conner MW, et al. Protein farnesyltransferase inhibitors block the growth of ras-dependent tumors in nude mice. *Proc Natl Acad Sci USA* 1994; 91: 9141–9145.
16. Sun J, Qian Y, Hamilton AD, Sebti SM. Ras CAAX peptidomimetic FTI 276 selectively blocks tumour growth in nude mice of a human lung carcinoma with K-Ras mutation and p53 deletion. *Cancer Res* 1995; 55:4243–4247,
17. Nagasu T, Yoshimatsu K, Rowell C, Lewis MD, Garcia AM. Inhibition of human tumor xenograft growth by treatment with the farnesyl transferase inhibitor B956. *Cancer Res* 1995; 55:5310–5314.
18. Bishop WR, Bond R, Petrin J, Wang L, Patton R, Doll R, et al. Novel tricyclic inhibitors of farnesyl transferase. *J Biol Chem* 1995; 270:30,611–30,618.
19. Billah MM, Chapman RW, Egan RW, Gilchrest H, Piwinski JJ, Sherwood J, et al. SCH 37370: a potent, orally active, dual antagonist of PAF and histamine. *J Pharmacol Exp Ther* 1990; 252:1090–1096.
20. James GL, Brown MS, Cobb MH, Goldstein JL. Benzodiazepine peptidomimetic BZA-5B interrupts the MAP kinase pathway in H-Ras-transformed Rat-1 cells, but not in untransformed cells. *J Biol Chem* 1994; 269:27,705–27,714.
21. Cox AD, Garcia AM, Westwick JK, Kowalczyk JJ, Lewis MD, Brenner DA, Der CJ. The CAAX peptidomimetic compound B581 specifically blocks farnesylated, but not geranylgeranylated or myristylated, oncogenic Ras signaling and transformation. *J Biol Chem* 1994; 269:19,203–19,206.
22. Dalton MB, Fantle KS, Bechtold HA, DeMaio L, Evans RM, Krystosek A, Sinensky M. The farnesyl protein transferase inhibitor BZA-5B blocks farnesylation of nuclear lamins and $p21^{ras}$ but does not affect their function or localization. *Cancer Res* 1995; 55:3295–3304.
23. Maltese WA. Posttranslational modification of proteins by isoprenoids in mammalian cells. *FASEB J* 1990; 4:3319–3328.
24. Glomset J, Gelb M, Farnsworth C. Prenyl proteins in eukaryotic cells: a new type of membrane anchor. *Trends Biochem Sci* 1990; 15:139–142.
25. Casey PJ, Thissen JA, Moomaw J. Enzymatic modification of proteins with a geranylgeranyl isoprenoid. *Proc Natl Acad Sci USA* 1991; 88:8631–8635.
26. Yokoyama K, Gelb MH. Purification of a mammalian protein geranylgeranyltransferase. Formation and catalytic properties of an enzyme-geranylgeranyl pyrophosphate complex. *J Biol Chem* 1993; 268: 4055–4060.
27. Seabra MG, Goldstein JL, Sudhof TC, Brown MS. Rab geranylgeranyl transferase. A multisubunit enzyme that prenylates GTP-binding proteins terminating in Cys-X-Cys or Cys-Cys. *J Biol Chem* 1992; 267:14,497–14,503.
28. Seabra MG, Reiss Y, Casey PJ, Brown MS, Goldstein JL. Protein farnesyltransferase and geranylgeranyltransferase share a common α subunit. *Cell* 1991; 65:429–434.
29. Khosravi-Far R, Clark G, Abe K, Cox A, McLain T, Lutz RJ, et al. Ras (CXXX) and Rab (CC/CXC) prenylation signal sequences are unique and functionally distinct. *J Biol Chem* 1992; 267:24,363–24,368.
30. Kinsella BT, Erdman RA, Maltese WA. Posttranslational modification of H-Ras p21 by farnesyl versus geranylgeranyl isoprenoids is determined by the COOH-terminal amino acid. *Proc Natl Acad Sci USA* 1991; 88:8934–8938.
31. Reiss Y, Stradley SJ, Gierasch LM, Brown MS, Goldstein JL. Sequence requirement for peptide recognition by rat brain $p21^{ras}$ protein farnesyltransferase. *Proc Natl Acad Sci USA* 1991; 88:732–736.
32. Yokoyama K, Goodwin GW, Ghomashchi F, Glomset JA, Gelb MH. A protein geranylgeranyltransferase from bovine brain: implications for protein prenylation specificity. *Proc Natl Acad Sci USA* 1991; 88:5302–5306.
33. Moomaw JF, Casey PJ. Mammalian protein geranylgeranyl transferase: subunit composition and metal requirements. *J Biol Chem* 1992; 267:17,438–17,443.
34. Reiss Y, Brown MS, Goldstein JL. Divalent cation and prenyl pyrophosphate specificities of the protein farnesyltransferase from rat brain is a zinc metalloenzyme. *J Biol Chem* 1992; 267:6403–6408.
35. Zhang FL, Fu H-W, Casey PJ, Bishop WR. Substitution of cadmium for zinc in farnesyl protein transferase alters its substrate specificity. *Biochemistry* 1996; 35:8166–8171.
36. Lowy DR, Willumsen BM. Function and regulation of Ras. *Annu Rev Biochem* 1993; 62:851–891.
37. Bos JL. ras oncogenes in human cancer: a review. *Cancer Res* 1989; 49:4682–4689.
38. Jackson JH, Cochrane CG, Bourne JR, Solski PA, Buss JE, Der CJ. Farnesol modification of Kirsten-ras exon 4B protein is essential for transformation. *Proc Natl Acad Sci USA* 1990; 87:3042–3046.

39. Kato K, Cox AD, Hisaka MM, Graham SM, Buss JE, Der CJ. Isoprenoid addition to Ras protein is the critical modification for its membrane association and transforming activity. *Proc Natl Acad Sci USA* 1992; 89:6403–6407.
40. James GL, Goldstein JL, Brown MS. Polylysine and CVIM sequences of K-RasB dictate specificity of prenylation and confer resistance to benzodiaepine peptidomimetic in vitro. *J Biol Chem* 1995; 270: 6221–6226.
41. Zhang F, Bond R, Wang L, Windsor W, Kirschmeier P, Carr D, Bishop WR. Characterization of H, K and N-Ras as in vitro substrates for isoprenyl protein transferases. *J Biol Chem* 1997; 272:10,232–10,239.
42. Kalman VK, Erdman RA, Maltese WA, Robishaw JD. Regions outside of the CAAX motif influence the specificity of prenylation of G protein gamma subunits. *J Biol Chem* 1995; 270:14,835–14,841.
43. Cox AD, Graham SM, Solski PA, Buss JE, Der CJ. The carboxyl-terminal CXXX sequence of Gi alpha, but not Rab5 or Rab11, supports Ras processing and transforming activity. *J Biol Chem* 1993; 268: 11,548–11,552.
44. James G, Goldstein JL, Brown MS. Resistance of K-RasBV12 proteins to farnesyltransferase inhibitors in Rat1 cells. *Proc Natl Acad Sci USA* 1996; 93:4454–4458.
45. Whyte DB, Kirschmeir P, Hockenberry TN, Nunez-Oliva I, James L, Cantino JJ, et al. K- and N-Ras are geranylgeranylated in cells treated with farnesyl protein transferase inhibitors. *J Biol Chem* 1997; 272:14,459–14,464.
46. Rowell CA, Kowalczyk JJ, Lewis MD, Garcia AM. Direct demonstration of geranylgeranylation and farnesylation of Ki-Ras in vivo. *J Biol Chem* 1997; 272:14,093–14,097.
47. Der CJ, Cox AD. Farnesyltransferase inhibitors and cancer treatment: targeting simply ras? *Biochim Biophys Acta Rev Cancer* 1997; 1333:F51–F71.
48. Njoroge FG, Doll RJ, Vibulbhan B, Alvarez CS, Bishop WR, Petrin J, et al. Discovery of novel non-peptide tricyclic inhibitors of ras farnesyl protein transferase. *Bioorg Med Chem* 1997; 5:101–113.
49. Mallams AK, Njoroge FG, Doll RJ, Snow ME, Kaminski JJ, Rossman RR, et al. Antitumor 8-chlorobenzocyclohepta-pyridines: a new class of selective, nonpeptidic, nonsulfhydryl inhibitors of Ras farnesylation. *Bioorg Med Chem* 1997; 5:93–99.
50. Njoroge G, Vibulbhan B, Rane DF, Bishop WR, Petrin J, Patton R, et al. Structure-activity relationship of 3-substituted N-pyrinylacetyl-4-(8-chloro-5,6-dihydro-11H-benzo[5,6]cyclohepta-[1,2-b]pyridinyl-11-ylidene) piperidine inhibitors of farnesyl protein transferase: design and synthesis of in vivo active antitumor compound 38. *J Med Chem* 1997; 40:4290–4301.
51. Njoroge FG, Vibulbhan B, Pinto P, Bishop WR, Bryant MS, Nomeir AA, et al. Potent, selective, and orally bioavailable tricyclic pyridyl acetamide N-oxide inhibitors of farnesyl protein transferase with enhanced in vivo antitumor activity. *J Med Chem* 1998; 41:1561–1567.
52. Mallams AK, Rossman RR, Doll RJ, Girijavallabhan V, Ganguly AK, Petrin J, et al. Inhibitors of farnesyl protein transferase. 4-amido 4-carbomyl and 4-carboxamido derivaties of 1-(8-chloro-6,11-dihydro-5H-benzo[5,6]cyclohepta[1,2-b]pyridin-11-yl)piperazine and 1-(3-bromo-8-chloro-6,11-dihydro-5H-benzo[5,6]cyclohepta[1,2-b]pyridin-11-yl)piperazine. *J Med Chem* 1998; 41:877–893.
53. Liu M, Bryant MS, Chen J, Lee S, Yaremko B, Li Z, et al. Effects of SCH 59228, an orally bioavailable farnesyl protein transferase inhibitor, on the growth of oncogene-transformed fibroblasts and a human colon carcinoma xenograft in nude mice. *Cancer Chemo Pharm* 1999; 43:50–58.
54. Liu M, Bryant MS, Chen J, Lee S, Yaremko B, Lipari P, et al. Antitumor activity of SCH 66336, an orally bioavailable tricyclic inhibitor of farnesyl protein transferase inhibitor, in human tumor xenograft models and wap-ras transgenic mice. *Cancer Res* 1998; 58:4947–4956.
55. Sepp-Lorenzino L, Ma Z, Rands E, Kohl NE, Gibbs JB, Oliff A, Rosen N. A peptidomimetic inhibitor of farnesyl:protein transferase blocks the anchorage-dependent and -independent growth of human tumor cell lines. *Cancer Res* 1995; 55:5302–5309.
56. Bryant MS, Korfmacher WA, Wang S, Nardo C, Nomeir AA, Lin C-C. Pharmacokinetic screening for selection of new drug discovery candidates is greatly enhanced through the use of liquid chromatography-atmospheric pressure ionization tandem mass spectrometry. *J Chroma* 1997; 777:61–66.
57. Bryant M. S, Liu M, Korfmacher WA, Nardo C, Wang S, Chen KJ, et al. LC-APCI/MS/MS analysis of serum and tumor samples from mice treated with a potent antitumor compound (SCH 59228; Proc. 45th ASMS Conf. mass spectrom. allied. topics. Palm Springs, CA, June 1–5, 1997.
58. Moasser MM, Sepp-Lorenzino L, Kohl NE, Oliff A, Balog A, Su D-S, et al. Farnesyl transferase inhibitors cause enhanced mitotic sensitivity to taxol and epothilones. *Proc Natl Acad Sci USA* 1998; 95:1369–1374.

59. Andres A-C, Schonenberger C-A, Groner B, Hennighausen L, LeMeur M, Gerlinger P. Ha-ras oncogene expression directed by a milk protein gene promoter: tissue specificity, hormonal regulation, and tumor induction in transgenic mice. *Proc Natl Acad Sci USA* 1987; 84:1299–1303.
60. Nielsen LL, Discafani CM, Gurnani M, Tyler RD. Histopathology of salivary and mammary gland tumors in transgenic mice expressing a human Ha-ras oncogene. *Cancer Res* 1991; 51:3762–3767.
61. Nielsen LL, Gurnani M, Tyler RD. Evaluation of the wap-ras transgenic mouse as a model system for testing anticancer drugs. *Cancer Res* 1992; 52:3733–3738.
62. Lebowitz PF, Sakamuro D, Prendergast GC. Farnesyl transferase inhibitors induce apoptosis of ras-transformed cells denied substratum attachment. *Cancer Res* 1997; 57:708–713.
63. Barrington RE, Subler MA, Rands E, Omer CA, Miller PJ, Hundley JE, et al. A farnesyltransferase inhibitor induces tumor regression in transgenic mice harboring multiple oncogenic mutations by mediating alterations in both cell cycle control and apoptosis. *Mol Cell Biol* 1998; 18:85–92.
64. James GL, Goldstein JL, Pathak RK, Anderson RGW, Brown MS. PxF, a prenylated protein of peroxisomes. *J Biol Chem* 1994; 269:14,182–14,190.
65. Farnsworth CC, Wolda SL, Gelb MH, Glomset JA. Human lamin B contains a farnesylated cysteine residue. *J Biol Chem* 1989; 264:20,422–20,429.
66. Holtz D, Tanaka RA, Hartwig J, McKeon F. The CaaX motif of lamin A functions in conjunction with the nuclear localization signal to target assembly of the nuclear envelope. *Cell* 1989; 59:969–977.
67. Beck LA, Hosick TJ, Sinensky M. Isoprenylation is required for the processing of the lamin A precursor. *J Cell Biol* 1990; 110:1489–1499.
68. Sinensky M, Fantle K, Trujillo M, McLain T, Kupfer A, Dalton M. The processing pathway of prelamin A. *J Cell Sci* 1994; 107:61–67.
69. Weber K, Plessmann U, Traub P. Maturation of nuclear lamin A involves specific carboxy-terminal trimming, which removes the polyisoprenylation site from the precursor; implications for the structure of the nuclear lamina. *FEBS Lett* 1989; 257:411–414.
70. Lutz RJ, Trujillo MA, Denham KS, Wenger L, Sinensky M. Nucleoplasmic localization of prelamin A: implications for prenylation-dependent lamin A assembly into the nuclear lamina. *Proc Natl Acad Sci USA* 1992; 89:3000–3004.
71. Kilic F, Dalton MB, Burrell SK, Mayer JP, Patterson SD, Sinensky M. In vitro assay and characterization of the farnesylation-dependent prelamin A endoprotease. *J Biol Chem* 1997; 272:5298–5304.
72. Firmbach-Kraft I, Stick R. Analysis of nuclear lamin isoprenylation in Xenopus oocytes: isoprenylation of lamin B3 precedes its uptake into the nucleus. *J Cell Biol* 1995; 129:17–24.
73. Sinensky M, Fantle K, Dalton M. An antibody which specifically recognizes prelamin A but not mature lamin A: application to detection of blocks in farnesylation-dependent protein processing. *Cancer Res* 1994; 54:3229–3232.
74. Lebowitz PF, Davide JP, Prendergast GC. Evidence that farnesyltransferase inhibitors suppress Ras transformation by interfering with Rho activity. *Mol Cell Biol* 1995; 15:6613–6622.
75. Lebowitz PF, Casey PJ, Prendergast GC, Thissen JA. Farnesyltransferase inhibitors alter the prenylation and growth-stimulating function of RhoB. *J Biol Chem* 1997; 272:15,591–15,594.
76. Cates CA, Michael RL, Stayrook KR, Harvey KA, Burke YD, Randall SK, et al. Prenylation of oncogenic human PTP(CAAX) protein tyrosine phosphatases. *Cancer Lett* 1996; 110:49–55.
77. Park H-W, Boduluri SR, Moomaw JF, Casey PJ, Beese LS. Crystal structure of protein farnesyltransferase at 2.25 Angstrom resolution. *Science* 1997; 275:1800–1804.

8 Histidylbenzylglycinamides

A Novel Class of Farnesyl Diphosphate-Competitive Peptidic Farnesyltransferase Inhibitors

Judith S. Sebolt-Leopold, PHD,
Daniele M. Leonard, PHD,
and W. R. Leopold, PHD

CONTENTS

1. INTRODUCTION

Therapeutic interest in farnesyltransferase inhibitors (FTIs) historically grew out of a desire to target the *ras* oncogene. Roughly 30% of all human tumors possess a mutation in one of the *ras* proto-oncogenes. These mutations result in guanosine triphosphate (GTP) becoming permanently bound to Ras, resulting in uncontrolled cellular proliferation. Some tumor types, namely pancreatic and colon cancers, are especially refractory to conventional chemotherapy and exhibit a particularly high incidence of activating *ras* mutations (approx 90 and 50%, respectively). Interest in farnesyltransferase (FTase) arose because this enzyme represented a potential target for interfering with Ras function. Ras must be anchored to the plasma membrane to act as a transducer of signaling events. Because it has no transmembrane domain, Ras is post-translationally modifed by FTase, thereby acquiring a farnesyl group and consequently the requisite hydrophobicity for membrane attachment. The rationale for FTase as an anticancer drug target has been comprehensively reviewed elsewhere *(1–3)*.

During the past year, FTIs have entered into the early stages of clinical testing *(4)*. Without a clinically approved predecessor, a number of critical issues require resolution before the ideal properties of an FTI can be determined. For example, what degree of selectivity for FTase relative to other prenylation enzymes, namely geranylgeranyltransferases,

From: *Farnesyltransferase Inhibitors in Cancer Therapy*
Edited by: S. M. Sebti and A. D. Hamilton © Humana Press Inc., Totowa, NJ

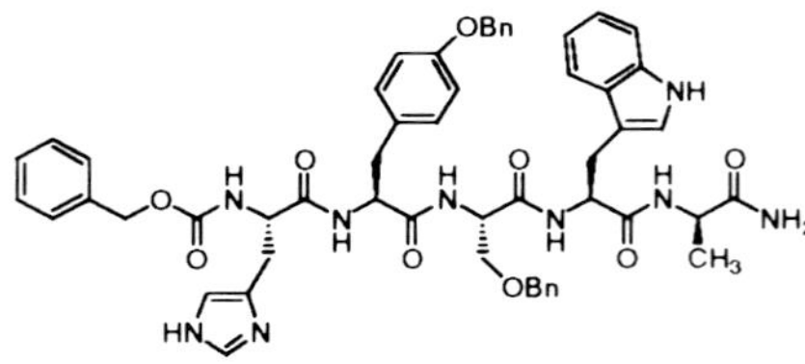

PD 083176

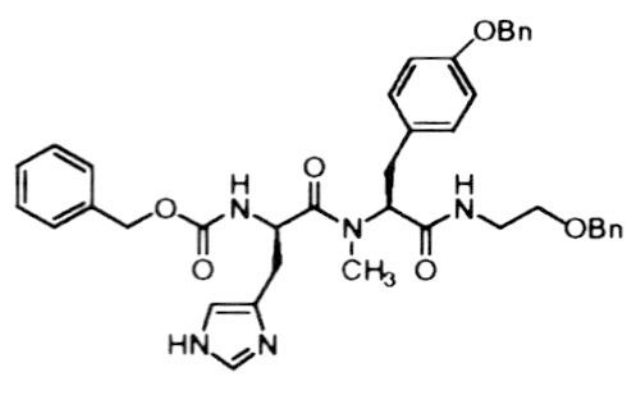

PD 154309

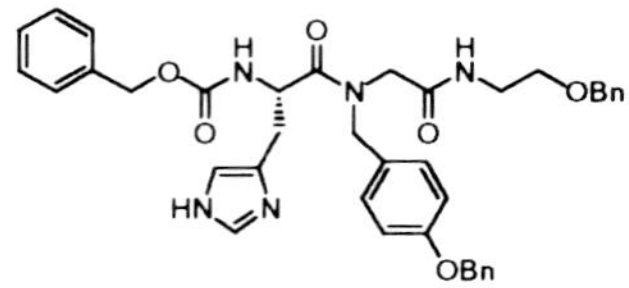

PD 152440

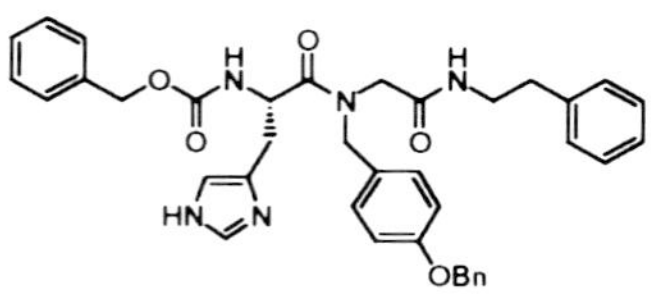

PD 161956

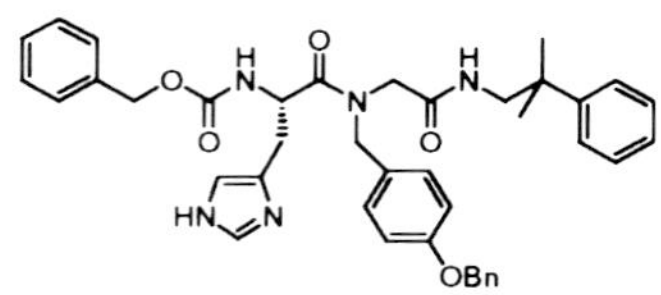

PD 169451

Fig. 1. Chemical structures of key compounds.

needs to be achieved? The finding that K-Ras is geranylgeranylated when farnesylation is shut down does not make this a straightforward question *(5)*. The existence of normal cellular farnesylated proteins, such as G-proteins involved in visual signal transduction and endogenous Ras, also raises the issue of whether FTIs will prove toxic based on mechanistic considerations. Mounting preclinical evidence suggests that this will not be the case *(6,7)*.

The vast majority of research devoted to the development of efficacious FTI drug candidates has purposely focused on the CAAX binding region of FTase. The existing bias against farnesyl diphosphate (FPP) analogs stems largely from their potential for toxicity based on the ubiquitous role played by FPP in cellular metabolism. However, FPP-competitive inhibitors need not be FPP analogs. The studies to be described below provide evidence that one class of FPP-competitive FTIs, the histidylbenzylglycinamides, exhibit promising preclinical activity against multiple human tumor xenograft models.

2. EVOLUTION OF THE HISTIDYLBENZYLGLYCINAMIDE SERIES

Mass screening of the Parke-Davis chemical library resulted in the identification of PD 083176 as a potent inhibitor of rat brain FTase, as evidenced by an IC_{50} of 20 n*M*. As shown in Fig. 1, the structure of this protected pentapeptide, which does not contain a cysteine residue, is uniquely different from the CAAX-based tetrapeptides of comparable potency reported by Brown and Goldstein *(8)*. Presumably because of the high degree of hydrophobicity of PD 083176, this inhibitor proved to be cell-impermeable and incapable of inhibiting cellular farnesylation at concentrations as high as 250 μ*M*. A drug discovery effort focusing on the development of truncated analogs of PD 083176 was initiated in part because of the excellent potency of PD 083176 against FTase coupled with a reasonable degree of selectivity for this enzyme. (PD 083176 inhibits purified geranylgeranyltransferase I (GGTase I) with an IC_{50} of 1.25 μ*M*.) Microinjection experiments provided further impetus for pursuit of the PD 083176 series; Xenopus oocytes injected with PD 083176 exhibited only a 32% maturation frequency in response to insulin, an event dependent on Ras function *(9)*. In contrast, insulin treatment elicited a 68% positive response in the dimethyl sulfoxide (DMSO)-treated control group. These experiments provided evidence that members of the PD 083176 series could effectively block Ras function once cell permeability problems were overcome.

2.1. Novel Biochemical Attributes of PD 083176 Series

Although peptidic in nature, PD 083176 surprisingly was found to be competitive with FPP and noncompetitive with the protein substrate *(10)*. Furthermore, this compound was found to serve as a more effective inhibitor of FTase when assayed in a buffer system containing phosphate *(10)*. The nature of synergistic inhibition of FTase by compounds from this chemical series and various anions was further explored, focusing on the inhibitor Cbz-His-Tyr-Ser(OBn)-Trp-NH_2 *(11)*. This compound (PD 156157) exhibited an IC_{50} of 6.1 μ*M* against purified FTase when assayed in HEPES buffer, compared to significantly enhanced inhibition (IC_{50} = 0.16 μ*M*) when assayed in HEPES buffer supplemented with 5 m*M* phosphate. A comparison of various anions for their relative synergistic effects revealed that phosphate was the most effective anion. Furthermore, covalent attachment of a phosphate group to the hydroxyl group of the tyrosine residue of PD 156157 resulted in significant enhancement of inhibitory potency, as evidenced by an IC_{50} of 0.003 μ*M* (compared to 0.16 μ*M* for PD 156157); inhibition by this phosphate-containing compound was independent of the presence of phosphate in the buffer system.

Kinetic evidence revealed that the anions were noncompetitive with respect to FPP. Therefore, it is reasonable to assume that phosphate binds to two different forms of the enzyme. A model has been proposed whereby phosphate can bind to the free form of FTase because the pyrophosphate pocket is unoccupied; after release of pyrophosphate from the E:farnesylated peptide:PPi enzyme complex, the E:farnesylated peptide enzyme complex would have a pyrophosphate binding pocket accessible to phosphate anion. It has been proposed that PD 156157 acts like the farnesylated product (F-CAAX), allowing phosphate to bind to the enzyme:inhibitor form of the enzyme *(11)*. This mechanistic scheme assumes that pyrophosphate leaves first and the release of F-CAAX is rate-limiting. The observed synergy between PD 156157 and phosphate ion could be explained on the basis of E:F-CAAX being a kinetically long-lasting species. This novel attribute of the PD 083176 series has not been observed for other FPP-competititve FTase inhibitors as exemplified by data obtained with α-hydroxyfarnesyl phosphonic acid *(11)*.

2.2. SAR of PD 083176 Series

A structure-activity relationship study around PD 083176 was carried out to identify the critical features for activity and to optimize affinity for FTase. The N-terminal histidine residue, protected with a benzyloxycarbonyl (Cbz) group, was found to be optimal

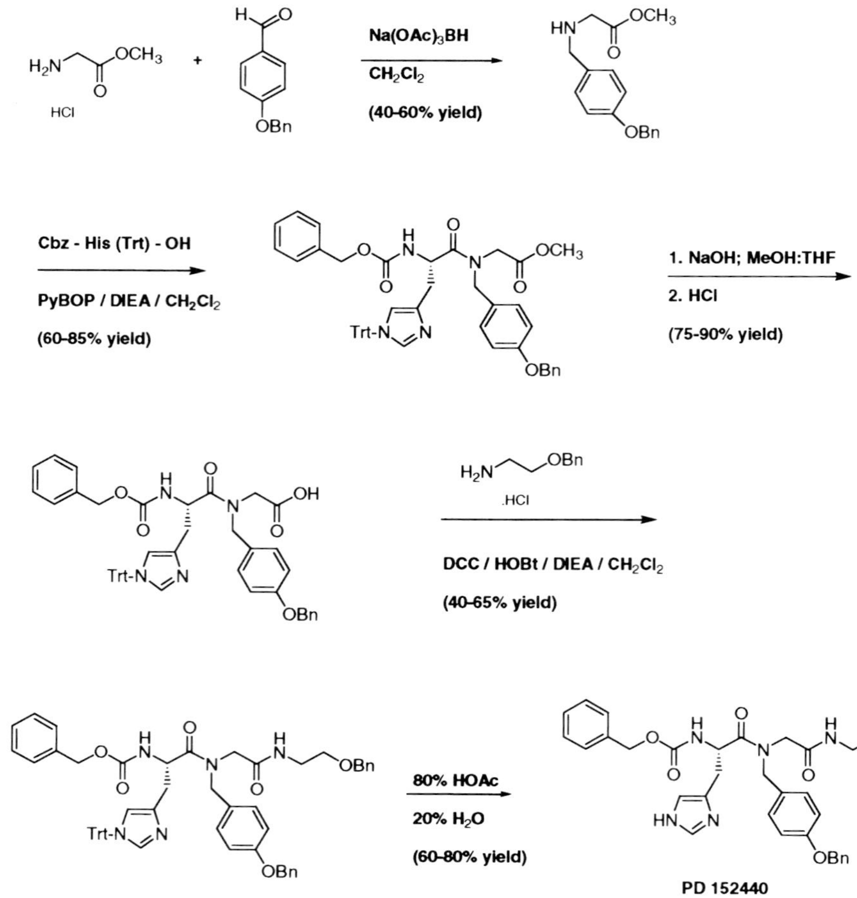

Fig. 2. Synthetic route leading to PD 152440 and analogs.

for inhibitory activity against FTase. The Tyr(OBn) residue was found to be more tolerant to a variety of modifications. In particular, replacement with a phenylalanine residue resulted in a pentapeptide with twofold enhanced potency relative to PD 083176. The Ser(OBn) residue could be replaced by Thr(OBn) giving a compound of similar potency to the parent compound (IC_{50} = 0.015 μ*M*). Cysteine substitution for Ser(OBn) gave rise to a significantly more active FTase inhibitor with an IC_{50} of 0.004 μ*M*. Limited structure-activity relationship (SAR) studies at the tryptophan and D-alanine residues were carried out, because it was found that these two residues could be truncated to a tripeptide with cellular activity as described below.

2.3. Truncated Tripeptide Analogs in the PD 083176 Series

Evaluation of the tripeptide Cbz-His-Tyr(OBn)-Ser(OBn)-NH_2 revealed that although this inhibitor was 18-fold less potent than the corresponding pentapeptide against purified FTase (IC_{50} = 0.37 μ*M*), it was effective at inhibiting Ras farnesylation in H-ras-transfected NIH 3T3 cells at a concentration of 20 μ*M*. Efforts were then carried out to improve inhibitory potency. Subsequently, the Cbz-histidine N-terminal moiety was shown to be important for activity, although inversion of the stereochemistry at the histidine residue was tolerated. In the tripeptide, the Tyr(OBn) residue was shown to be critical, whereas the aromatic OBn group on the serine residue was optimal for activity. Modifications were then carried out focusing on the amide bonds and the C-terminal Ser(OBn) residue of the tripeptide. Replacement of Ser(OBn) by O-benzylethylamine, followed by either α-*N*-methylation of the Tyr(OBn) residue, or transposition of the side chain of Tyr(OBn) to its α-nitrogen, led to compounds PD 154309 and PD 152440, respectively (Fig. 1). Both compounds were shown to have increased cellular activity as evidenced by inhibition of cellular Ras processing at a concentration of 1 μ*M*. Continued SAR studies of PD 152440 were carried out. As will be shown later in this chapter, PD 152440 proved to be nonselective for FTase relative to GGTase I. However, selectivity for FTase was achieved as more potent inhibitors described in the next section were developed.

2.4. Emergence of Histidylbenzylglycinamides from the Tripeptide Series

The chemistry for the synthesis of PD 152440 and analogs is illustrated in Fig. 2. The glycine methyl ester hydrochloride salt is reductively aminated with 4-(benzyloxy)benzaldehyde using sodium triacetoxyborohydride to give the N-substituted glycine residue methyl ester. Acetylation of this product with Cbz-His(trityl) *(12)* was then carried out followed by hydrolysis of the methyl ester. The resulting acid was then coupled to O-benzylethylamine giving rise to the trityl protected modified dipeptide. Treatment with acid to remove the trityl group gave the desired analog.

SAR studies were then carried out focusing on the C-terminus of PD 152440 *(13)*. Replacement of the O-benzyl group with a phenyl group resulted in PD 161956, which is 10-fold more potent than the parent compound (PD 152440) at inhibiting cellular farnesylation, being effective at a concentration of 0.1 μ*M*. This modification also resulted in a greater degree of selectivity against FTase (100-fold) as evidenced by IC_{50}s of 0.062 and 6.6 μ*M* against purified FTase and GGTase I, respectively.

Subsequent analog development focusing on PD 161956 was carried out. It was determined that the ethylene spacer between the amide moiety and the phenyl group of the C-terminus moiety was optimal for activity; a shorter spacer such as a methylene group resulted in a compound with an IC_{50} of 9.9 μ*M*, whereas extending the spacer to a propyl or butyl group gave rise to compounds with decreased activity (IC_{50}s of 3.3 and 0.92 μ*M*, respectively). Substitution on the phenyl ring was also studied, whereupon it was determined that the unsubstituted ring resulted in the best activity.

Substitution of the ethylene spacer at the C-terminus was then carried out. At the α-position, a (R)-methyl group led to an increase in activity (IC_{50} = 0.009 μ*M*); however, in the S configuration, it was twofold less active than the parent compound. Disubstitution also led to a less active analog. The ethylene spacer was then substituted at the β-position. The (R/S)-methyl, -ethyl, and -propyl substituents gave potent inhibitors of FTase (IC_{50}s of 0.007, 0.005, and 0.025 μ*M*, respectively). However, the (R/S)-methyl and -ethyl analogs were more potent in cells, being effective at 0.05 μ*M*, compared to 0.2 μ*M* for the propyl-containing analog. Resolution of the mono-methyl analog to its respective diastereomers was carried out. The R isomer analog was active against FTase with an IC_{50} of 0.005 μ*M* and was effective in cells at a concentration of 0.05 μ*M*, whereas the S isomer was slightly less potent (IC_{50} = 0.016 μ*M* and effective in cells at 0.2 μ*M*).

To avoid the addition of a chiral center, disubstitution at the β-position was then carried out, resulting in PD 169451 (Fig. 2), which was strongly potent against purified

Table 1
Evolution of the Histidylbenzylglycinamides

Compound	*Soft agar IC_{50} (μM)*[a]	*Assessment*
PD 083176	> 100	Potent but not cell permeable.
PD 152440	14.2	Cell permeable but not selective and too toxic in vivo.
PD 161956	4.3	50× selective for FTase, but not sufficiently potent or selective. However, first indication of in vivo activity for this series.
PD 169451	0.18	Significant improvement in selectivity, potency, and degree of in vivo efficacy (i.p. and s.c. routes). No efficacy upon oral dosing.[a]

[a]H-ras transfected 3T3 cells.

Fig. 3. Lack of selectivity of PD 152440 for farnesyltransferase. NIH 3T3 fibroblasts expressing oncogenic H-ras (Ras-F) or Ras-CVLL (Ras-GG) were kindly provided by Channing Der (University of North Carolina, Chapel Hill). Cells were treated overnight with the indicated concentration of PD 152440 prior to lysis, immunoprecipitation of Ras, and Western blot analysis of prenylation status. IC_{50} against purified enzymes: FTase = 0.26 μ*M*; GGTase I = 0.60 μ*M*.

FTase (IC_{50} = 0.004 μ*M*) and in cells at 0.05 μ*M*. Cyclopropyl and cyclobutyl substitutions resulted in increased activity, whereas the cyclopentyl substitution as well as the diethyl substitution resulted in a decrease in activity *(13)*.

PD 169451, like the pentapeptide PD 083176, is competitive with FPP, with a K_i value of 0.74 n*M*. It was also shown to be 4500-fold selective for FTase over GGTase I. PD 169451 has effectively served as a prototype inhibitor from the histidylbenzylglycinamide series, exhibiting a promising preclinical profile.

3. BIOLOGICAL EVALUATION OF THE HISTIDYLBENZYLGLYCINAMIDES

3.1. Lessons Learned from Predecessors to PD 169451

Table 1 summarizes the biological properties of the key compounds that led to the synthesis of PD 169451. Efforts to improve the cellular permeability of this chemical series resulted in PD 152440, which inhibited cellular farnesylation at 5 μ*M*, but proved to be nonselective against FTase relative to other prenylation enzymes (*see* Fig. 3). Because an increasing number of literature reports indicated that K-Ras was an efficient substrate for geranylgeranylation when challenged with an FTI, we proceeded with the in vivo evaluation of PD 152440 despite its lack of selectivity. Treatment of H-ras transfected

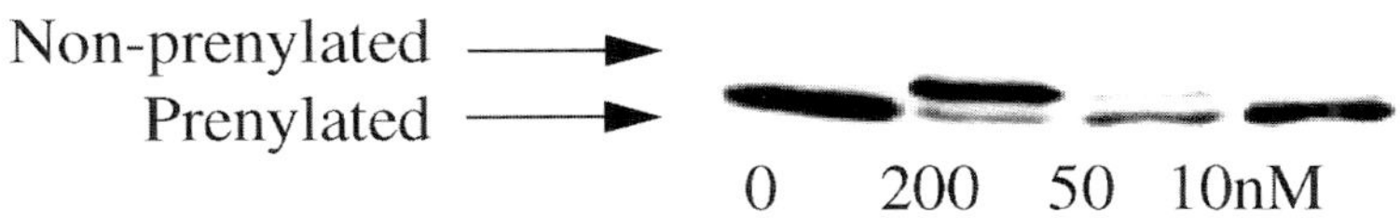

Fig. 4. Cellular potency of PD 169451. H-ras transformed 3T3 fibroblasts were treated overnight with the indicated concentration of PD 169451. Inhibition of Ras prenylation was evaluated as described in the legend to Fig. 3.

fibroblast tumors grown in nude mice failed to respond to treatment with this inhibitor; in a life-span assay, not only was there no evidence of antitumor activity, but the compound proved to be toxic at the two highest doses tested (120 and 75 mg/kg/injection) (data not shown). It is not clear whether the observed toxicity resulted from inhibition of geranylgeranylation. However, further efforts in this series were then directed toward the development of analogs with increased potency as well as selectivity against FTase.

Replacement of the O-benzyl group of PD 152440 with a phenyl ring resulted in PD 161956, which accomplished both increased potency (active against cellular farnesylation at 0.1 μ*M*) as well as a 50-fold gain in selectivity against FTase relative to GGTase I. Although PD 161956 gave the first indication of in vivo antitumor activity for this chemical series, the degree of activity was modest (data not shown). Further efforts to improve potency and selectivity continued.

PD 169451, characterized by dialkylation of the ethylene spacer between the amide group and the phenyl ring of PD 161956, represented a big advance in the attainment of enhanced selectivity, potency, and in vivo efficacy. The remainder of this chapter will be devoted to describing the biological profile of this inhibitor, which has served as a useful prototype from the histidylbenzylglycinamide series.

3.2. Preclinical Profile of PD 169451

A high degree of selectivity against FTase (4500-fold) relative to inhibition of GGTase I was demonstrated against the purified enzymes, with an IC_{50} of 4 n*M* against the target enzyme. Subsequent experiments at the cellular level confirmed this high degree of selectivity, as well as revealing 50 n*M* potency at inhibition of Ras farnesylation (Fig. 4). In vivo evaluation of PD 169451 against tumors derived from fibroblasts transfected with H-ras revealed a higher degree of activity than that observed against tumors derived from Ras-CVLL transfected fibroblasts (Fig. 5), a result consistent with the in vitro selectivity observed for PD 169451 against FTase.

It is generally assumed that significantly higher plasma levels for a FTI will need to be achieved relative to its IC_{50} against purified enzyme, because relatively small amounts of membrane-bound Ras could be sufficient to drive tumor cell proliferation. Figure 6 illustrates that a single subcutaneously delivered dose of 125 mg/kg is sufficient for achieving roughly 10 μ*M* levels of PD 169451; furthermore, this high plasma level is sustained for approx 20 h. Experiments then turned to demonstrating that this FTI possessed broad spectrum in vivo activity against a panel of human tumor xenografts.

As shown in Fig. 7, PD 169451 at 10 μ*M* proved to be effective at inhibiting anchorage-independent growth of multiple human tumor cell lines expressing mutant K-Ras, encompassing those of breast, lung, and pancreatic origin. In all of these experiments, cells did

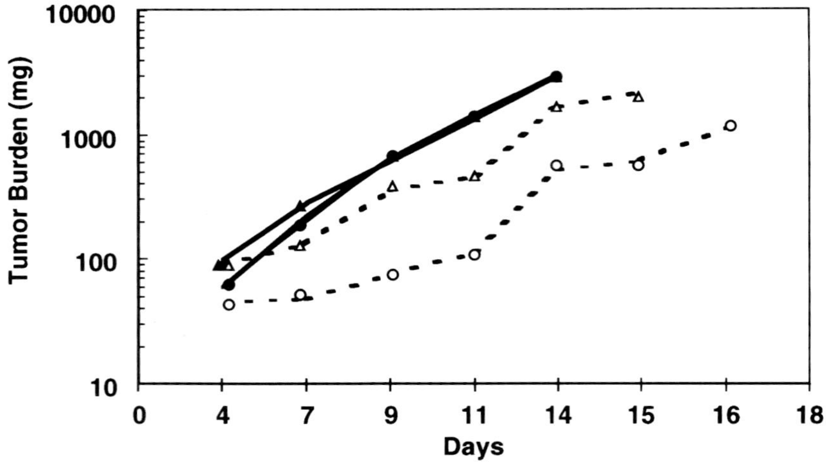

Fig. 5. In vivo selectivity of PD 169451 for farnesyltransferase. Nude mice bearing tumors derived from fibroblasts transfected with either H-Ras (●,○) or Ras-CVLL (▲,△) (implanted subcutaneously) were treated with vehicle (cremophor/ethanol/water; 10/10/80) [●,▲] or PD 169451 (125 mg/kg) [○,△] once daily. Treatment was administered subcutaneously at a distal site from the tumor.

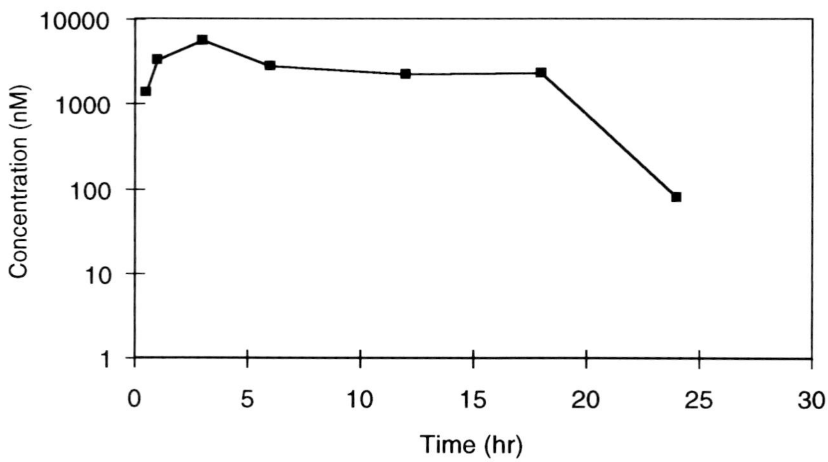

Fig. 6. Plasma concentration of PD 169451 as a function of time after dosing. Mice were treated with a single subcutaneous dose of 125 mg/kg PD 169451. The area under the curve (AUC) in this experiment was determined to be 54296 nmol/h. (Data kindly provided by Chetan Lathia, Parke-Davis.)

not appear to be killed by treatment with PD 169451, but rather were suppressed in their ability to grow in soft agar, a finding suggestive of a cytostatic mechanism of action.

Consistent with the results obtained in soft agar, a diverse array of human tumors proved to be sensitive to in vivo treatment with PD 169451. Representative data are summarized in Figs. 8 and 9 for a tumor model that expresses mutant K-Ras (H460 nonsmall cell lung carcinoma) as well as a model that is wild-type with respect to *ras* (MCF-7 breast carcinoma). Against both of these models, tumor growth was significantly inhibited for the duration of subcutaneous treatment (15 d); upon cessation of treatment, tumor growth resumed, a finding that again is consistent with a cytostatic mechanism. Table 2 summarizes the in vivo data observed for a panel of human tumor xenografts, encompassing a variety of tumor sites that differ also in their *ras* mutation status. Only the SK-OV-3

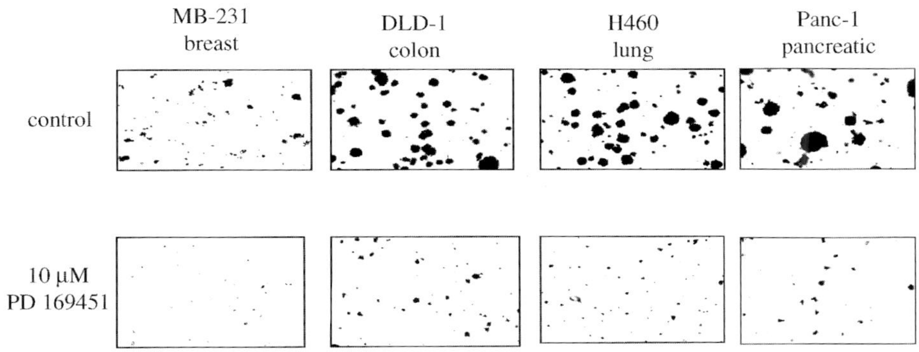

Fig. 7. Inhibition of anchorage-independent growth of human tumor cell lines expressing mutant K-Ras. The indicated cell line was plated in soft agar in the presence or absence of 10 μ*M* PD 169451 as previously described *(13)*. Colony formation was evaluated after 14–21 d of incubation.

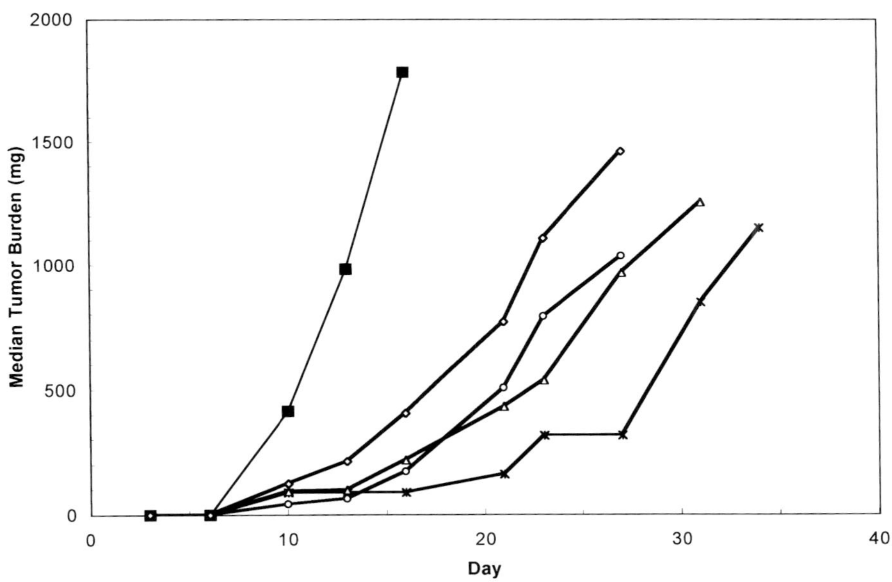

Fig. 8. Activity of PD 169451 against H460 nonsmall cell lung carcinoma xenografts. Mice bearing s.c. H460 tumors were treated with PD 169451 subcutaneously q. 12 h on d 3–7, 10–14, 17–21. Animals were injected with vehicle (described in legend to Fig. 5) [■] or PD 169451 at dose levels of 200, 125, 78, and 48 mg/kg/injection [✱], [○], [△], and [◇], respectively.

ovarian model failed to respond to treatment with PD 169451; generally, 100% tumor growth inhibition was observed across this panel of tumors.

4. FUTURE DIRECTIONS

The data outlined in the previous section supports a cytostatic mechanism of action for FTIs in the PD 083176 series. For single agent therapy, the potential usefulness of such an agent is therefore likely to depend on attainment of an orally active formulation. Although PD 169451 possesses significant antitumor activity when dosed parenterally, it is inactive upon oral administration (data not shown). Current efforts are therefore focusing on improvements in the bioavailability of these inhibitors so as to allow oral dosing.

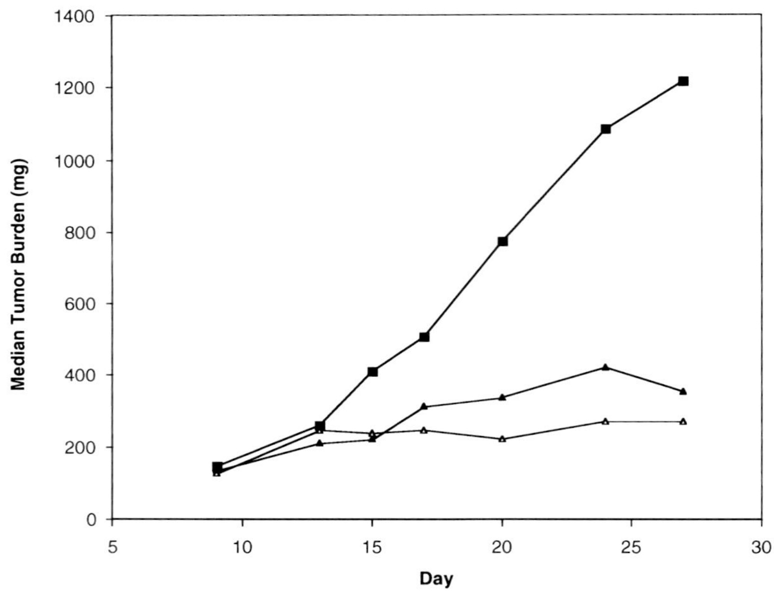

Fig. 9. Activity of PD 169451 against MCF-7 breast carcinoma xenografts. Mice bearing s.c. MCF-7 tumors were treated subcutaneously q. 12 h daily on d 9–23 with either vehicle (described in legend to Fig. 5) [■] or PD 169451 at dose levels of 100 or 50 mg/kg/injection, [△] and [▲], respectively.

Table 2
In Vivo Activity of PD 169451
Against a Panel of Human Tumor Xenografts[a]

Tumor model	*Tumor type*	*Ras mutation*	*% Tumor growth inhibition*[b]
A549	NSCLC[c]	K	140
H460	NSCLC	K	100
HT-29	Colon	Wt	108
MCF-7	Mammary	Wt	130
MCF-7	Mammary	Wt	100
Ras-F	NIH-3T3	H	43
SK-OV-3	Ovarian	Unknown	0
SW620	Colon	K	66

[a] PD 169451 was dosed at its maximum tolerated dose.
[b] %TGI = [(T-C) / # of days of Rx] × 100
[c] NSCLC = nonsmall cell lung cancer.

In addition, various combination treatment regimens are being explored in an attempt to optimize activity of compounds in this series. Combination strategies are beginning to receive increasing attention from researchers in the FTase field. Synergistic effects of FTIs with both radiation and taxol have been reported *(14,15)*. In considering various combination strategies, it is likely that not all FTIs will behave similarly. For example, because PD 169451 and related compounds are competitive with FPP, it may prove beneficial to combine them with HMG CoA reductase inhibitors. The latter agents, which

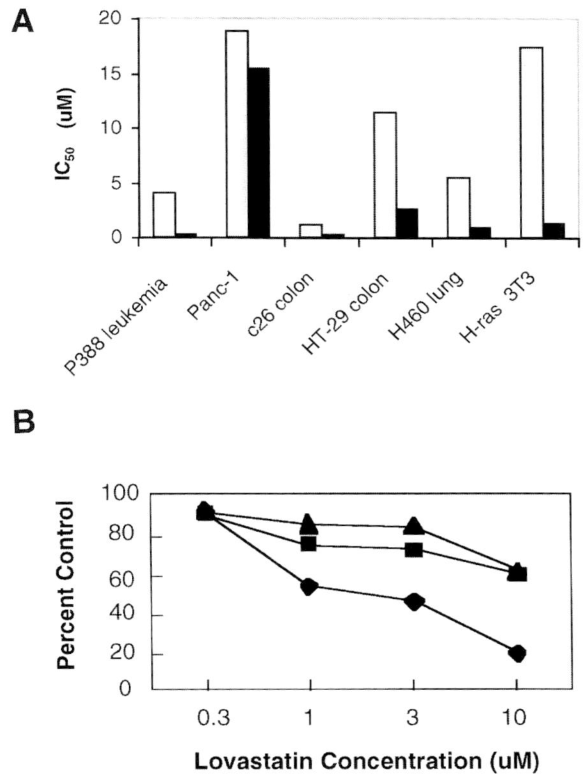

Fig. 10. Synergistic growth inhibition of tumor cells by a combination of PD 169451 and lovastatin. (**A**) A panel of tumor cell lines encompassing tumors of both murine and human origin were treated for 72 h in the continuous presence (shaded bars) or absence (unshaded bars) of 1 μ*M* PD 169451 plus varying concentrations of lovastatin. Cell number was quantitated after incubation and the data expressed as IC_{50} for lovastatin treatment; combination data was corrected for inhibition of growth induced by PD 169451 alone, which generally resulted in a 40–60% reduction of growth relative to the DMSO-treated controls. (**B**) H-ras transfected 3T3 fibroblasts were treated continuously for 72 h with either agent alone or co-incubated with both agents simultaneously. The concentration of each of the FTIs was 0.1 μ*M*, which resulted in negligible growth inhibition (< 10%) when used alone. Symbols: DMSO, [▲]; PD 169451, [◆]; L-745,631, [■].

would be expected to lower endogenous FPP levels, may create significant metabolic liabilities for tumor cells that are also impaired in their ability to utilize FPP because of the presence of an FPP-competitive compound. Figure 10A illustrates the enhancement in lovastatin-induced growth inhibition in a panel of tumor cell lines in response to treatment with PD 169451. Furthermore, synergistic growth inhibition by lovastatin in combination with CAAX-competitive FTIs was not observed; as shown in Fig. 10B, L-745,631 *(16)*, which exhibits comparable potency to PD 169451 at inhibition of cellular farnesylation, does not elicit the same effect. In vivo validation of this approach is currently under study.

5. SUMMARY

A number of pharmacological challenges have been overcome to address the limitations of PD 083176, a potent cell-impermeable farnesyltransferase inhibitor. Because of its unique nonthiol structure and biochemical profile, a series of analogs have been developed to improve cellular potency and subsequently in vivo efficacy. Emerging from these studies was the histidylbenzylglycinamide series of FTIs. These compounds remain peptidic in nature, yet are FPP-competitive and exhibit inhibition of FTase that is synergistic with phosphate. Exhibiting >500-fold selectivity for FTase relative to other prenylation enzymes, the histidylbenzylglycinamides also possess nanomolar potency against cellular farnesylation. Their broad spectrum in vivo activity against a panel of human tumor xenografts is independent of *ras* status. Collectively, data point to a novel biochemical profile for these inhibitors resulting in promising biological activity. The future clinical potential of this series of FTIs is likely to depend on the attainment of improved bioavailability and/or the judicious design of combination regimens that take into account their unique features.

REFERENCES

1. Sebolt-Leopold JS. A case for ras targeted agents as antineoplastics, in: *Cancer Therapeutics: Experimental and Clinical Agents* (Teicher B, ed.), Humana Press, Totowa, NJ, 1995, pp. 395–415.
2. Leonard DM. Ras farnesyltransferase: a new therapeutic target. *J Med Chem* 1997; 40:2971–2990.
3. Cox AD, Der CJ. Farnesyltransferase inhibitors and cancer treatment: targeting simply Ras? *Biochim Biophys Acta* 1997; 1333:F51–F71.
4. Venet M, Angibarud P, Sanz G, Poignet H, End D, Bowden C. Synthesis and in-vitro structure-activity relationships of imidazolyl-2-quinolinones as farnesyl protein transferase inhibitors (FTI). *Proc Amer Assoc Cancer Res* 1998; 39:318.
5. James GL, Goldstein JL, Brown MS. Polylysine and CVIM sequences of KrasB dictate specificity of prenylation and confer resistance to benzodiazepine peptidomimetic in vitro. *J Biol Chem* 1995; 270: 6221–5226.
6 Sun J, Qian Y, Hamilton AD, Sebti SM. Ras CAAX peptidomimetic FTI 276 selectively blocks tumour growth in nude mice of a human lung carcinoma with k-ras mutation and p53 deletion. *Cancer Res* 1995; 55:4243–4247.
7. Kohl NE, Omer CA, Conner MW, Anthony NJ, Davide JP, Desolms SJ, et al. Inhibition of farnesyltransferase induces regresssion of mammary and salivary carcinomas in ras transgenic mice. *Nature Med* 1995; 1:792–797.
8. Goldstein JL, Brown MS, Stradley SJ, Reiss Y, Gierasch LM. Nonfarnesylated tetrapeptide inhibitors of protein farnesyltransferase. *J Biol Chem* 1991; 266:15,575–15,578.
9. Leonard DM, Shuler KR, Poulter CJ, Eaton SR, Sawyer TK, Hodges JC, et al. Structure-activity relationships of cysteine-lacking pentapeptide derivatives that inhibit ras farnesyltransferase. *J Med Chem* 1997; 40:192–200.
10. Scholten JD, Zimmerman K, Oxender GM, Sebolt-Leopold JS, Gowan R, Leonard D, Hupe DJ. Inhibitors of farnesyl:protein transferase—a possible cancer chemotherapeutic. *Bioorg Med Chem* 1996; 4: 1537–1543.
11. Scholten JD, Zimmerman KK, Oxender MG, Leonard D, Sebolt-Leopold J, Gowan R, Hupe DJ. Synergy between anions and farnesyldiphosphate competitive inhibitors of farnesyl:protein transferase. *J Biol Chem* 1997; 272:18,077–18,081.
12. Hudspeth JP, Kaltenbronn JS, Repine JT, Roark WH, Stier MA. U.S. Patent No. 4,735,933, 1988.
13. McNamara DJ, Dobrusin E, Leonard DM, Shuler KR, Kaltenbronn, JS, Quin J III, et al. C-Terminal modifications of histidyl-N-benzylglycinamides to give improved inhibition of ras farnesyltransferase, cellular activity, anticancer activity in mice. *J Med Chem* 1997; 40:3319–3322.
14. Moasser MM, Sepp-Lorenzino L, Kohl NE, Oliff A, Balog A, Su DS, et al. Farnesyl transferase inhibitors cause enhanced mitotic sensitivity to taxol and epothilones. *Proc Natl Acad Sci USA* 1998; 95:1369–1374.
15. Bernhard EJ, Kao G, Cox AD, Sebti SM, Hamilton AD, Muschel RJ, McKenna WG. The farnesyltransferase inhibitor FTI-277 radiosensitizes H-ras-transformed rat embryo fibroblasts. *Cancer Res* 1996; 56:1727–1730.
16. Williams TM, Ciccarone TM, MacTough SC, Bock RL, Conner MW, Davide JP, et al. 2-Substituted piperazines as constrained amino acids. Application to the synthesis of potent, non carboxylic acid inhibitors of farnesyltransferase. *J Med Chem* 1996; 39:1345–1348.

9

From Random Screening of Chemical Libraries to the Optimization of FPP-Competitive Inhibitors of Farnesyltransferase

Patrick Mailliet, MD, *Abdel Laoui*, PHD, *Jean-Dominique Bourzat*, PHD, *Marc Capet*, PHD, *Michel Chevé*, PHD, *Alain Commerçon*, PHD, *Norbert Dereu*, PHD, *Alain LeBrun*, PHD, *Jean-Paul Martin*, MD, *Jean-François Peyronel*, PHD, *Christophe Salagnad*, MD, *Fabienne Thompson*, MD, *Martine Zucco*, PHD, *Jean-Dominique Guitton*, PHD, *Guy Pantel*, MD, *Marie-Christine Bissery*, PHD, *Clive Brealey*, PHD, *Jacques Lavayre*, PHD, *Yves Lelièvre*, PHD, *Jean-François Riou*, PHD, *Patricia Vrignaud*, PHD, *Marc Duchesne*, PHD, and *François Lavelle*, PHD

CONTENTS

From: *Farnesyltransferase Inhibitors in Cancer Therapy*
Edited by: S. M. Sebti and A. D. Hamilton © Humana Press Inc., Totowa, NJ

1. INTRODUCTION

Together with gene alterations of the p53 tumor suppressor gene, mutations of the ras genes represent the most frequent gene modification in human cancers. Ras mutations are found in at least 90% of pancreas, 50% of colon, and 30% of both lung and thyroid cancers *(1,2)*.

P21 Ras proteins (Ha-, N-, Ki4A-, and Ki4B-) act as molecular switches between their inactive cytosolic GDP-bound form and their activated GTP-bound form. Ras proteins are anchored at the inner face of the plasma membrane. This localization is a result of a series of four successive steps of posttranslational modifications occurring on the C-terminal tetrapeptide of the protein, referred to as the CAAX-box (C = cysteine, A = any aliphatic amino acid, usually X = serine, methionine, or glutamine). The posttranslational steps include farnesylation of the cysteine, catalyzed by protein farnesyltransferase (FTase), cleavage of the three last amino acids, methylation of the resulting carboxylic acid, and, finally, anchorage into the plasma membrane. FTase transfers a C15 farnesyl residue to the cysteine thiol of the CAAX-box *(3)*. Because farnesylation of mutated Ras proteins is required for cell transformation *(4)*, inhibition of FTase emerged as an attractive target for therapeutic intervention in the field of cancer *(5–9)*.

2. MATERIAL AND METHODS

2.1. Enzyme Purification

Human FTase and geranylgeranyl transferase (GGTase) were purified according to Reiss *(3)*. The 100,000 G supernatant fraction from THP-1 lysate was fractionated by ammonium sulfate. The solubilized dialyzed 30–60% fraction was purified on a fast-flow Q Sepharose anionic gel exchanger. The 0.18 *N* NaCl gradient eluted fraction contained the GGTase, free from FTase. The 0.23 *N* NaCl eluted fraction was further purified on an affinity column of TKCVIM-CH-Sepharose, thus providing FTase, free from GGTase.

2.2. Precipation Assays with TriChloroAcetic Acid (TCA)

Purified human FTase was preincubated for 10 min with various concentrations of inhibitors. p21Ha-Ras protein (3 μ*M*) was added, and the enzymatic reaction initiated by the addition of 0.075 μ*M* of tritiated farnesylpyrophosphate (FPP) in a 60 μL final volume. After incubation at 37°C for 30 min, the reaction was stopped by the addition of 30% TCA in methanol. The precipitation was completed by the addition of 1% sodium dodecyl sulfate (SDS) at 0°C. The final precipitate, filtered by a Skatron device on a glass filter, was counted in a scintillation counter after scintillate impregnation. Similar TCA assay was performed using p21 Ki-Ras, protein, the experimental concentration being adjusted according to the p21 Ki-Ras Km value.

2.3. Scintillation Proximity Assays (SPA)

Purified human FTase was incubated at 37°C for 1 h with 50 n*M* of a Ki-Ras related peptide, Biot-(bA)3-S-K-D-G-(K)6-S-K-T-K-C-V-I-M, 120 n*M* of FPP in 100 μL final volume. Then 150 μL of a suspension of streptavidin PVT beads were added at pH 4.0, thus blocking the farnesylation reaction and—thanks to the biotin/streptavidin interaction—giving a scintillation that was measured on a scintillation counter. Similarly, 100 n*M* of a lamin B related peptide, Biot-Y-R-A-S-N-R-S-C-A-I-M, was used in a lamin B/SPA assay. The same assay conditions were also used for the Ki-Ras peptide and the purified GGTase.

2.4. Competition Assays

Kinetic experiments were performed, according to the previously described TCA assay conditions, with increasing concentrations of FPP, or p21 Ha-Ras protein, around its Km value and with an excess of p21 Ha-Ras protein (Km × 5), or FPP (Km × 2).

2.5. Ras Processing in THAC Cells

THAC cells (Chinese Hamster fibroblasts, CCL39 cells transformed with activated Ha-Ras) were treated with various concentrations of inhibitors for 24 h and lysed in Triton X-114 buffer (20 m*M* Tris-HCl, 1% Triton X-114, 5 m*M* $MgCl_2$, 5 m*M* DTT, pH 7.4). The farnesylated Ras protein was separated from the nonfarnesylated Ras protein by phase extraction with Triton X-114. The proteins were fractionated by 14% SDS-polyacrylamide gel electrophoresis (PAGE) and transferred onto a polyvinylene membrane. The filters were incubated with specific anti-Ras monoclonal antibodies (panRas Ab3) and then with radiolabeled (^{125}I) protein A. The Ras-specific bands were cut off and counted in a scintillation counter. The ratio between radioactivity corresponding to farnesylated and nonfarnesylated Ras allowed the determination of Ras inhibition.

2.6. Colony Formation Assay Conditions

Human cell lines harboring Ras-mutations were provided by the American Type Culture Collection, (Rockville, MD). Cell lines were grown as monolayers in Dulbecco's Modified Eagle Medium (DMEM) containing 2 m*M* L-glutamine, 200 U/mL penicillin, 200 µg/mL streptomycin and supplemented with 10% (v/v) heat-inactivated fetal calf serum (FCS). Cells in exponential growth were trypsinized, washed with phosphate-buffered saline (PBS) solution, and diluted to a final concentration of 5000 cells/mL in complete medium. Drugs to be tested or control solvent were added to the cell suspension (2.5 mL) under a volume of 50 µL and 0.4 mL of 2.4% Noble Difco agar maintained at 45°C was added and mixed. The mixture was immediately poured into Petri dishes and allowed to stand for 5 min at 4°C. The numbers of cellular clones (>60 cells) were measured after a 12-d incubation at 37°C under 5% CO_2 atmosphere. Each drug was tested at 10, 1, 0.1, 0.01, and 0.001 µg/mL (final concentration in agar) in duplicate. Results are expressed in percent of inhibition of clonogenicity relative to untreated controls. The 50% inhibitory concentrations (IC_{50}) were graphically determined from semi-logarithmic plots of the mean value determined for each concentration.

2.7. Pharmacokinetics

RPR 130401 was suspended in water containing 0.5% tween 80 and 0.5% methyl cellulose and administered to female B6D2F1 mice at 250 mg/kg. Plasma was collected 5, 15, 30 min, 1, 2, 3, 4, 6, 8, 12, and 24 h. Plasma levels were determined by HPLC with U.V. detection (LC-MS/MS, the limit of quantification was 0.05 µg/mL).

2.8. In Vivo Antitumor Activity

Swiss-nu female mice were bred at Iffa-Credo (France). They had free access to food and water, and were maintained under pathogen-free conditions. Mice were over 18 g at the beginning of therapy.

Human colon carcinoma HCT 116, a Ki-Ras activated human tumor, was obtained from the American Type Culture Collection *(10)*. It was established in nude mice (subcutaneous [s.c.] implant of 1.4 10^7 cells) and maintained by s.c. serial passage once every 4 wk in Swiss-nu mice.

The methods of chemotherapy were previously reported *(11–14)*. Briefly, mice were implanted s.c. bilaterally with tumor fragments with a 12-gage trocar. The tumor size at the start of therapy ranged from 80–180 mg in the various treatment and control groups. RPR 130401 was suspended in water containing 1% Tween 80 and 0.5% methyl cellulose and administered orally (p.o.) twice daily for 19 consecutive days. Mice were checked daily and adverse clinical reactions were noted. Each group of mice was weighed as a whole once daily until the weight nadir was reached. Then groups were weighed once or twice weekly until the end of the experiment.

The s.c. tumors were measured with a caliper two to three times a week until the tumor reached 2000 mg. Solid tumor weights were estimated from two dimensional tumor measurements: Tumor weight (mg) = [length (mm) × width2 (mm2)] / 2). The tumor growth delay (T-C), used as an end point for estimating the experimental antitumor activity, is based on the median time (days) required for the treated group (T) and the control group (C) to reach a predetermined size of tumor. A T-C value superior or equal to the total duration of the treatment period indicates at least a cytostatic effect (no tumor growth during treatments). The tumor doubling time (Td) is estimated from the best fit straight line from a log linear growth. The $\log_{10}$ cell kill net was calculated from the following formula: [(T-C) – treatment duration] / (3.32 × Td). A positive log cell kill net indicates at least a cytostatic activity.

3. STRATEGIES FOR THE DISCOVERY OF INHIBITORS

FTase is a bisubstrate (FPP and p21-Ras protein) enzyme. Among various possible strategies, the inhibition of its activity may be achieved with compounds able to mimic and (or) to compete with either FPP or p21-Ras proteins.

3.1. Rationale Design of "CAAX" Mimic p21-Ras Competitive Inhibitors

The discovery, that some p21 Ras CAAX tetrapeptides *(3)* inhibit FTase, stimulated intensive rational drug design of CAAX mimics leading to peptidic *(15–17)*, pseudopeptidic and peptidomimetic inhibitors *(18–25)*. Although some of these compounds were found to be active both in cellular and in animal models *(26–29)*, only recently have some of these reached early clinical trials.

Our contribution to the design of p21-Ras-competitive inhibitors of FTase started with molecular modeling studies of CVFM peptide analogs (Fig. 1; **I/1**) using molecular modeling dynamics and energy minimization *(16)*. These studies demonstrated a direct correlation between FTase inhibitory activity and the proclivity of the inhibitors to adopt an extended conformation with CaCys-CaMet distance >7.5 Å. This prompted us to search for hydrophobic scaffolds able to orientate the cysteine and methionine units according to a pharmacophore model derived from the potent prototype Cys-(N-Me)Val-Tic-Met *(22)* (Fig. 1; **I/2**). We intentionally restricted this search to scaffolds that were unable to access low-energy turn conformations. The 1,5-naphthyl scaffold proved to be worthy of further investigation (Fig. 2; **I/3**) *(25)*. It satisfies the distance requirement between Cys and Met units and also provides a rigid template that directs these units

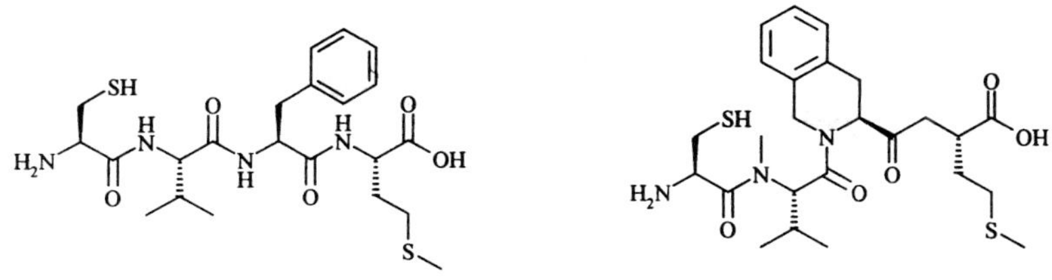

I/1 I/2

human Ftase / Ha-Ras
IC_{50} = 1.0 μM 0.005 μM

Fig. 1. Structure and FTase inhibition by of CVFM peptide **I/1** and C-(NMe)V-Tic-M peptidomimetic **I/2**.

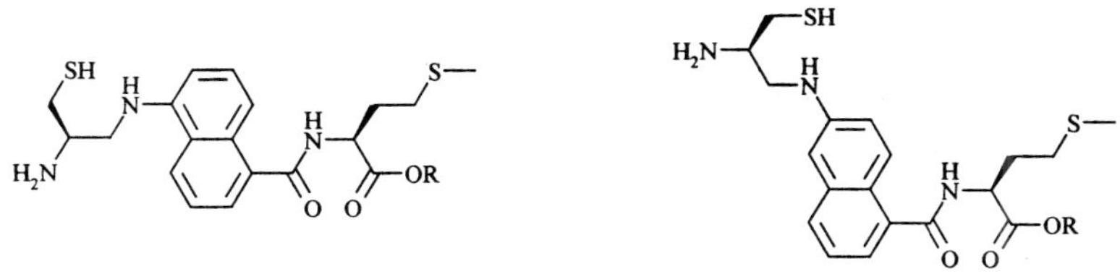

I/3a R= H
I/3b R= Me

I/4a R= H
I/4b R= Me

Fig. 2. Structure of "naphthalene based" CAAX-mimic inhibitors **I/3** and **I/4**.

unambiguously into an extended conformation. Further examination of naphthalene-based scaffolds suggested that the isomeric 1,6-naphthyl scaffold *(25)* (Fig. 2; **I/4**) would also be capable of maintaining the approximate distance and conformational constraints between Cys and Met projected in the earlier analysis *(23)*. In the FTase enzyme assay the 1,6-naphthyl series is three- to fourfold more potent than the 1,5 series, and the carboxylic acids are roughly 50-fold more potent than the corresponding esters (Table 1). Usually Ki-Ras protein (or related peptide) is 10-fold less sensitive to the "naphthyl-inhibitors" than Ha-Ras protein, possibly reflecting the 50-fold higher affinity of Ki-Ras for FTase as compared to Ha-Ras *(30)*.

Historically one of the major challenges in the research on "CAAX" peptides and peptidomimetics has been the translation of potent isolated-enzyme inhibitory activity to activity in cells. Considerable progress has been made in this area owing to ester prodrugs on the carboxylate. Application of this prodrug strategy led to a dramatic improvement of the cellular activity of the 1,6-series, as demonstrated by compound **I/4b**, which inhibits both the farnesylation of Ha-Ras in THAC cells (Table 1) and the anchorage-independent cell-growth human cancer cell lines harboring Ki-Ras mutations such as HCT 116 colon carcinoma, in the 10–20 μ*M* range (Table 1). However, such a level of cell-growth inhibition in vitro is itself insufficient to produce cell-growth inhibition in vivo, as demonstrated by the lack of activity observed with compound **I/4b** against HCT 116 human xenografts implanted in nude mice (data not shown). Further optimization of these types of conformationaly extended naphthalene-based inhibitors is currently in progress in our group through three main routes: optimization of the scaffold itself, replacement of the methionine by an aromatic surrogate, and replacement of the cysteine by a more versatile zinc-chelating moiety.

Table 1
Structure of Enzymatic and Cellular Inhibition of Naphthalene-Based CAAX-Mimic Inhibitors I/3–4[a]

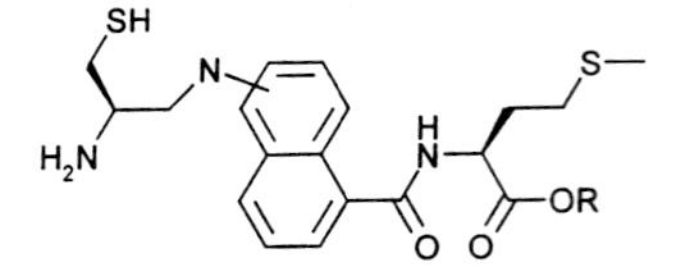

Compound		enzymatic inhibition (human FTase)		% inhibition of Ha-Ras	HCT 116 growth inhibition
		Ha-Ras IC50 μM	Ki-Ras IC50 μM	process at 10 μM	IC50 μM
1-5 series	**I/3a**	0.0056	n.d.	0	>> 20
	I/3b	0.325	n.d.	20	>> 20
1-6 series	**I/4a**	0.0018	0.0175	10	>> 20
	I/4b	0.08	0.5	> 50	20

[a]For experimental details, *see* Section 2.

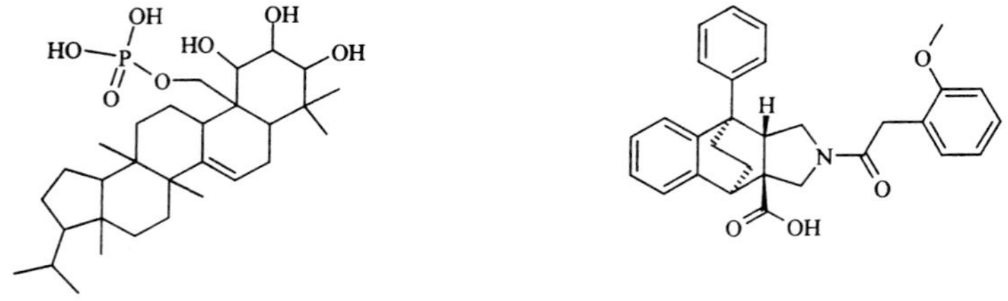

RPR 113228
(IC_{50} = 0.83 μM)

RPR 104213
(IC_{50} = 0.2-0.3 μM)

Fig. 3. Chemical structures and FTase/lamin B inhibitions by RPR 113228 and RPR 104213 (+/–) **II/1**.

3.2. Random Screening of Libraries

Beside drug design, intensive screening of natural products *(30–32)* and chemical libraries *(33–38)* also led to the discovery of potent p21-competitive and FPP-competitive inhibitors of FTase. Recently, two p21-competitive FTase inhibitors R115777 *(37)* and SCH 66336 *(38)* entered clinical trials.

Such a general screening strategy has been used in our group by using the Scintillation Proximity Assay (SPA) technology with a lamin B-related peptide as protein substrate. This led us to the discovery of various hits, including RPR 113228 *(32)* and RPR 104213 (Fig. 3), as micromolar/submicromolar, mainly FPP-competitive inhibitors of FTase. Because RPR 104213 was the most potent FTase inhibitor and also exhibited significant inhibition of Ha-Ras processing in intact cells (THAC hamster fibroblasts), we focused on this particular series of benzo[f]perhydroisoindole (BPHI) derivatives for further optimization. Starting from RPR 104213, this optimization process led us successively to RPR 115135, a potent inhibitor of FTase in cells, then to RPR 130401, which displays in vivo activity in animal models (Fig. 4).

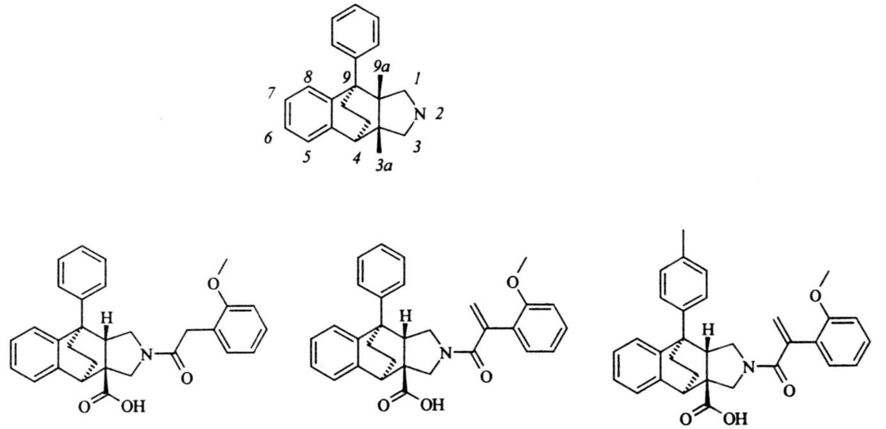

IC_{50} (μM)	(+/-) II/1 RPR 104213	(+) II/2 RPR 115135	(+) II/25 RPR 130401
FTase/KiRas	0.2	0.05	0.008
HCT116 cloning	> 10	0.72	0.045
HCT116 *vivo*	n.d.	almost inactive	cytostatic activity

Fig. 4. Numbering of BPHI: chemical structures and biological activities of RPR 104213 **(+/–) II/1**, RPR 115135 **(+) II/2**, and RPR 130401 **(+) II/25**.

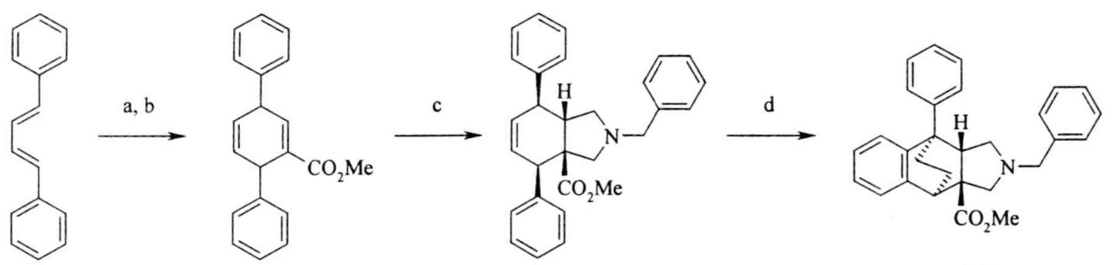

II/A

Scheme 1. Synthesis of methyl 4,9-ethano-9-phenyl-2-phenylmethyl-2,3,3a,4,9,9a-hexahydrobenzo [f]isoindole-3a-carboxylate-3aRS,4SR,9SR,9aRS **II/A**. (a) Propiolic acid / 0.1 Eq hydroquinone / toluene (rfx.). (b) H_2SO_4 / MeOH (rfx.). (c) $Me_3Si\text{-}CH_2\text{-}N(Bn)\text{-}CH_2\text{-}O\text{-}n$Bu/TFA (cat.) / CH_2Cl_2 (r.t.). (d) 10 Eq. CF_3SO_3H / CH_2CL_2 (0° to r.t.).

4. CHEMISTRY OF BENZO[F]PERHYDROISOINDOLES

The general chemical structure and numbering of the BPHI skeleton are outlined in Fig. 4.

4.1. Modifications at Positions 2 and 3: First Chemical Pathway (Schemes 1 and 2)

As outlined in Scheme 1, a straightforward synthesis of the "9-phenyl-BPHI"-polycyclic skeleton has been developed starting from trans,trans-1,4-diphenyl-1,3-butadiene. Diels-Alder reaction between trans,trans-1,4-diphenyl-1,3-butadiene and propiolic acid, followed by esterification, afforded methyl 2,5-diphenyl-1,3-cyclohexadiene-carboxylate in 67% yield. Subsequent 1,3-dipolar cycloaddition with *in situ*-generated azomethine ylid dipole *(39)* gave methyl 2-benzyl-4,7-diphenyl-2,3,3a,4,7,7a-hexahydro-1H-isoindol-3a-carboxylate in 68% yield. When treated with an excess of triflic acid in dichlorometh-

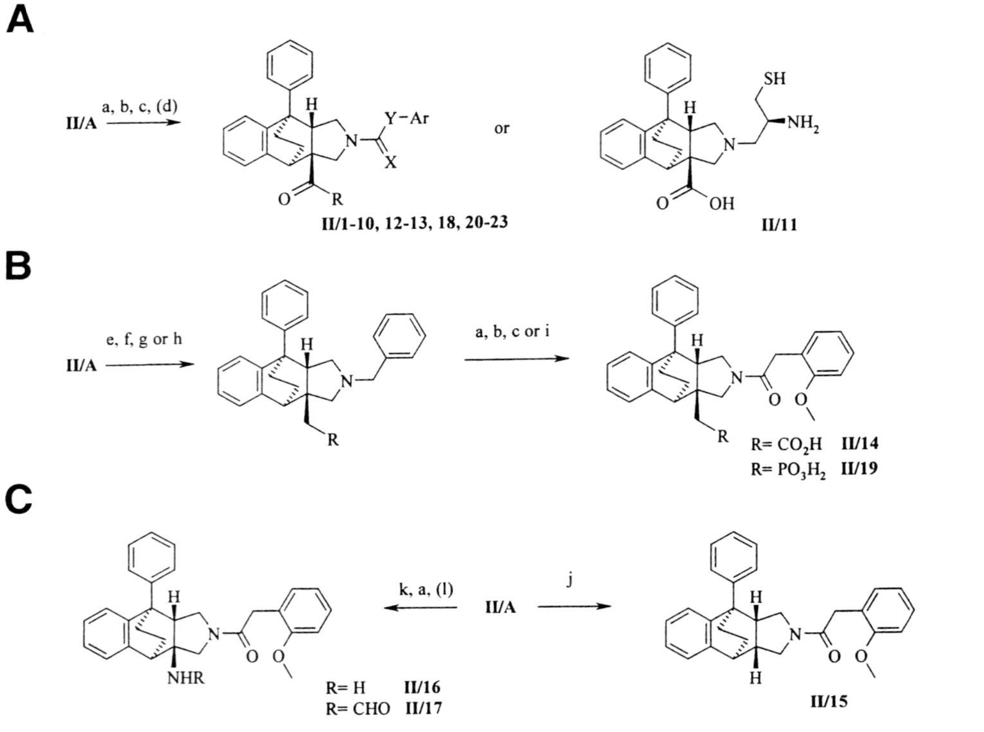

Scheme 2. Chemical modification of BPHI at position 2 and 3a. (**A**) Synthesis of carboxylic acid and derivatives **II/1–13**, **18**, **20–23**. (**B**) Synthesis of compound **II/14**, **19**. (**C**) Synthesis of compounds **II/15**, **16**, and **17**. (a) NH_4^+ HCO_2^- / Pd-C/MeOH (rfx.); or H_2 / $Pd(OH)_2$ / EtOH (r.t.). (b) Ar-CH_2-COOH / EDCI / HOBT (r.t.); or ArNCO / DMAP / THF (r.t.); or Ar-$(CH_2)_2$-Ots / TEA / toulene (rfx.). (c) Aq. NaOH (0.1–1 *N*) / EtOH (r.t. to rfx.). (d) Im_2CO / CH_2Cl_2; then 0.5 *M* NH_3 in ethanol, or amine / EDCI / HOBT / CH_2Cl_2, or $CsCO_3$ / alcohol / DMF. (e) LAH / THF (rfx.). (f) Tf_2O / TEA / dioxane (r.t.). (g)KCN / DMSO (100°); H_2SO_4 / MeOH (rfx.). (h) Triethyl phosphite (140°). (i) Me_3SiI / CCl_4; then 0.1 *N* aq. $Na_2S_2O_3$. (j) Oxalyl chloride then 2-mercaptopyridine *N*-oxyde / toluene (rfx.); Bu_3SnH / toluene (rfx.). (k) $SOCl_2$ / Toluene (rfx.); NaN_3 / Et_2O (-10° to r.t.) then heating in toluene (75–90°); BnOH / toluene (rfx.); H_2 / $Pd(OH)_2$ / MeOH. (l) HCOOH / Ac_2O (50°).

ane at room temperature, it underwent migration of the double-bond, protonation to the carbocation, and intramolecular Friedel-Crafts reaction providing exclusively methyl 2-benzyl-9-phenyl-2,3,3a,4,9,9a-hexahydro-1H-benzo[f]isoindole-3a-carboxylate-3aRS, 4SR,9SR,9aRS (**II-A**) in 60% yield.

Racemic compounds **II/1-10–13**, **18**, **20–23** (Table 1) have been prepared by usual sequences of debenzylation, coupling reaction at position 2, saponification and eventually amide-coupling at position 3a (Scheme 2A). Lithium-aluminum hydride reduction of II-A, followed by activation and substitution of the hydroxymethyl to a cyanomethyl group, and further saponification, gave the homologous racemic carboxylic acid **II-14** (Scheme 2B). Similarly, the 3a-hydroxymethyl derivative, upon reaction with triethylphosphite, led to the phosphonic acid **II-19**.

Radical Barton decarboxylation *(40)* of **II-1** allowed the synthesis of the racemic 3a-hydrogeno analog **II-15** in low yield (Scheme 2C). Curtius rearrangement of acyl-azide derivative of compound **II-1** afforded racemic amine **II-16** in 50% yield (Scheme 2C).

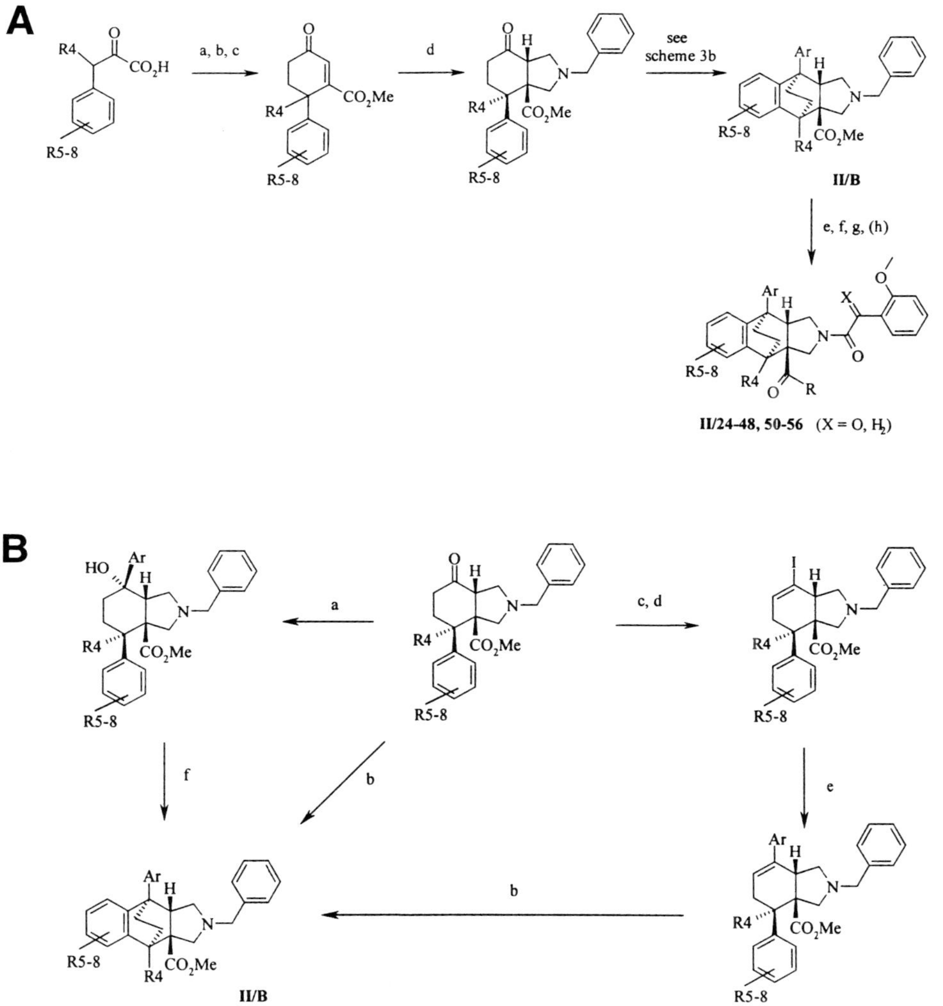

Scheme 3. Modifications of BPHI at positions 4, 5, 6, 7, 8 and 9. (**A**) Syntheses of intermediate carboxylates **II/B** and syntheses of compounds BPHI **II/24–48**, **50–56**. (a) Methyl vinyl ketone / KOH / MeOH (rfx.) (b) p. TsOH toluene (rfx.). (c) MeI / DBU / acetone (rfx.). (d) $Me_3Si\text{-}CH_2^-$ N(Bn)-Ch_2-O-*n*Bu / TFA (cat.) / CH_2Cl_2 (r.t.). (d) $NH_4^+HCO_2^-$ / Pd-C / MeOH (rfx.) or VocCl / Ch_2Cl_2 then HCl / MeOH. (e) Oxalyl chloride / (o.methoxyphenyl)acetic acid / E_3N / CH_2Cl_2. (f) NaOH 0.1–1 *N* in MeOH or dioxane (rfx.). (g) Im_2CO / CH_2Cl_2 then 0.5 *M* NH_3 or amine / EDCI / HOBT / CH_2Cl_2. (**B**) Methods for the introduction of the 9-aryl ring and subsequent cyclization to compound **II/B**. (a) ArMgX ($CeCl_3$) or ArLi / Et_2O / toluene (0° to rfx.). (b) Excess CF_3SO_3H (CH_2Cl_2) (0° to rfx.). (c) Hydrazine / KOH / MeOH (rfx.). (d) iodine / TEA / MeOH (rfx.). (e) $ArB(O_2H)$ / $Pd(PPh_3)_4$ / aq. Na_2CO_3 / toluene (rfx.). (f) ArH / CF_3 SO_3H (CH_2Cl_2) (0° to rfx.).

4.2. Modifications at Positions 4, 5, 6, 7, 8, and 9 (Schemes 3 and 4)

The synthesis developed in Scheme 1 allowed, almost exclusively, chemical modifications at position 2 and 3a. Therefore we investigated a more versatile synthesis, providing modifications at positions 4, 5, 6, 7, and 8 of the BPHI polycyclic skeleton. Moreover, this chemical route also led to a wide range of modifications of the phenyl ring at position 9. As outlined in Scheme 3A, when treated with methyl vinyl ketone in alkaline medium,

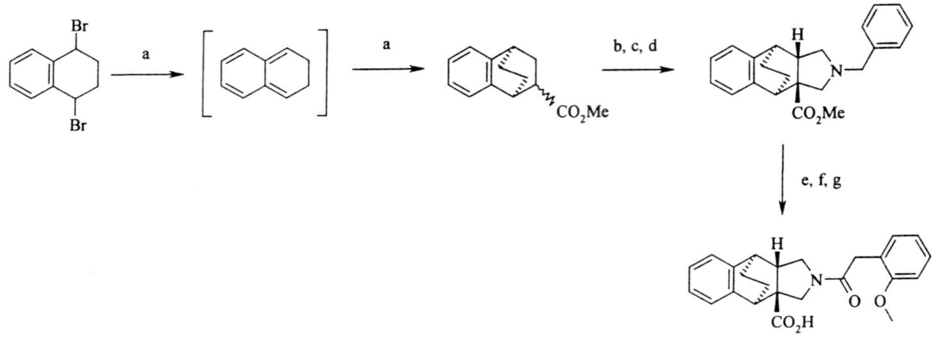

Scheme 4. Modifications at position 9: suppression of the phenyl ring, synthesis of compound **11/49**. (a) Zn-Cu amalgam/methyl acrylate / CH_2Cl_2 (rfx.). (b) PhSeCl/LDA/THF (–70° to 0°). (c) H_2O_2/ THF (–40° to r.t.). (d) Me_3Si-CH_2-N(Bn)-CH_2-O-nBu / TFA (cat.) / CH_2Cl_2 (r.t.). (e) $NH_4^+HCO_2^-$/ Pd-C / MeOH (rfx.). (e) (o.methoxyphenyl)acetic acid / EDCI / HOBT / CH_2Cl_2. (f) NaOH *N* in MeOH (rfx.).

substituted phenylpyruvic acids underwent annelation. Further dehydration and esterification gave methyl 6-aryl-cyclohex-1-en-3-one-carboxylates in 60–90% yields *(41)*. Dipolar cycloaddition was then performed in 70–90% yield *(39)*. The "9-aromatic ring" was subsequently introduced at the keto position and finally intramolecular Friedel-Craft reaction gave methyl 9-aryl-4,9-ethano-2-phenylmethyl-2,3,3a,4,9,9a-hexahydro-1H-benzo[f]isoindole-3a-carboxylates (**II-B**) in 20–90% yields, potentially substituted at positions 4, 5, 6, 7, or 8.

As outlined in Scheme 3B, depending on the nature and the reactivity of the entering "9-aromatic ring," three different strategies have been developed for its introduction and subsequent cyclization:

1. Grignard or cerium-modified Grignard *(42)* reaction of the ketone with arylmagnesium halides, or in few cases of aryllithium reagents, gave an alcohol. When protonated in triflic acid, this alcohol gave rise to a carbocation that underwent Friedel-Craft reaction.
2. According to the method developed by Barton *(43)*, the ketone was first transformed into an vinyl-iodo derivative, which was further coupled with arylboronic acid through the Suzuki reaction. The resulting vinyl aryl compound underwent similar Friedel-Craft cyclization through protonation of the double bond,
3. According to a method developed in our group, in triflic acid, electron-rich aromatic rings reacted with the ketone through intermolecular Friedel-Craft reaction to give an alcohol that immediately cycled through intramolecular Friedel-Craft reaction.

From methyl 9-aryl-4,9-ethano-2-phenylmethyl-2,3,3a,4,9,9a-hexahydro-1H-benzo [f]isoindole-3a-carboxylates (**II-B**), racemic compounds **II/24–48, 50–56** (Table 2) have been prepared by usual sequences of debenzylation, coupling reaction at position 2, saponification and eventually amide-coupling at position 3a (Scheme 3A).

From 2,3-dihydronaphthalene *(44)*, easily generated *in situ* by reaction of 1,4-dibromotetraline *(45)* with activated Zn-Cu couple, through successive Diels-Alder and dipolar-Achiwa cycloaddition reactions as key steps, we synthesized the racemic analogous compound **II/49**, which lacks the phenyl ring at position 9, in an overall yield of <2% (Scheme 4).

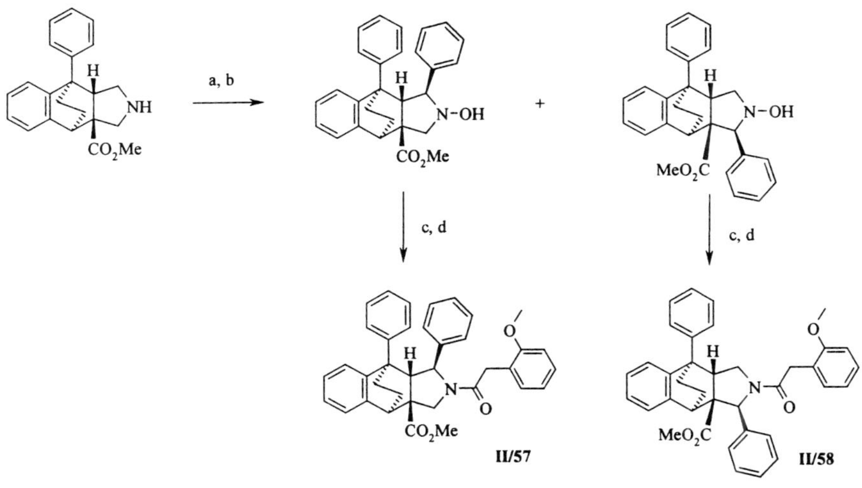

Scheme 5. Modifications of BPHI at positions 1 or 3, synthesis of compounds **II/57–58**. (a) 2-(phenylsulfonyl)-3-phenyloxaziridine / $CHCl_3$ (rt). (b) PhLi / THF (0° to rfx.). (c) H_2 / Pd-C / MeOH. (d) (o.methoxyphenyl)acetic acid / EDCI / HOBT / CH_2Cl_2.

4.3. Modifications at Positions 1 and 3 on the Pyrrolidine Ring (Scheme 5)

The oxidation of methyl 2H-4,9-ethano-9-phenyl-2,3,3a,4,9,9a-hexahydro-1H-benzo [f]isoindole-3a-carboxylate by 2-(phenylsulfonyl)-3-phenyloxaziridine *(46)*, followed by easy chromatographic separation of the resulting nitrones and by subsequent reaction of phenyllithium (or alkyllithium reagents) on the less hindered face of the molecule, led to the preparation of BPHI stereoselectively substituted at positions 1 and 3 by alkyl or aryl groups such as phenyl derivatives **II/57–58**. (Scheme 5).

4.4. Chiral Resolution of BPHI

When needed for further biological evaluations, enantiomers of active BPHI were easily obtained through chiral HPLC resolution on silica-gel bearing (2,3-dinitrobenzoyl)-S-phenylalanine residues as chiral resolving agents *(47,48)*.

5. ENZYMATIC FTASE INHIBITION

More than 400 compounds have been synthesized in the BPHI series. The strategy used to evaluate the FTase inhibition is primarily based on their potency both to inhibit the farnesylation of Ras proteins or related peptides by human FTase and to prevent Ras-processing in intact cells. For historical reasons, owing to the availability of Ras protein, the first 200 compounds have been evaluated using Ha-Ras as protein substrate. We then evaluated the ability of inhibitors to prevent the farnesylation of a Ki-Ras related peptide, which is more relevant to the clinical situation. Usually BPHI inhibit the farnesylation of both substrates with near or equal potencies.

5.1. Enzymatic Structure Activity Relationships

The FTase inhibitory potency of **(±) II/1** (IC_{50} = 0.2–0.31 μM) in SPA/lamin B assay was confirmed in a TCA/ Ha-Ras assay (IC_{50} = 0.31 μM) as well as a SPA/Ki-Ras assay (IC_{50} = 0.2–0.3 μM). This lack of protein substrate selectivity is a general feature within the BPHI series and is likely to be related to their FPP-competitive mode of inhibition (*see* Section 5.3.). Moreover these equipotent enzymatic inhibitions of Ha-Ras or Ki-Ras farnesylation allows us to analyze enzymatic SAR data independently of the precise assay or substrate conditions.

5.1.1. Stereochemical Requirements

Chiral resolution of (±) **II/1** emphasized the dextrogyre enantiomer **(+) II/1** as being at least 100-fold more potent than the levogyre enantiomer **(–) II/1**, with IC_{50} values of 0.14 and 45 μ*M*, respectively. At least for 3a-carboxylic acids, this dextrogyre-specific activity seems to be a general feature within the series, as exemplified by compounds **(+) II/2**, **(+) II/25**, and **(+) II/53**, which are 200–5000 times more potent than the corresponding levogyre enantiomers.

X-ray analysis of crystals from 4,9-ethano-2-[2-(2-methoxyphenyl)propen-2-oyl]-9-phenyl-2,3,3a,4,9,9a-hexahydro-1H-benzo[f]isoindol-3a-N-[1-(R)-phenylethyl]carboxamide RPR 117101 **(+) II/22**, resulting from the condensation of RPR 115135 **(+) II/2** with 1-(R)-phenylethylamine, demonstrated that the bioactive dextrogyre series of BPHI corresponds to the 3aS,4R,9R,9aS absolute configuration, with the 4,9-ethano bridge back and the 3a-carboxylic acid in front *(49)*. Moreover, in these crystals, the phenyl ring at position 9 and the phenyl ring of the side chain at position 2 are stacking together. Such a π-stacking conformation is also seen as the preferred one according to molecular modeling and energy minimization studies. This bioactive conformation will be assessed further by ongoing analysis of co-crystals obtained with RPR 130401 **(+) II/25** and human FTase.

With (3-pyridyl)methylamides, the enzymatic stereoselection is much less crucial, because dextrogyre enantiomers **(+) II/47** or **II/48** are only 4–8 times more potent than their levogyre enantiomers. This discrepancy in the stereochemical requirements may reflect differences in the binding of carboxylic acids and (3-pyridyl)methylamides at position 3a.

5.1.2. S.A.R. on the Side-Chain at Position 2 (Table 2A)

The nature and the length of the side chain at position 2 also play a key role on the inhibitory potency. Either removal of the side chain at position 2 or its replacement by a 4-methoxy-butanoyl moiety leads to inactive compounds (data not shown), thus demonstrating the need for an aromatic ring on this side chain. The nature of the linking group is also crucial. Replacement of the arylacetyl moiety by an arylethyl group **II/3**, an arylcarbonylamino II/4 or an arylcarbonyloxy group (data not shown), or its shortening to a benzoyl group **II/5**, or its lengthening to an arylpropanoyl group (**II/6**) can significantly decrease FTase inhibition.

The position of the substituent on the phenyl ring of the side chain is also crucial. Simply shifting the methoxy group from the ortho to the meta **II/7** or the para position (data not shown) also reduces activity. The precise role of the ortho methoxy group is not clear, as demonstrated by the drop in activity observed when it is removed **II/8** or replaced by electron-withdrawing groups (data not shown), while some electron-donating groups usually lead to two- to fivefold less active analogues **II/9–10**. The introduction of a methylene group **II/2**, at the benzylic position, is the most favorable modification, thus leading to 2- to 3-fold increase in inhibition. More generally substitutions by alkyl or

Table 2A
Structure of Enzymatic and Cellular FTase Inhibitions of BPHI Modified At Position 2 II/1-11

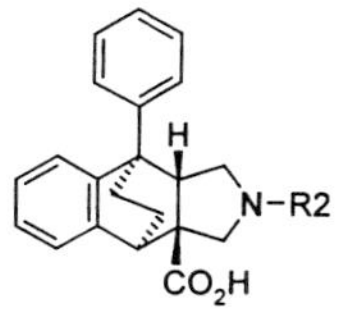

Compound	Chemical structure R2	Enzymatic FTase inhibition Ha-Ras/ TCA (1)	Ki-Ras/SPA (1)	Ras process. inhibition (2)
(±) II/1	C(O)-CH2-(2-MeO-Ph)	0.31	0.2-0.4	8-10
(+) II/1	"	0.135	0.14	4
(-) II/1	"	45	n.d.	>> 10
(+) II/2	C(O)-C(CH2)-(2-MeO-Ph)	0.02-0.04	0.025	1
(-) II/2	"	100	100	>> 10
(±) II/3	CH2-CH2-(2-MeO-Ph)	52.5	100	>> 10
(±) II/4	C(O)-NH-(2-MeO-Ph)	9.2	n.d.	>> 10
(±) II/5	C(O)-(2-MeO-Ph)	20	n.d.	>> 10
(±) II/6	C(O)-(CH2)2-(2-MeO-Ph)	5	2.5	>> 10
(±) II/7	C(O)-(CH2)2-(3-MeO-Ph)	100	100	>> 10
(±) II/8	C(O)-CH2-Ph	10-40	6	> 10
(±) II/9	C(O)-CH2-(2-OH-Ph)	1.8	n.d.	> 10
(±) II/10	C(O)-CH2-(2-Me-Ph)	1.15	1.5	> 10
(±) II/11	(S)-CH2-CH(NH2)-CH2-SH	1.8	1.7	>> 10

(1) IC_{50} values (μM) as means of two or more determinations. For experimental details, *see* Section 2.3.
(2) IC_{50} values (μM) of Ras proteins in THAC cells (Ha-Ras). >10 means 10–50% inhibition at 10 μM; >>10 means <10% inhibition at 10 μM. For experimental details, *see* Section 2.5.

alkenyl residues at the benzylic position of the side chain are also usually tolerated (data not shown).

We introduced more than 100 different side chains at position 2. The more active BPHI always corresponds to phenylacetyl derivatives, favorably substituted by an methoxy group at the ortho position, with the only surprising exception being a cysteinyl function such as in **II-11**. The cysteinyl exception may suggest that the side chain at position 2 is likely to be involved, even indirectly, in metal (magnesium or zinc) binding.

Table 2B
Structure of Enzymatic and Cellular FTase Inhibitions of BPHI Modified at Position 3a II/12–23

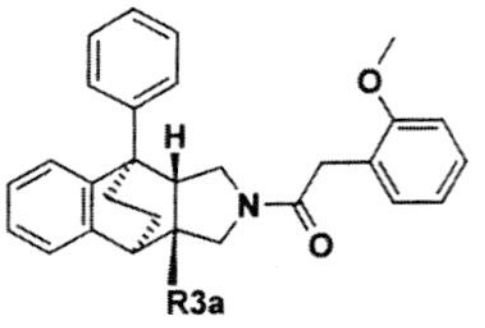

Compound	Chemical structure R3a	Enzymatic FTase inhibition Ha-Ras/TCA (1)	Ki-Ras/SPA (1)	Ras process. inhibition (2)
(±) II/1	COOH	0.31	0.2-0.4	8-10
(±) II/12	COOMe	0.4-1	1.7-3.3	> 10
(±) II/13	CONH2	0.33	0.35	10
(±) II/14	CH2-COOH	5.6	1.7-3.3	>> 10
(±) II/15	H	6.7	40	>> 10
(±) II/16	NH2	24.5	25	>> 10
(±) II/17	NH-C(O)H	6.95	n.d.	>> 10
(±) II/18	C(O)-NHOH	0.27	0.5	> 10
(±) II/19	CH2-P(O)(OH)2	2.6	1.0	>> 10
(±) II/20	C(O)-NH-NH2	5.5	5.05	> 10
(±) II/21	C(O)-NH-CH2-COOH	0.95	1.94	> 10
(±) II/22	(R)-C(O)-NH-CH(Me)-Ph	0.19	0.11	10
(+) II/23	C(O)-CH2-(3-Py)	0.15	0.21	1.5

(1) IC_{50} values (μM) as means of two or more determinations. For experimental details, *see* Section 2.3.
(2) IC_{50} values (μM) of Ras proteins in THAC cells (Ha-Ras). >10 means 10–50% inhibition at 10 μM; >>10 means <10% inhibition at 10 μM. For experimental details, *see* Section 2.5.

5.1.3. S.A.R. at Position 3a (Table 2B)

We also investigated the role of the carboxylic function at position 3a. Methyl ester **II/12** or carboxamide **II/13** exhibit rather similar potency to the carboxylic acid **II/1**. However, whereas higher esters (ethyl, isopropyl, or benzyl) are at least 10 times less potent, substituted amides **II/22–23** or aminoacid derivatives **II/21** are usually active. The homologation of the carboxylic acid **II/14**, or its suppression **II/15**, or its replacement by an amino **II/15** or an aminocarbonyl group **II/16** always lead to a significant loss of activity. Interestingly, activity can be restored by the introduction of the carboxamic acid **II/18** or, to lesser extent, the phosphonic acid **II/19**. As compared to position 2, functional requirements at position 3 are far less easy to determine. At least one hydrogen-bonding acceptor function is likely to be necessary, suggesting possible hydrogen-bonding interaction between the substituent at position 3a and the enzyme.

5.1.4. S.A.R at Position 9 (Table 2C)

Like the aromatic ring at position 2, the removal of the aromatic ring at position 9 completely abolishes activity **II/49**. However, unlike the aromatic ring at position 2, a very wide range of modifications on the phenyl at position 9 are allowed, often leading to more potent inhibitors.

More specifically, para **II/24**, meta **II/26**, and to a lesser extent ortho substitutions **II/27** with a methyl group are favorable. Positive para and meta substitutions include various functions such as methoxy or methylthio groups **II/28–29**, higher alkyls **II/30**, halogens **II/33–34**, trifluoromethyl **II/35**, or methoxy groups **II/37**, dimethylamino groups (data not shown). Electron-donating substituents usually appear to be slightly more propitious than electron-withdrawing ones. Only two limitations of these positive substitution effects can be noticed. The first one corresponds to steric hindrance as demonstrated by a modest drop of activity of 4-t.butyl derivative **II/31** and more dramatic drop corresponding to 4-phenyl substitution **II/32**. The second one is related to the polar nature of the substituent, as exemplified by 4-hydroxy analog **II/36**. 3-4-Disubstitutions are also usually favorable **II/38**, **40**, with similar limitations as shown by 3,4-dimethoxy compound **II/39**. 3,5- or 2,4-disubstitutions also lead to potent inhibitors **II/44–45**, whereas 3,4,5-trisubstitutions give less potent inhibitors, in accordance with the previously noticed steric constraints (data not shown).

The phenyl ring at position 9 may be replaced by heterocyles such as thiophene (data not shown) or more favorably by heterocycles corresponding to 3,4-disubstituted phenyl such as dihydrobenzofurane, benzodioxane or benzothiophene **II/41–43**.

S.A.R. at positions 2 and 3a, previously described in the 9-phenyl series, are mostly retrieved in the 9-substituted-phenyl series: methylene substitution at the benzylic position of the side chain at position 2 leads to a twofold improvement of the activity and maintains the biological stereoselection of the dextrogyre enantiomer, as exemplified by RPR 130401 **(+) II/25**. Methyl esters **II/46**, hydroxamic acids (data not shown), hydrazides, and amides (data not shown) are active; (3-pyridyl)methyl amides **II/47–48** being more specifically interesting both for their inhibitory potency and the previously described activity of the two enantiomers.

Substituent effects on the phenyl ring at position 9 (favorable substitutions with alkyl, alkyloxy residues, or halogen atoms, steric hindrance, unfavorable substitution with polar functions) suggest that this ring may participate mainly to a tight lipohilic interaction with the enzyme.

5.1.5. S.A.R at other Positions (Table 2D)

Substitutions at position 4 by either a methyl group **II/50** or an hydroxy function **II/51** and substitution at position 5 by a methoxy group **II/52** lead to a slight (1.5–2-fold) enhancement of the inhibitory potency in the 9-phenyl ring. This enhancement should be related, at least for hydroxy and methoxy functions, to a possible participation of the oxygen electron pair in the probable hydrogen-bonding acceptor interaction occurring with the carbonyl group of the 3a-substituent. When adapted to the 9-(4-methyl-phenyl) series, these modifications do not improve activity **(+) II/53**.

Table 2C
Structure of Enzymatic and Cellular FTase Inhibitions of BPHI Modified at Position 9 II/24–49

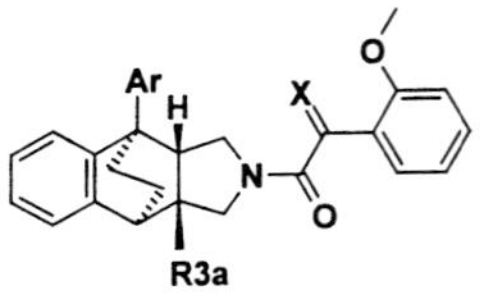

Compound	Chemical structure			Enzymatic FTase inhibition		Ras process.
	Ar	X	R3a	Ha-Ras/TCA (1)	Ki-Ras/SPA (1)	inhibition
(±) II/1	phenyl	H2	COOH	0.31	0.2-0.4	8-10
(±) II/24	4-Me-phenyl	"	"	0.046	0.035	1
(±) II/25	"	CH2	"	0.025	0.02-0.04	0.35
(+) II/25	"	"	"	n.d.	0.008 (3)	0.2
(-) II/25	"	"	"	"	1.8	25
(±) II/26	3-Me-phenyl	H2	"	0.05	0.053	1.2
(±) II/27	2-Me-phenyl	"	"	n.d.	0.099	n.d.
(±) II/28	4-MeO-phenyl	"	"	0.047	0.04	2.5
(±) II/29	4-MeS-phenyl	"	"	0.05	0.062	8
(±) II/30	4-Et-phenyl	CH2	"	n.d.	0.064	0.1
(±) II/31	4-t.Bu-phenyl	"	"	"	0.396	10
(±) II/32	4-Ph-phenyl	"	"	"	12	n.d.
(±) II/33	4-Cl-phenyl	"	"	"	0.02	1.0
(±) II/34	4-Br-phenyl	"	"	"	0.059	1.0
(±) II/35	4-F3C-phenyl	"	"	"	0.083	1.0
(±) II/36	4-HO-phenyl	"	"	"	1.13	>> 10
(±) II/37	4-F3CO-phenyl	"	"	"	0.097	n.d.

(±) II/38	3,4-diMe-phenyl	"	"	"	0.045	0.3
(±) II/39	3 4-diMeO-phenyl	H2	"	"	10	n.d.
(±) II/40	3,4-diCl-phenyl	CH2	"	"	0.068	0.1
(±) II/41	6-benzodioxanyl	"	"	"	0.106	1.0
(±) II/42	-(2,3-dihydrobenzofuryl)	"	"	"	0.036	n.d
(±) II/43	5-benzothienyl	"	"	"	0.011	"
(±) II/44	3,5-diMe-phenyl	"	"	"	0.103	5.4
(±) II/45	2,4-diCl-phenyl	"	"	"	0.012	n.d.
(±) II/46	4-Me-phenyl	"	COOMe	0.35	0.305	10
(+) II/ 47	"	"	CONH-CH2-(3-Py)	n.d.	0.055	0.8
(-) II/47	"	"	"	"	0.213	3.0
(+) II/48	6-benzodioxanyl	"	"	"	0.01	0.2
(-) II/48	"	"	"	"	0.084	1.4
(±) II/49	H	CH2	COOH	85.5	100	>> 10

(1) IC_{50} values (μ*M*) as means of two or more determinations. For experimental details, *see* Section 2.3.
(2) IC_{50} values (μ*M*) of Ras proteins in THAC cells (Ha-Ras). >10 means 10–50% inhibition at 10 μ*M*; >>10 means <10% inhibition at 10 μ*M*. For experimental details, *see* Section 2.5.
(3) As an internal reference compound in FTase/Ki-Ras/SPA assay, RPR 130401 **(+) II/25** has been extensively evaluated and always retrieved in the 0.002–0.02 μ*M* range.

Substitutions at positions 6 or 7 appear to exert only few effects on the FTase inhibition, as exemplified by methoxy derivatives **II/54–55**, thus suggesting that this part of the BPHI molecule is not interacting directly with the enzyme.

Substitutions at positions 8 significantly decrease inhibitory potency as exemplified by **II/56**. This should be related to modifications in the orientation of the phenyl ring at position 9 induced by substituents at position 8. Similarly, stereoselective introduction of a phenyl ring at position 1 **II/57** also dramatically diminishes activity. A similar drop in activity is retrieved with a methyl group in place of phenyl at position 1 (data not shown). Taken together, these results suggest that substituents at position 1 may disturb the correct orientation of the phenyl ring at position 9 and more certainly prevent π-stacking interactions between phenyl rings at position 2 and 9. The stereoselective introduction of a phenyl ring **II/58** and even a methyl group at position 3 also decreases activity (data not shown). One plausible explanation might be related to steric hindrance preventing hydrogen-bonding acceptor interaction of the carbonyl function at position 3a with the enzyme.

Table 2D
Structure of Enzymatic and Cellular
FTase Inhibitions of BPHI Modified At Other Positions II/50–57

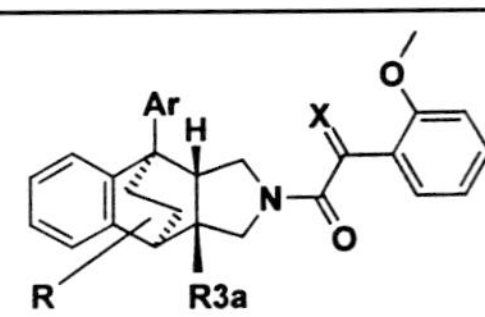

Compound	Chemical structure				Enzymatic FTase inhibition		Ras process.
	R	Ar	X	R3a	Ha-Ras / TCA (1)	Ki-Ras / SPA (1)	inhibition (2)
(±) **II/1**	no	Ph	H2	COOH	0.31	0.2-0.4	8-10
(±) **II/50**	4-Me	"	"	"	n.d.	0.086	2.5
(±) **II/51**	4-HO	"	"	"	0.17	0.25	> 10
(±) **II/52**	5-MeO	"	"	"	n.d.	0.099	10
(+) **II/53**	"	4-Me-Ph	CH2	"	"	0.018	0.3
(-) **II/53**	"			"	"	100	n.d.
(±) **II/54**	6-MeO	Ph	H2	"	"	0.29	10
(±) **II/55**	7-MeO	"	"	"	2.85	2.5	> 10
(±) **II/56**	8-MeO	"	"	"	"	12	>> 10
(±) **II/57**	1-Ph	"	"	"	39	"	>> 10
(±) **II/58**	3-Ph	"	"	"	51	"	>> 10

(1) IC_{50} values (μ*M*) as means of two or more determinations For experimental details, *see* Section 2.3.
(2) IC_{50} values (μ*M*) of Ras proteins in THAC cells (Ha-Ras). >10 means 10–50% inhibition at 10 μ*M*; >>10 means <10% inhibition at 10 μ*M*. For experimental details, *see* Section 2.5.

5.2. Mode of FTase Inhibition (Fig. 5)

As shown by Fig. 5, kinetic experiments performed with RPR 115135 **(+) II/2** indicate a competitive profile in the BPHI series with regards to FPP and a noncompetitive profile with the p21-Ha-Ras protein. Similar kinetic experiments, performed with either RPR 130401 **(+) II/25** or the cysteinyl analog **(+) II/11**, confirm the FPP-competitive mode of binding in the BPHI series. However kinetic experiments with (3-pyridyl)methyl-amides **(+) II/47** and **(–) II/47** are more difficult to analyze and do not exactly correspond to FPP-competitors. Additional support for this mode of binding is provided by significant differences in circular dichroism (CD) spectra of FTase observed with compound **(+) II/2** and with the p21-Ras competitive pentapeptide inhibitor K-C-V-Tic-M *(16)*. Ongoing analysis of co-crystallization studies of RPR 130401 **(+) II/25** and RPR 201542 **(+) II/48** with human FTase will lead to understanding the exact mode of binding of this series of inhibitors into the FPP-binding site of the enzyme (unpublished data).

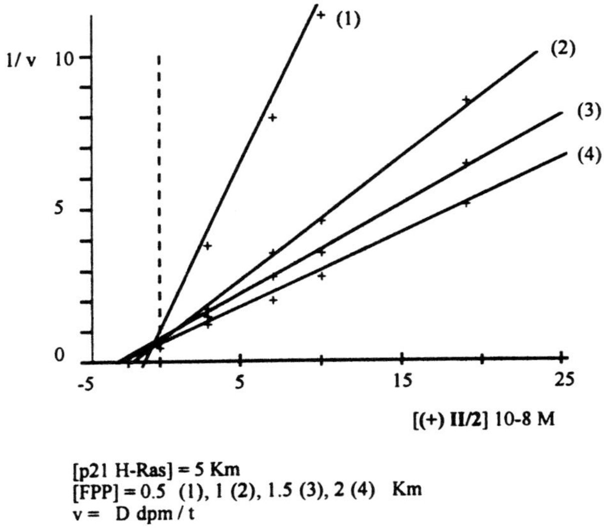

Fig. 5. Dixon representation of the kinetics of FTase inhibition by RPR 115135 (+) **II/2** at various concentrations of FPP. Assay by TCA precipitation as described in ref. *(3)*. For experimental details, *see* Section 2.4.

Table 3
Substrate and Enzyme Selectivities of BPHI

	FTase				*GGTase*	
Compound	*Lamin B/SPA* [a]	*Ha-Ras/TCA* [b]	*Ki-Ras/SPA* [a]	*Ki-Ras/TCA* [b]	*Ki-Ras/SPA* [a]	*SS* [a]
(±) **II/11**	0.2–0.3	0.31	0.2–0.4	0.75	>10	>10
(+) **II/2**	0.07	0.02–0.04	0.05	0.025	5.9	>10
(±) **II/25**	n.d.	0.051	0.02–0.04	0.052	>10	ND
(±) **12**	0.35	0.33	0.35	1.7	>10	ND

[a] IC_{50} (μ*M*) values as means of two or more determinations. For experimental details, *see* Sections 2.1 and 2.2.

[b] IC_{50} (μ*M*) values for FTase-catalyzed transfer of [^{3}H]FPP to human H-Ras protein, as means of two or more determinations. >10 means <50% inhibition at 10 μ*M*.

ND, not determined.

From previously described S.A.R. data and molecular modeling studies with the recently published FTase structures *(50–53)*, one can hypothesize that the BPHI scaffold is positioned within the deep part of the FPP-binding pocket: a) the conformation is probably close to the isolated crystalline structure of the BPHI (π-stacking interaction between aromatic rings at positions 2 and 9); b) the interactions should be of a lipophilic nature, especially through the aromatic rings. Additional interactions could arise from the methoxy group of the phenyl ring at position 2, with the zinc atom, and from the carbonyl group at position 3a, through hydrogen-bonding interactions with the main chain of the enzyme.

5.3. Substrate and Enzyme Selectivities (Table 3)

As a consequence of their FPP-competitive mode of binding to FTase, BPHI are inhibitors of the farnesylation of Ha-Ras, Ki-Ras, and lamin B, giving similar results using either purified proteins (TCA assays) or related peptides (SPA assays). This lack of protein substrate specificity should be of particular interest for the development of this series of FTase inhibitors, because farnesylated proteins other than Ras, such as Rho (B, E), Rheb, or protein–tyrosine phosphatases (PRL-1, IP3 5-PTPase) participate in transformation by Ras-dependent or -independent pathways *(54,55)*. The recent discovery of farnesylated oncogenic protein tyrosine phosphatases harboring a CAAX-box *(56,57)* reinforced the potential interest of BPHI as FTase inhibitors, irrespective of the FTase substrate.

BPHI are highly selective (more than 100-fold) for FTase with regards to GGTase. This is an unexpected result because GGTase can accommodate and transfer both FPP and GGPP to CAAX-proteins *(58,59)*. BPHI also appear to be selective for FTase with respect to squalene synthase (SS), which also uses FPP as substrate. (For a recent review of squalene synthase and its inhibitors, *see* ref. *60*). This enzyme selectivity remains to be confirmed by additional studies with other FPP-using enzymes such as geranyl geranyl diphosphate synthase, which transforms FPP into GGPP *(61)*.

5.4. Cellular Structure Activity Relationships (Tables 2A–D)

Ras processing was determined in THAC cells (hamster fibroblasts transformed by Ha-Ras) by evaluating the ratio between cytosolic and membrane-anchored p21 Ha-Ras protein (*see* Section 2.5.). The S.A.R. results, observed with the isolated enzyme, are still maintained in cells with only few exceptions of compounds that strongly inhibit FTase but probably do not penetrate cells, such as hydroxamic acid **II/18** or phosphonic acid **II/19** at position 3a, or cysteinyl derivative **II/11** at position 2. Furthermore, carboxylic acids, esters, or amides at position 3a inhibit FTase in cells, usually with IC_{50} values 10–50-fold higher than with the enzymatic ones.

Moreover, favorable substitutions (e.g., introduction of a methylene at the benzylic position of the side chain at position 2, substitution with a methoxy group at position 5 or substitutions at the para and meta positions of the 9-phenyl ring) almost always lead to a greater enhancement of the FTase inhibition in intact cells as compared to isolated enzymes. Thus the positive effect of the additional methylene on the side chain at position 2 roughly translates from 2–3-fold against isolated enzyme to 5–10-fold in cells. Similarly, favorable substitutions on the phenyl ring at position 9 (such as ethyl, bromo, dimethyl or dichloro **II/30**, **34**, **38**, **40**), which only induce a 1.5–3-fold improvement against enzyme, gives a more than 10-fold enhancement in the cellular FTase inhibition. This jump in cellular activity is likely to be related to an enhanced cellular uptake.

In summary, the more potent BPHI inhibit the farnesylation of Ki-Ras with purified enzyme in the 0.008–0.2 μ*M* range, and block the farnesylation of Ha-Ras within intact cells in the 0.1–0.2 μ*M* range over a 24-h drug treatment.

6. ANTITUMOR PROPERTIES

6.1. Inhibition of Anchorage-Independent Cell Growth (Tables 4–7)

Active compounds (e.g., compounds with FTase inhibition IC_{50} values of <0.5 or 10 μ*M* against purified enzyme or in THAC cells, respectively) have been systematically evaluated in a colony formation assay, using the Ki-Ras mutated HCT 116 colon carcinoma cell line (Table 4). The more potent inhibitors have been further evaluated in a panel of Ki-Ras mutated human cell lines (Table 5). Finally, the most representative leads such as RPR 115135 **(+) II/2** (Table 6) and RPR 130401 **(+) II/25** (Table 7) have also been evaluated in a panel of non-Ki-Ras mutated cell lines.

Table 4
Inhibition of Anchorage-Independent Cell-Growth of HCT 116 Human Colon Carcinoma with BPHI[a]

Compound	*IC_{50} μM*	*Compound*	*IC_{50} μM*
(±) **II/1**	7.5	(±) **II/38**	0.5
(+) **II/2**	0.72	(±) **II/40**	0.42
(+) **II/23**	0.24	(±) **II/41**	1.85
(±) **II/24**	0.41	(±) **II/42**	0.42
(±) **II/25**	0.1	(±) **II/43**	0.39
(+) **II/25**	0.045	(±) **II/44**	9
(–) **II/25**	10	(±) **II/45**	0.41
(±) **II/26**	0.54	(±) **II/46**	0.84
(±) **II/27**	3.0	(+) **II/47**	0.54
(±) **II/28**	ND	(–) **II/47**	2.9
(±) **II/29**	0.54	(+) **II/48**	0.053
(±) **II/30**	0.098	(–) **II/49**	1.7
(±) **II/31**	1.12	(±) **II/50**	3
(±) **II/33**	0.64	(±) **II/52**	4
(±) **II/34**	0.125	(+) **II/53**	0.069
(±) **II/35**	0.098	(±) **II/55**	5
(±) **II/37**	6.2		

[a] IC_{50} values as means of two or more determinations. For experimental details, *see* Section 2.6.

6.1.1. HCT 116 Human Colon Carcinoma: S.A.R. (Table 4)

Compound **(±) II/1** inhibits the colony formation in soft agar of the Ki-Ras mutated human colon carcinoma cell line HCT 116 with an IC_{50} value of 7.5 μ*M*, consistent with its IC_{50} value for the inhibition of Ha-Ras processing in THAC cells (8–10 μ*M*). Such a similar level of cellular activity was further confirmed with either RPR 115135 **(+) II/2**, which inhibits Ha-Ras processing in THAC cells and HCT 116 colony formation at 1 μ*M* and 0.72 μ*M*, respectively, or RPR 130401 **(+) II/25**, with respective inhibition IC_{50} values of 0.2 μ*M* (in the 0.1–0.25 μ*M* range) for Ras processing and of 0.045 (in the 0.02–0.12 μ*M* range) for HCT 116 cell-growth. Such a difference between the IC_{50} ranges could be explained by differences in the duration of treatment in the two assays (24 h for cellular Ras processing and 12 d for colony formation).

The main S.A.R., previously observed against FTase both in enzymatic and cellular assays, are reestablished in HCT 116 cell cloning inhibition assays: the introduction of a methylene at the benzylic position of the side chain at position 2 usually increases potency by 5–10-fold. Similar features were observed with substitutions on the phenyl ring at position 9, as exemplified by compounds **(+) II/25**, **(±) II/30**, **(±) II/34**, or **(±) II/35**. The positive effect of the methoxy group at position 5 of the BPHI skeleton is also seen with **(+) II/53**, as well as the effect of carboxylic acids or amides, more specifically (3-pyridyl)methylamides **(+) II/23**, **48**, at position 3a.

6.1.2. Panel of Ki-Ras Human Mutated Cell Lines (Table 5)

The more potent BPHI have been further evaluated in a panel of human Ki-Ras mutated cancer cell lines, including colon SW 620 and LoVo, lung H 460 and A 549, and pancreas MIA PaCa-2 tumors. With the noticeable exception of lung A 459, which is less sensitive

Table 5
Inhibition of Anchorage-Independent Cell-Growth of Various Human Cancer Cells Lines Harboring Ki-Ras Mutations with Selected BPHI [a]

Compound	*HCT 116 (colon)*	*SW 620 (colon)*	*LoVo (colon)*	*H460 (lung)*	*A 549 (lung)*	*MIA PaCa-2 (pancreas)*
(+) **II/2**	0.72	0.95	ND	0.95	4.8	1.38
(+) **II/25**	0.045	0.09	0.015	0.11	0.4	0.19
(±) **II/30**	0.098	ND	ND	0.89	2.2	0.78
(+) **II/48**	0.053	ND	ND	ND	0.52	0.072
(+) **II/53**	0.069	ND	ND	ND	0.57	0.067

[a] IC_{50} values (μ*M*) as means of two or more determinations. For experimental details, *see* Section 2.6. ND, not determined.

to all tested BPHI, IC_{50} values of the various cell lines to a specific inhibitor are always in the same range, micromolar for RPR 115135 (+) **II/2** and 0.05–0.2 micromolar for BPHIs substituted on the phenyl ring at position 9. This indicates that the sensitivity is not generally tissue-related. However, some subtle differences were sometimes observed: an enhanced sensitivity to (+) **II/48** and (+) **II/53** and a relative resistance to (+) **II/35** as compared to RPR 130401 (+) **II/25** was noted in pancreas MIA PaCa-2.

6.1.3. Sensitivity of Ha-Ras Mutated, N-Ras Mutated, and w.t.-Ras Cell Lines (Tables 6 and 7)

Elucidation of the Ras signaling pathway also emphasized the potential role of wild-type Ras in tumors related to upstream oncoproteins, mainly receptor- or nonreceptor-protein-tyrosine kinases, such as EGF-R, ErbB-2 (HER-2), Bcr/Abl, src *(62,63)*. Therefore, RPR 115135 was further evaluated for growth inhibition of various human cell lines with either Ha-, N-, or Ki-Ras mutations or without Ras mutation (Table 6). As can be expected from its FPP-competitive properties, no significant differences have been detected in the ability to inhibit cell growth of either Ha-Ras (T24), N-Ras (HL60 or THP1), or Ki-Ras (HCT 116, SW 620, H 460, A 549, or MIA PaCa-2; *see* Table 5) mutated cell lines from various origins.

More interestingly, RPR 115135 also blocks the cell growth of wild-type Ras cell lines (SKBr-3) in the same low-micromolar range. This lack of correlation between cell-growth inhibition and Ras-status has already been described for CAAX-peptidomimetics B956 *(27)* and L-744,832 *(64)*. The ability to block non-Ras transformed cell lines as well as Ras-transformed cell lines was confirmed by looking at NIH3T3 fibroblasts transformed by various oncogenes: similar level of growth inhibition was observed with fibroblasts transformed by Ras oncogenes (Ha- and Ki-), oncogenes located upstream (HER2, Src) or downstream (v-Raf) of Ras, or apparently unrelated to Ras (v-mos) (Table 6).

Similarly, RPR 130401 has been evaluated in a large panel of human cell lines with Ras mutations (Ki- or N-) or with wild-type or undetermined Ras status, including a set of solid tumors from various origins (colon, lung, pancreas, breast, prostate, liver) and leukemia. From these assays, sensitivity to RPR 130401 is not correlated to Ras mutation status and not subject to tissue/species selectivities, as demonstrated by the comparison of results between human and rodent tumor cell lines. Additional experiments are ongoing for documenting these points.

Table 6

Inhibition of Colony Formation of Human Cancer Lines or NHI3T3 Fibroblasts Transformed by Various Oncogenes by RPR 115135 (+) II/2[a]

					NIH3T3					
Cell line	*T24*	*HL60*	*THP1*	*SKBr3*	*Ki-Ras*	*Ha-Ras*	*HER2*	*Src*	*v-Raf*	*v-mos*
IC_{50} µ*M*	0.1	1	0.5	0.5	1–5	0.5–1	1–5	1–5	1	1–5
Ras type	Ha[b]	N[b]	N[b]	w.t.	Ki[b]	Ha[b]	w.t.	w.t.	w.t.	w.t.
Origin	Bladder	Leukemia	Leukemia	Breast	Fibroblast	Fibroblast	Fibroblast	Fibroblast	Fibroblast	Fibroblast

[a] IC_{50} values as a mean of two or more determinations. For experimental details, *see* Section 2.6.
[b] Ras mutation.

Table 7

Inhibition of Colony Formation of Non-Ki-ras Human Cancer Lines and of Mouse and Rat Cancer Cell Lines by RPR 130401 (+) II/25[a]

Cell line	*HCT 116*	*HL60*	*HT29*	*PC 3*	*Calc C 18*	*HLF*	*3LL*	*C 6*
IC_{50} µ*M*	0.045	0.08	0.04	0.19	0.06	0.12	0.3	0.03
Ras type	Ki[b]	N[b]	w.t.	w.t.	ND	ND	Ki[b]	ND
Other mutation	p53	p53 (-/-)				p53		PDGF
Origin	Human Colon	Human Leukemia	Human Colon	Human Prostate	Human Breast	Human Liver	Mouse Lung	Rat Glioblastoma

[a] IC_{50} values as a mean of two or more determinations. For experimental details, *see* Section 2.6.
[b] Ras mutation.
ND, not determined.

6.2. Mechanistic Aspects of Cell Growth Inhibition

6.2.1. Cytostatic Cellular Activity

Two types of cloning experiments have been performed on HCT 116 cell lines with RPR 130401. In the first set of experiments HCT 116 cells were grown in agar in the presence of RPR 130401 (0.1, 1, or 10 μ*M*) for 4–12 consecutive days. Formation of clones were inhibited in a dose- and time-dependent manner. In the second set of experiments, HCT 116 cells have been grown in the presence of RPR 130401 for 4 consecutive days. The inhibitor was washed out from the culture medium and cells were allowed to grow for 12 additional days. In the latter case, clones grew in a normal way with no diminution of their number, thus clearly demonstrating that the cellular effect of RPR 130401 is essentially cytostatic (data not shown).

In other experiments, the formation of apoptotic cells was evaluated in a panel of cell lines (HCT 116, A 549, H 460, SW 620, MIA PaCa-2, and HL 60) in conditions that corresponds to a cytostatic effect on HCT 116 cells (4 d incubation with 20 μ*M* RPR 130401). Only 3–15% of apoptotic cells were found, with the only exception being the HL 60 cell line (55%). This confirms once again the main cytostatic nature of cell-growth inhibition.

Preliminary experiments on the cell cycle have been performed with RPR 115135 to document the cytostatic properties. First results indicate that RPR 115135 apparently induces a G0-G1 arrest in the HCT 116 cell line upon serum starvation *(65)*. These cell-cycle effects seem to be different from those observed with a p21-competitive inhibitor such as FTI-277, which rather induces a G2-M enrichment in A 549 Ki-Ras mutated cell line. This apparent discrepancy may be related to different regulation pathways in different cell lines *(66)*.

6.2.2. Cellular Combination with Chemotherapeutic Agents

Combination of signal transduction inhibitors with cytotoxic agents has recently emerged as a promising opportunity for future therapy. This has been documented by combination studies of paclitaxel and epothilones with the peptidomimetic p21-competitive inhibitor L-744.832 *(67)*. In addition, combination-therapy experiments done with another representative of the class of signal transduction modulators, such as the cyclin-dependent kinase inhibitor flavopiridol, demonstrated that the schedule of administration of the two drugs is important *(68)*.

We first tested the combination between RPR 130401 and camptothecin. When simultaneously given, the two drugs antagonized. On the other hand, when camptothecin was given first for 24 h, followed by washing and subsequent treatment with RPR 130401 for 72 h, the effects of the two drugs were found to be additive (Table 8). These in vitro studies now need to be extended in vivo.

Then, we tested in vitro combination with either 5-FU and docetaxel. In both cases, additive effects of these drugs with RPR 130401 were observed when the drugs were given together (data not shown). Additional studies are in progress with other chemotherapeutic agents, with HMG-Co reductase inhibitors able to lower the FPP cellular pool and also with various signal transduction inhibitors acting upstream or downstream in the Ras pathway.

6.3. In Vivo Efficacy of RPR130401

RPR 115135 has been tested with negative results *(69,70)*. RPR 130401 was found to exert a much more potent cytostatic effect in vitro than RPR 115135 and was therefore

Table 8
Percentage of Cloning Inhibition of HCT 116 Cell Line by the In Vitro Sequential Combination of Camptothecin and RPR 130401 (+) II/25

RPR 130401	*Camptothecin (μg/mL)*		
(μg/mL)	*0*	*0.01*	*0.1*
0	0	0	46
0.03	7	0	49
0.1	13	0	57
0.3	17.5	9	62
1	21	17	65
3	31.5	34	73

For experimental details, *see* Section 6.2.2.

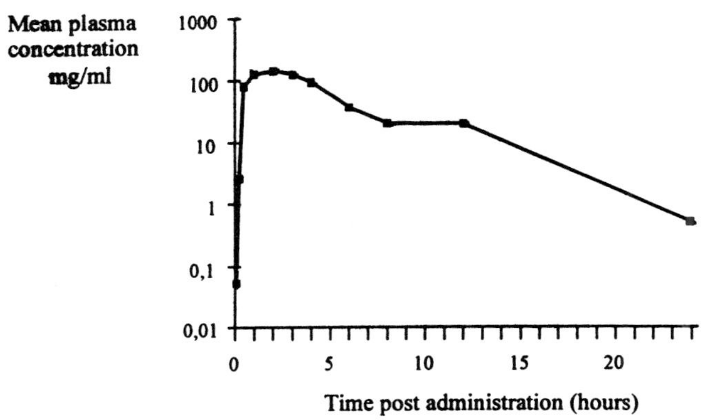

Fig. 6. RPR 130401 **(+) II/25** plasma level after oral administration at 250 mg/kg in mice. For experimental details, *see* Section 2.7.

a valuable candidate for evaluating its in vivo antitumor activity. The observed cytostatic effect supposes that the compound has: 1) an acceptable toxicological profile making repeated and high dosage administrations possible and 2) a bioavailability allowing suitable plasma concentrations after oral administration.

When administered at 400 mg/kg p.o. for 5 consecutive days in mice, RPR 130401 appears to be well-tolerated: no overt lethality/toxicity was detected in contrast to classical chemotherapeutic agents. No anatomopathological abnormalities were detected in necropsied animals at the end of the observation period (2–3 wk).

Concerning oral pharmacokinetics, a good bioavailability was observed: as indicated in Fig. 6 and Table 9, a single administration of 250 mg/kg p.o. 130401 gives plasma concentrations largely above the active cellular IC_{50}. Drug levels in plasma were very high (mean Cp_{max} 142 μg/mL plasma at 2-h post-dose and 0.5 μg/mL at 24 h post-dose). Furthermore RPR 130401 has been shown to achieve substantial levels in tumor bulk tissue (data not shown).

Preliminary LC-MS/MS identification of RPR 130401 metabolites in extracts of ex vivo mouse plasma samples and in vitro rat liver slices suggested hydroxylation of the 2-methoxyphenyl moiety and direct glucuronidation of the 3a carboxylic acid.

Table 9
Pharmacokinetics of RPR 130401 (+) II/25 after a Single Oral Administration at 250 mg/kg in Female B6D2F1Mice

C max (μg/mL)	*T max (h)*	*AUC (0-24 h) (h/μg/mL)*	*T 1/2 (h)*
142.1	2	820.3	2.8

For experimental details, *see* Section 2.7.

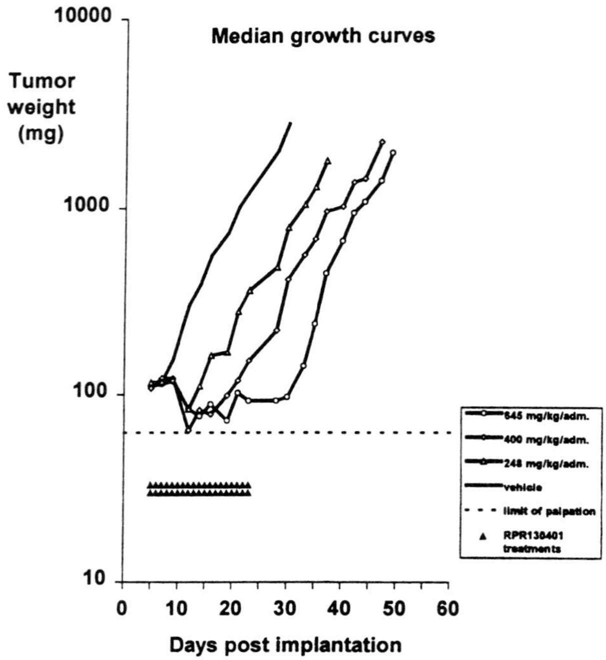

Fig. 7. Oral antitumor activity of RPR 130401 **(+) II/25** against HCT 116 colon carcinoma xenogratfed in nude mice. For experimental details, *see* Section 2.8.

Mice bearing human colon carcinoma HCT 116 were randomized according to their tumor weight on the first day of therapy. The schedule of administration was twice a day from days 5–23 post-tumor implantation (i.e., 19 consecutive days) at 645, 400, and 248 mg/kg/administration. The highest dosage administered was well-tolerated with no body loss and no adverse effects. Of interest, a cytostatic activity was achieved with a tumor growth delay of 22.8 d for a treatment duration of 19 d and a positive $\log_{10}$ cell kill net of 0.3 (*see* Fig. 7 and Table 10). This indicates that no tumor growth occurred during treatments. There was a clear dose-response effect in the tumor growth delays at the two dosages below, although no complete cytostatic effect was obtained (negative $\log_{10}$ cell kill net). Further in vivo evaluations are being performed with RPR 130401, as a single agent, against tumors exhibiting various forms of Ras as well as combination therapies with cytotoxic agents.

7. CONCLUSION

The discovery of new anticancer agents is carried out using two complementary directions: a classical one focused on the targets of the conventional chemotherapy, and a

Table 10
Oral Antitumor Activity of RPR 130401 (+) II/25
Against HCT 116 Colon Carcinoma Xenogratfed on Nude Mice

Dosage (mg/kg/adm.)	*Schedule (d)*	*Total dose (mg/kg)*	*Drug death*	*T-C (d)*	*Log cell kill net*
645	5–23 (2×/d)	24,510	0/6	22.3	0.3
400	"	15,200	0/6	17.3	-0.1
248	"	9424	0/6	11.7	-0.5

Treatment duration: 19 d. Tumor doubling time: 4.5 d. Tumor size at start of therapy: 87–172 mg. Time to reach 1000 mg in solvent-treated mice: 20.9 d. For experimental details, *see* Section 2.8.

prospective one based on the advances of basic research. Intracellular signal transduction is certainly the area that has been and still continues to be the most widely explored in pharmaceutical companies and academic institutions *(71)*.

RPR 130401 is the emerging part of a 5-yr research program targeting FTase inhibitors in the pharmaceutical industry. This compound adds to the list of other nonpeptidomimetic inhibitors of FTase such as R 115777 *(37)* and SCH 66336 *(38)* which are presently in Phase I clinical trials. These three compounds bear strong similarities at the preclinical stage. All have excellent tolerance and exert cytostatic/antiproliferative activities against human tumor xenografts following daily long-term administration by the oral route. Moreover, BPHI including RPR 130401, the most active compound, are, to the best of our knowledge, representatives of a new class of FTase inhibitors. BPHI are competitive inhibitors with the FPP substrate and are noncompetitive with respect to the farnesylated proteins. This unique feature confers BPHI identical inhibitory properties regardless of the nature of the farnesylated targets in biochemical and cellular models. Because we still do not know the exact nature of the farnesylated proteins important for the transformation and for the maintenance of tumorigenic properties, BPHI might be key compounds for answering the final and important therapeutic question: "Are FTase inhibitors clinically active antitumor agents?"

Similar properties were also reported for another class of signal transduction modulators, the tyrosine kinase inhibitors of the Epidermal Growth Factor Receptor, which are also in clinical trials *(71)*. Finally, all these signal transduction modulators have very different preclinical properties than those of the conventional chemotherapeutic agents. Such innovative features open exciting new avenues for treating cancer, but also raise issues about the methodologies used for selecting and evaluating such compounds. Everyone in the field looks forward to hearing about the results of the ongoing FTI clinical trials in monotherapy and in association with other therapeutic modalities *(72)*.

ACKNOWLEDGMENTS

We thank A. Bello, P.F. Berne, J.P. Guilloteau, A. Hasnain, R. Jenkins, S. Maignan, S. Michiels, V. Mikol, F. Parker, M. Robin, P. Sédivy, A. Selingue, L. Tahraoui, B. Tocqué, J.Y. Crenne, N. Michot, and M. Vuilhorgne for their active contributions to the accomplishment of this work. We also thank N. Baudouy, S. Beaudoin, L. Bertin, S. Blais, R. Boulay, G. Baudry, S. Caillon, S. d'Heilly, S. Galiné, C. Gobert, S. Gontier, N. Grevet, G. Leclerc, M. LeRoux, L. Maestrali, C. Martinez, S. Monget, S. Perron, O. Petitgenet, A. Renaud, B. Sanz, D. Sévère and C. Souder for their expert technical assistance. The financial support of the Bioavenir program is also gratefully acknowledged.

REFERENCES

1. Barbacid, M. Ras genes. *Ann. Rev. Biochem.* 1987; 56:779–827.
2. Bos JL. ras oncogenes in human cancer: a review. *Cancer Res* 1989; 49:4682–4689.
3. Reiss Y, Goldstein JL, Seabra MC, Casey PC, Brown MS. Inhibition of purified p21ras farnesyl: protein transferase by Cys-AAX tetrapeptides. *Cell* 1990; 62:81–88.
4. Kato K, Cox AD, Hisaka MM, Graham SM, Buss JE, Der CJ. Isoprenoid addition to Ras protein is the critical modification for its membrane association and transforming activity. *Proc Natl Acad Sci USA* 1992; 89:6403–6407.
5. Gibbs JB. Ras C-terminal processing enzymes: new drug targets? *Cell* 1991; 65:1–4.
6. Gibbs JB, Oliff A, Kohl NE. Farnesyltransferase inhibitors: Ras research yields a potential cancer therapeutic. *Cell* 1994; 77:175–178.
7. Buss JE, Marsters JC Jr. Farnesyl transferase inhibitors: the successes and surprises of a new class of potential cancer chemotherapeutics. *Chem Biol* 1995; 2:787–791.
8. Leonard DM. Ras farnesyltransferase: a new therapeutic target. *J Med Chem* 1997; 40:2971–2993.
9. Lerner EC, Hamilton AD, Sebti SM. Inhibition of Ras prenylation: a signaling target for novel anticancer drug design. *Anti-Cancer Drug Design* 1997; 12:229–238.
10. Brattain MG, Fine WD, Khaled FM, Thompson J, Brattain DE. Heterogeneity of malignant cells from a human colonic carcinoma. *Cancer Res* 1981; 41:1751–1756.
11. Corbett TH, Roberts BJ, Trader MW, Laster WR Jr, Griswold DP Jr, Schabel FM Jr. Response of transplantable tumors of mice to anthracendione derivatives alone and in combination with clinically useful agents. *Cancer Treat Rep* 1982; 66:1187–1200.
12. Corbett TH, Leopold WR, Dykes DJ, Roberts BJ, Griswold DP Jr, Schabel FM Jr. Toxicity and anticancer activity of a new triazine antifolate (NSC 127755). *Cancer Res* 1982; 42:1701–1715.
13. Schabel FM Jr, Griswold DP Jr, Corbett TH, Laster WR Jr, Mayo JG, Llyod HH. Testing on the therapeutic activity against advanced solid tumors of man, in: (Busch H, DeVita V Jr. eds.), *Cancer Drug Development,* Part B, Vol. 17, Academic, New York, 1979, pp. 3–51.
14. Schabel FM Jr, Griswold DP Jr, Corbett TH, Laster WR Jr, Mayo JG, Lloyd HH. Quantitative evaluation of anticancer agents in experimental animals. *Pharmacol Ther* 1977; 1:411–435.
15. Goldstein JL, Brown MS, Stradley SJ, Reiss Y, Gierasch LM. Nonfarnesylated tetrapeptide inhibitors of protein farnesyltransferase. *J Biol Chem* 1991; 266:15,575–15,578.
16. Clerc F-F, Guitton J-D, Fromage N, Lelièvre Y, Duchesne M, Tocqué B, et al. Constrained analogs of KCVFM with improved inhibitory properties against farnesyl transferase. *Bioorg Med Chem Lett* 1995; 5:1779–1784.
17. Hunt JT, Lee VG, Leftheris K, Seizinger B, Carboni J, Mabus J, et al. Potent cell active, non-thiol tetrapeptide inhibitors of farnesyltransferase. *J Med Chem* 1996; 39:353–358.
18. James GL, Goldstein JL, Brown MS, Rawson TE, Somers TC, McDowell RS, et al. Benzodiazepine peptidomimetics: potent inhibitors of Ras farnesylation in animal cells. *Science* 1993; 260:1937–1942.
19. Nigam M, Seong C-M, Qian Y, Hamilton AD, Sebti SM. Potent inhibition of human tumor p21ras farnesyltransferase by A1A2-lacking p21ras CA1A2X peptidomimetics. *J Biol Chem* 1993; 268:20,695–20,698.
20. Graham SL, deSolms SJ, Giuliani EA, Kohl NE, Mosser SD, Oliff AI, et al. Pseudopeptide inhibitors of Ras farnesyl-protein transferase. *J Med Chem* 1994; 37:725–732.
21. Lerner EC, Qian Y, Blaskovich MA, Fossum RD, Vogt A, Sun J, et al. Ras CAAX peptidomimetic FTI-277 selectively blocks oncogenic Ras signaling by inducing cytoplasmic accumulation of inactive Ras-Raf complexes. *J Biol Chem* 1995; 270:26,802–26,806.
22. Byk G, Duchesne M, Parker F, Lelievre Y, Guitton JD, Clerc FF, et al. Local constrained shifty pseudopeptides inhibitors of ras-farnesyl transferase. *Bioorg Med Chem Lett* 1995; 5:2677–2682.
23. Qian Y, Vogt A, Sebti SM, Hamilton AD. Design and synthesis of non-peptide Ras CAAX mimetics as potent farnesyltransferase inhibitors. *J Med Chem* 1996; 39:217–223.
24. Williams TM, Ciccarone TM, MacTough SC, Bock RL, Conner MW, Davide JP, et al. 2-Substituted piperazines as constrained amino acids. Application to the synthesis of potent, non carboxylic acid inhib-itors of farnesyltransferase. *J Med Chem* 1996; 39:1345–1348.
25. Burns CJ, Guitton J-D, Baudoin B, Lelièvre Y, Duchesne M, Parker F, et al. Novel conformationally extended naphthalene-based inhibitors of farnesyltransferase. *J Med Chem* 1997; 40:1763–1767.
26. Kohl NE, Omer CA, Conner MW, Anthony NJ, Davide JP, deSolms SJ, et al. Inhibition of farnesyltransferase induces regression of mammary and salivary carcinomas in ras transgenic mice. *Nature Med* 1995; 1:792–797.

27. Nagasu T, Yoshimatsu K, Rowell C, Lewis MD, Garcia AM. Inhibition of human tumor xenograft growth by treatment with the farnesyl transferase inhibitor B956. *Cancer Res* 1995; 55:5310–5314.
28. Leftheris K, Kline T, Vite GD, Cho YH, Bhide RS, Patel DV, et al. Development of highly potent inhibitors of Ras farnesyltransferase possessing cellular and in vivo activity. *J Med Chem* 1996; 39:224–236.
29. James GL, Goldstein JL, Brown MS. Polylysine and CVIM sequences of K-RasB dictate specificity of prenylation and confer resistance to benzodiazepine peptidomimetic in vitro. *J Biol Chem* 1995; 270: 6221–6226.
30. Singh SB, Zink DL, Liesch JM, Goetz MA, Jenkins RG, Nallin-Omstead M, et al. Isolation and structure of chaetomellic acids A and B from chaetomella acutiseta: farnesyl pyrophosphate mimic inhibitors of Ras farnesyl-protein transferase. *Tetrahedron* 1993; 49:5917–5926.
31. Hara M, Akasaka K, Akaniga S, Okabe M, Nakano H, Gomez R, et al. Identification of Ras farnesyltransferase inhibitors by microbial screening. *Proc Natl Acad Sci USA* 1993; 90:2281–2285.
32. van der Pyl D, Cans P, Debernard JJ, Herman F, Lelièvre Y, Tahraoui L, et al. RPR 113228, a novel farnesyl-protein transferase inhibitor produced by chrysporium lobatum. *J Antibiot* 1995; 48:736,737.
33. Bishop WR, Bond R, Petrin J, Wang L, Patton R, Doll R, et al. Novel tricyclic inhibitors of farnesyl protein transferase. *J Biol Chem* 1995; 270:30,611–30,618.
34. Wallace A, Koblan KS, Hamilton K, Marquis-Omer DJ, Miller PJ, Mosser SD, et al. Selection of potent inhibitors of farnesyl-protein transferase from a synthetic tetrapeptide combinatorial library. *J Biol Chem* 1996; 271:31,306–31,311.
35. McNamara DJ, Dobrusin E, Leonard DM, Shuler KR, Kaltenbronn JS, Quin J III, et al. C-terminal modifications of histidyl-N-benzylglycinamides to give improved inhibition of Ras farnesyltransferase, cellular activity, and anticancer activity in mice. *J Med Chem* 1997; 40:3319–3322.
36. Aoyama T, Satoh T, Yonemoto M, Shibata J, Nonoshita K, Arai S, et al. A new class of highly potent far-nesyl diphosphate-competitive inhibitors of farnesyltransferase. *J Med Chem* 1998; 41:143–147.
37. Sktzat S, Angibaud P, Venet M, Sanz G, Bowden C, End D. R115777, a novel imidazole farnesyl transferase inhibitor (FTI) with potent oral antitumor activity. *Proc Am Assoc Cancer Res Meeting* 1998; 39:317, # 2169.
38. Liu M, Lee S, Yaremko B, Chen J, Dell J, Nielsen L, et al. SCH 66336, an orally bioavailable tricyclic farnesyl protein transferase inhibitor demonstrates broad and potent in-vivo antitumor activity. *Proc Am Assoc Cancer Res Meeting* 1998; 39:270, #1843.
39. Terao Y, Kotaki H, Imai N, Achiwa K. A definite evidence on the ambivalent azomethine ylide intermediate in trifluoroacetic acid- and fluoride anion-promoted 1,3-Cyclo-additions involving the silicon-carbon bond cleavage. *Chem Pharm Bull* 1985; 33:896–898.
40. Barton DHR, Crich D, Motherwell WB. The invention of new radical chain reactions. Part VIII. Radical chemistry of thiohydroxamic esters: a new method for the generation of carbon radicals from carboxylic acids. *Tetrahedron* 1985; 41:3901–3924.
41. Ziegler FE, Condon ME. Gibbane synthons via hexahydrofluorenones. An intramolecular Reformatsky reaction. *J Org Chem* 1971; 36:3707–3715.
42. Imamoto T, Takiyama N, Nakamura K. Cerium chloride-promoted nucleophilic addition of grignard reagents to ketones: an efficient method for the synthesis of tertiary alcohols. *Tetrahedron Lett* 1985; 26:4763–4766.
43. Barton DHR, O'Brien RE, Sternhell S. A new reaction of hydrazones. *J Chem Soc* 1962; 470–476.
44. Anderson AG Jr, Anderson RG. Studies related to pyracene. An improved synthesis. *J Org Chem* 1957; 22:1197–1200.
45. Ito Y, Nakatsuka M, Saegusa T. Syntheses of polycyclic ring systems based on the new generation of o-quinodimethanes. *J Am Chem Soc* 1982; 104:7609–7622.
46. Zajac WW Jr, Walters TR, Darcy MG. Oxidation of amines with 2-sulfonyloxyaziridines (Davis' reagents). *J Org Chem* 1988; 53:5856–5860.
47. Commerçon A, LeBrun A, Mailliet P, Peyronel JF, Sounigo F, Truchon A, Zucco M (Rhône-Poulenc-Rorer SA). New farnesyl transferase inhibitors, their preparation and pharmaceutical compositions which contain them. WO 97/03050.
48. Bourzat JD, Commerçon A, Dereu N, Mailliet P, Sounigo-Thompson F, Martin JP, Capet M, Chevé, M (Rhône-Poulenc Rorer SA). New farnesyl transferase inhibitors their preparation and compositions which contains them. WO 98/29390.
49. Mornon JP. (Unpublished data) (Laboratoire de mineralogie et de cristallographie Université Pierre et Marie Curie, Paris, France).

50. Park H-W, Boduluri SR, Moomaw JF, Casey PJ, Beese LB. Crystal structure of protein farnesyltransferase at 2.25 angstrom resolution. *Science* 1997; 275:1800–1804.
51. Park H-W, Beese L. Protein farnesyltransferase. *Curr Opin Struct Biol* 1997; 7:873–880.
52. Long SB, Casey PJ, Beese LS. Cocrystal structure of protein farnesyltransferase complexed with a farne-syl diphosphate substrate. *Biochemistry* 1998; 37:9612–9618.
53. Dunten P, Kammlott U, Crowther R, Weber D, Palermo R, Birktoft J. Protein farnesyltransferase: structure and implications for substrate binding. *Biochemistry* 1998; 37:7907–7912.
54. Cox AD, Der CJ. Farnesyltransferase inhibitors and cancer treatment: targeting simply ras? *Biochim Biophys Acta* 1997; 1333:F51–F71.
55. Khosravi-Far R, Campbell S, Rossman KL, Der CJ. Increasing complexity of Ras signal transduction: involvement of Rho family proteins. *Adv Cancer Res* 1998; 57–105.
56. Cates CA, Michael RL, Stayrook KR, Harvey KA, Burke YD, Randall SK, et al. Prenylation of oncogenic human PTPCAAX protein tyrosine phosphatases. *Cancer Lett* 1996; 49–55.
57. Crowell PL, Kuo CF, Randall SK, Crowell DN. Differential sensitivity of PTP-CAAX-transformed vs. untransformed cells to a farnesyl transferase inhibitor. *Proc Am Assoc Cancer Res Meeting* 1998; 39: 269, #1840.
58. Yokoyama K, Goddwin GW, Ghomashchi F, Glomset JA, Gelb MA. A protein geranylgeranyltransferase from bovine brain: implications for protein prenylation specificity. *Proc Natl Acad Sci USA* 1991; 88:5302–5306.
59. Armstrong SA, Hannah VC, Goldstein JL, Brown MS. CAAX geranylgeranyl transferase transfers farnesyl as efficiently as geranylgeranyl to RhoB. *J Biol Chem* 1995; 270:7864–7868.
60. Biller SA, Neuenschwander K, Ponpipom MM, Poulter CD. Squalene synthase inhibitors. *Curr Pharm Design* 1996; 2:1–40.
61. Tharshis LC, Proteau PJ, Kellogg BA, Sacchettini JC, Poulter CD. Regulation of product chain length by isoprenyl diphosphate synthases. *Proc Natl Acad Sci USA* 1996; 93:15,018–15,023.
62. Prendergast GC, Gibbs JB. Pathways of Ras function: connections to the actin cytoskeleton. *Adv Cancer Res* 1993; 62:19–69.
63. Khosravi-Far R, Der CJ. The Ras signal transduction pathway. *Cancer Metastasis Rev* 1994; 13:67–89.
64. Sepp-Lorenzino L, Ma Z, Rands E, Kohl NE, Gibbs JB, Oliff A, Rosen N. A peptidomimetic inhibitor of farnesyl: protein transferase blocks the anchorage-dependent and -independent growth of human tumor cell lines. *Cancer Res* 1995; 55:5302–5309.
65. Russo P, Reinhold W, Yu L, Ottoboni C, Kohn KW, Riou JF, et al. Cellular actions of farnesyl transferase inhibitor RPR 115135, in a human isogenic colon cancer cell line system. *Proc Am Assoc Cancer Res Meeting* 1998; 39:319, #2178.
66. Miguel K, Pradines A, Sun J, Qian Y, Hamilton AD, Sebti SM. GGTI-298 induces G0-G1 block and apoptosis whereas FTI-277 causes G2-M enrichment in A549 cells. *Cancer Res* 1997; 57:1846–1850.
67. Moasser MM, Sepp-Lorenzino L, Kohl NE, Oliff A, Balog A, Su D-S, et al. Farnesyl transferase inhibitors cause enhanced mitotic sensitivity to taxol and epothilones. *Proc Acad Natl Sci USA* 1998; 95: 1369–1374.
68. Bible KC, Kaufmann SH. Cytotoxic synergy between flavopiridol (NSC 649890, L86-8275) and various antineoplastic agents: the importance of the sequence of administration. *Cancer Res* 1997; 57: 3375–3380.
69. Mailliet P, Lelièvre Y, Chevé M, Fromage N, Lavelle F, Le-Brun A, et al. Synthesis and in vitro structure-activity relationships of a new promising series of non peptidic protein-farnesyl-transferase inhibitors. *Proc Am Assoc Cancer Res Meeting* 1997; 38:350, #2347.
70. Vrignaud P, Mailliet P, Bissery MC, Duchesne M, Lelièvre Y, Riou JF, et al. Structure activity relationships and in vivo evaluation in a new promising family of nonpeptidic farnesyltransferase inhibitors. *Proc Am Assoc Cancer Res Meeting* 1997; 38:350, #2348.
71. Lavelle F. American Association for Cancer Research 1997: Progress and new hope in the fight against cancer. *Exp Opin Inves Drugs* 1997; 6:771–775.
72. Lavelle F. American Association for Cancer Research 1998: Promises and prospects for the next century. *Exp Opin Inves Drugs* 1998; 7:1015–1021.

10 Genetic Analysis of FTase and GGTase I and Natural Product Farnesyltransferase Inhibitors

Fuyuhiko Tamanoi, PHD, *Keith Del Villar,* PHD, *Nicole Robinson,* PHD, *MeeRhan Kim,* PHD, *Jun Urano,* PHD, *and Wenli Yang,* PHD

CONTENTS

1. INTRODUCTION

Protein prenyltransferases are conserved from yeast to humans *(1,2)*. Because yeast and human enzymes are structurally and functionally similar, efforts have been made to use yeast as a genetic system to study protein prenyltransferases. Recently, the yeast system was used to obtain a number of farnesyltransferase (FTase) mutants that provided valuable information concerning residues of FTase important for substrate recognition and catalysis *(3,4)*. The yeast system's usefulness for the study of protein prenylation was initially established from the study of the yeast mating factor, **a**-factor *(5,6)*. This short peptide is modified by the addition of a farnesyl group. The structure determination of this peptide provided insights into the chemical nature of the modification as well as other processing events that accompany farnesylation. The study on the processing of yeast Ras proteins also provided critical information on farnesylation and related processing events, and subsequently led to the identification of the yeast genes encoding FTase *(1,7)*.

From: *Farnesyltransferase Inhibitors in Cancer Therapy*
Edited by: S. M. Sebti and A. D. Hamilton © Humana Press Inc., Totowa, NJ

Table 1
Yeast Prenyltransferase Genes[a]

	FTase		*GGTase I*	
	β	α	β	α
S. cerevisiae	*DPR1/RAM1* (431)	*RAM2* (316)	*CAL1/CDC43* (376)	*RAM2* (316)
S. pombe	*cpp1*$^+$ (382)	*cwp1*$^+$ (294)	*cwg2*$^+$ (355)	*cwp1*$^+$ (294)

[a] Numbers in parentheses represent amino acid residues encoded by prenyltransferase genes.

Identification of two yeast genes required for FTase activity provided hints that FTase consists of two distinct subunits. We have recently established a second genetic system, *Schizosaccharomyces pombe*, for the study of protein prenylation *(8,9)*. Since fission yeast genes encoding subunits of FTase and geranylgeranyltransferase I (GGTase I) have now been identified, the fission yeast can be exploited to study protein prenylation.

The study of farnesylation in yeast also contributed to the development of FTase inhibitors (FTIs). First, a yeast mutant was used to develop a microbial assay to screen for FTIs *(10–12)*. This was based on the characterization of a farnesylated protein, Ste18, which is the γ-subunit of a heterotrimeric G-protein involved in the mating factor signal transduction pathway. Application of the microbial screen to culture media from a large number of microbial organisms led to the identification of natural compound inhibitors of FTase *(11)*. Among the compounds identified, manumycin represents the most extensively characterized natural product. In addition, other screens using microbial sources led to a number of other natural product inhibitors. We present a compilation of these compounds and provide a brief description of the compounds that have been reported in recent years.

2. CHARACTERIZATION OF FTASE AND GGTASE I IN *SACCHAROMYCES CEREVISIAE*

2.1. Identification of S. cerevisiae *Genes Encoding FTase and GGTase I*

As shown in Table 1, *S. cerevisiae* FTase consists of two subunits that are encoded by *Dpr1/Ram1* and *Ram2*. These genes were identified from the study of the processing of yeast Ras proteins and identification of mutants defective in the processing *(13–16)*. The overall processing of the Ras2 protein leading to its membrane association can be divided into a series of modification events *(17–19)*. First, Ras2 is synthesized in the cytosol as a precursor form that ends with the C-terminal CAAX motif (C is cysteine; A is aliphatic amino acid; and X is the C-terminal amino acid, usually serine, cysteine, methionine, alanine, or glutamine). The precursor form is first modified by the addition of a farnesyl group to the cysteine in the CAAX motif. Then, the three C-terminal amino acids are proteolyzed and the C-terminus is carboxymethylated leading to the generation of an intermediate form. This intermediate form is then further modified by the addition of a palmitic acid to a cysteine directly upstream of the farnesylated cysteine. This mature form is stably associated with the plasma membrane. Mutants of yeast defective in the processing of Ras proteins have been identified *(20,21)*. These mutants accumulate precursor Ras proteins in the cytosol. The mutants were obtained by taking advantage of phenotypes caused by the expression of activated Ras, Ras2^{val19}. Yeast cells expressing Ras2^{val19} exhibit distinct phenotypes including increased heat-shock sensitivity and decreased glycogen accumulation *(22)*. Mutations that suppress the heat-shock sensitivity of the Ras2^{val19} cells were identified. Among these mutants, a type that exhibits temperature-sensitive growth and mating sterility specific to *MAT***a** cells was identified. Genetic characterization of these mutants led to the identification of two mutants, *dpr1/ram1* and *ram2 (14,16)*. Other alleles of *dpr1/ram1* were also obtained. The *ste16* allele was identified from the study of the biosynthesis of the mating peptide, **a**-factor *(23)*. Another allele of *dpr1/ram1*, called *sgp2*, suppresses the lethal phenotype of the disruption of *GPA1*, α-subunit of the heterotrimeric G-protein in yeast *(24)* (*see below*).

DPR1/RAM1 and *RAM2* were established as genes encoding the subunits of yeast FTase from the following three lines of observation. First, FTase activity was undetectable in the extracts of *dpr1/ram1* or *ram2* mutant *(14)*. Second, FTase activity was observed in *Escherichia coli* cells co-expressing *DPR1/RAM1* and *RAM2 (16)*. Finally, FTase purified from either wild-type yeast cells or cells overproducing *DPR1/RAM1* and *RAM2* contained two subunits corresponding to the sizes of Dpr1/Ram1 and Ram2 *(25)*. The Dpr1/Ram1 protein shares 37% identity with the β-subunit of the mammalian FTase, whereas the Ram2 protein shares 30% identity with the α-subunit of mammalian FTase.

The Ram2 subunit, in addition to acting as the α-subunit of FTase, is shared with GGTase I *(26,27)*. The β-subunit of GGTase I is encoded by *CAL1/CDC43* gene *(26–28)*. This β-subunit of GGTase I exhibits significant sequence similarity with the β-subunit of FTase, which reflects a related function. In fact, the activity of the two enzymes is very similar because both recognize similar C-terminal motifs in their peptide substrates (*see below*). *CAL1/CDC43* was originally identified as a gene involved in bud formation in yeast *(29)*. The bud formation process involves a number of proteins including two Ras-superfamily G-proteins, Cdc42 and Rsr1 *(30)*. These G-proteins terminate with the CAAL motif (C is cysteine, A is aliphatic amino acid, and the C-terminal amino acid is preferentially leucine), which is specifically modified by GGTase I.

Although both enzymes carry out similar enzymatic processes, FTase is not essential for growth, whereas GGTase I is essential. Thus, disruption of either *RAM2* or *CAL1/CDC43* is lethal. On the other hand, yeast with the *DPR1/RAM1* gene completely deleted are viable, however, the deletion mutants grow slowly at low temperatures and do not grow at high temperatures *(31)*. These mutants exhibit sterility specific to *MAT***a** cells owing to the processing deficiency of the mating pheromone **a**-factor *(20,21)*. In addition to Ras, **a**-factor and Ste18 protein, yeast FTase catalyzes farnesylation of other CAAX ending proteins. One of these is a Rho-family G-protein, Rho3 *(32)*. This protein is involved in the maintenance of cell polarity in yeast through its effects on the actin cytoskeleton and exocytosis *(33,34)*. Deficiency in Rho3 function results in the appearance of uniformly enlarged cells. These cells can be detected in the *dpr1/ram1* mutants. In addition, a search for the yeast genome has identified many more ORFs that encode proteins terminating with the CAAX motif.

2.2. Biochemical Properties

Biochemical characterization of *S. cerevisiae* FTase points to the similarity between the yeast and the mammalian enzymes. Like the mammalian enzyme, the yeast enzyme preferentially utilizes peptides ending with the CAAX motif over the CAAL motif and

FTase activity is inhibited by CAAX peptides *(25)*. Although both the mammalian and yeast enzymes form a complex with farnesyl diphosphate (FPP), the yeast enzyme requires the presence of Zn^{2+} for FPP binding whereas the mammalian enzyme exhibits Zn^{2+}-independent binding *(35)*. Through investigations including a combination of steady-state kinetic and equilibrium studies, the binding mechanism of substrates for the yeast enzyme has been deduced to be an ordered sequential mechanism in which FPP binds to the enzyme before peptide binding occurs *(25,36)*. Studies using a series of substrate analogs for FPP have suggested that farnesylation occurs by nucleophic substitution mechanism, with significant allyl cation character in the transition state *(37)*.

Reconstitution of active FTase from purified individual subunits has been accomplished *(38)*. The Dpr1/Ram1 subunit was purified as a fusion protein with glutathione S-transferase (GST) after expression in *Escherichia coli*. Alternatively, intein-chitin binding domain fusion could be used to obtain intact Dpr1/Ram1. Ram2 subunit was purified from *E. coli* as a fusion with the maltose binding protein (MBP). When these purified proteins were mixed, active FTase was reconstituted. The interaction between the Dpr1/Ram1 and Ram2 proteins was also detected by GST pull-down experiments using Dpr1/Ram1 fused with GST. This type of reconstitution has not been possible with the mammalian FTase enzymes; this is presumably owing to instability of individual subunits of the mammalian enzyme. Thus, the yeast FTase provides the opportunity to further investigate biochemical properties of FTase through the use of individually stable subunits.

2.3. Identification of FTase Mutants

Recent determination of the crystal structure of rat FTase provided insights into the structure of this enzyme *(39,40)*. One important issue concerning FTase is the identification of residues which are involved in the recognition of substrates, the CAAX motif peptide and FPP, which acts as a donor for the transferred farnesyl group. Yeast has enabled a genetic approach to be taken to investigate this question, and Table 2 summarizes a variety of yeast FTase mutants identified.

Mutants S159N (serine at position 159 is mutated to asparagine), Y362L, Y362I, Y362M, and Y366N were found by screening for FTase mutants with altered substrate specificity *(3,41)*. The overall aim of this screen was to identify residues of FTase critical for the recognition of the CAAX motif by taking advantage of the similarity between FTase and GGTase I. First, both proteins recognize similar motifs. CAAX (X is preferentially methionine, glutamine, cysteine, alanine, or serine) is recognized by FTase and CAAL (the C-terminal amino acid is preferentially leucine) by GGTase I *(26)*. Second, both enzymes are structurally similar; they share a common α-subunit and their β-subunits are approx 30% homologous *(28)*. In addition, FTase is able to recognize the CAAL motif, albeit at a low efficiency *(31)*. These discoveries led us to speculate that it is possible to identify FTase mutants that can efficiently recognize the CAAL motif on GGTase I substrates. Therefore, we set up a screen using a temperature sensitive mutant, *cal1*, which is defective for GGTase I. Mutant FTases were sought that would enable the *cal1* cells to grow at nonpermissive temperatures. Extensive screening led to the identification of amino-acid alterations occurring at one of three residues in FTase, S159, Y362, and Y366. Thus, alteration of a single amino acid can convert FTase to a form that is capable of efficiently recognizing GGTase I substrate. Further insights into residues of FTase critical for the recognition of the CAAX motif were obtained by changing residues S159

Table 2
Yeast FTase β Mutants

Mutants	*Properties*	*References*
Substrate affinity mutants		
G149E,D	Acquires the ability to recognize peptides ending with R,K	*(4)*
S159N	Increased affinity to CAAL motif	*(3,41)*
Y362LM,I	Increased affinity to CAAL motif	*(3)*
Y366N	Increased affinity to CAAL motif	*(3)*
I74D	Increased specificity for CIIS motif	*(43)*
206DDLF	Decreased geranylgeranylation	*(43)*
351FSKN	Decreased affinity for CIIL motif	*(43)*
D209A	15-fold higher K_m for peptide	*(42)*
E256A	130-fold higher K_m for FPP	*(42)*
Mutants affecting zinc coordination and catalytic activity		
D307A	Zinc coordination, decrease in k_{cat}	*(42)*
D307N	Zinc coordination, reduction in FTase activity	*(35)*
C309A	Zinc coordination, decrease in k_{cat}	*(35,42)*
H363A	Zinc coordination, decrease in k_{cat}	*(42)*
H363Q	Zinc coordination, reduction in FTase activity	*(35)*
R211Q	Decrease in k_{cat}	*(42)*
Y310F	Decrease in k_{cat}	*(42)*
H258N	Partial reduction in FTase activity	*(35)*
D360N	Partial reduction in FTase activity	*(35)*
H156Q	Partial reduction in FTase activity	*(35)*
G259V	*ram1-2* mutant	*(14,16)*
D209N	*ram1-1* mutant	*(14,16)*

and Y362 to all 20 possible amino acids *(3)*. This analysis demonstrated that only asparagine and aspartic acid substitutions at position 159 converted the substrate specificity of FTase to that of GGTase I. At position residue 362, only leucine, isoleucine, and methionine could cause the conversion. Because these residues that affected the substrate recognition have similar van der Waals volume, we speculated that it is the size of the amino acids at these residues that is critical for the altered substrate recognition in the mutant FTases *(3)*.

In another approach, Trueblood et al. *(4)* used the mating peptide **a**-factor to identify FTase mutants with altered substrate recognition. The C-terminus of **a**-factor contains the CVIA sequence and is farnesylated *(6)*. A strain of yeast expressing **a**-factor with an altered C-terminal motif, CAMQ, was used to identify FTase mutants that were much more defective in processing **a**-factor-CAMQ than the wild-type **a**-factor-CVIA. This led to the identification of a mutation *G149E* within the Dpr1/Ram1 protein. Further characterization of this mutant revealed that it also exhibited an increased efficiency for recognizing **a**-factors ending in CVIR or CVIK. In addition, the FTase β mutant G149E as well as G149D was able to recognize $Ras2^{val19}$ ending in CIIR or CIIK. On the other hand, FTase β G149R and G149K mutants were effective in farnesylating a $Ras2^{val19}$ protein ending with the CIID sequence. This evidence of a possible electrostatic interaction between residue 149 and the C-terminal amino acid of the CaaX motif is consistent with the idea that residue-149 of Dpr1/Ram1 is located in close proximity to the site of the binding of the protein substrate.

An alignment of amino-acid sequences for the β-subunits of FTase, GGTase I, and GGTase II reveals five regions of high similarity. Dolence et al. *(42)* substituted 13 of the conserved polar and charged residues in Dpr1/Ram1. Kinetic analysis led to the identification of five substitutions, R211Q, D307A, C309A, Y310F, and H363A, with substantially reduced k_{cat} values. In addition, this study identified the glutamine at residue-256 as being important in the recognition of FPP. Replacement of this glutamine with alanine resulted in a 130-fold increase of the K_m for FPP. Another mutant, D209A, showed a 15-fold increase in the K_m for peptide substrate. A similar approach by Kurth et al. *(35)* also identified E256, D307, C309, and H363 as residues critical for FTase activity.

Extensive amino acid alterations of Dpr1/Ram1 residues have also been carried out by Caplin et al. *(43)*. Comparison of the sequences of Dpr1/Ram1 and Cal1/Cdc43 showed that there are a limited number of residues that differ between the two proteins. These residues of Dpr1/Ram1 were changed to the corresponding ones in Cal1/Cdc43 in order to assess the significance of these residues specific to each protein. This analysis resulted in the identification of three types of FTase mutants that exhibited altered isoprenoid and CAAX preference. FTase containing the I74D of Dpr1/Ram1 farnesylated only Ras1-CIIS and not Ras1-CIIM or Ras1-CIIL. Furthermore, it geranylgeranylated all three substrates as well or better than the wild-type Ras1. FTase mutant 206DDLF, which contains amino-acid changes G206D/V208D/T210L/G212F of Dpr1/Ram1, farnesylated Ras1-CIIS, Ras1-CIIM and Ras1-CIIL at wild-type levels but could no longer geranylgeranylate the Ras1-CIIM or Ras1-CIIL substrates. Finally, FTase mutant 351FSKN, which contains amino-acid changes L351F/R352S/D353K/K354N of Dpr1/Ram1, farnesylated Ras1-CIIS, and Ras1-CIIM, but not Ras1-CIIL.

2.4. Comparison of the Genetic Data with the Structural Data

Insights into the significance of the Dpr1/Ram1 residues identified by yeast genetic studies as described previously have been obtained by the recent studies on the crystal structure of rat farnesyltransferase *(39,40,44)*. The FTase β-subunit is an α-helical barrel that contains a hydrophobic interior surface. This hydrophobic interior is conserved in rat and human enzymes. Because both substrates of this enzyme—FPP and CAAX peptides—are hydrophobic, the hydrophobic nature of the interior is consistent with the idea that the interior represents the active site. As described previously, yeast FTase β residues G149, S159, Y362, and Y366 are implicated in the recognition of the CAAX peptide. These residues correspond to G143, P152, Y361, and Y365 of the β-subunit of rat FTase. Interestingly, all four of these residues are located along one side of the hydrophobic pocket in the center of the β-subunit barrel *(40)*. P152 is near the side chain of R202 in the crystal structure, and it has been suggested that R202 forms a salt bridge with the C-terminal carboxylate of the peptide substrate *(40)*. E246 of the rat enzyme is located within the binding site for the diphosphate moiety of farnesyl diphosphate. The corresponding residue in the yeast FTase is E259 and the replacement of this residue by alanine results in a dramatic increase of the apparent K_m for farnesyl diphosphate *(42)*. Furthermore, residues D297, C299, and H362 of Dpr1/Ram1 were found to be critical for the catalytic activity of yeast FTase. These residues correspond to D307, C309, and H363 of the rat enzyme, which were shown by the X-ray crystallographic analysis to be involved in the coordination of a zinc ion *(39)*.

3. *S. POMBE*: A SECOND GENETIC SYSTEM FOR THE STUDY OF PROTEIN PRENYLATION

3.1. Identification of FTase and GGTase I Genes

We have recently identified *S. pombe* genes encoding FTase and GGTase I (Table 1) *(8,9)*. Our work carried out in collaboration with Dr. Pilar Perez was initiated by the isolation of the *cwg2*⁺ gene, which encodes the β-subunit of *S. pombe* GGTase I *(8)*. The *cwg2-1* mutant was originally identified as a mutant defective in cell-wall synthesis *(8)*. The mutant is temperature-sensitive for growth, which can be suppressed by the addition of sorbitol in the growth media. *S. pombe* cell wall synthesis involves (1–3)β-D-glucan synthase and the Rho1 protein. Rho1 regulates this process by stimulating the synthase activity in a GTP-dependent manner *(45)*. Because the Rho1 function depends on its geranylgeranylation, β-glucan synthase activity is defective in the *cwg2-1* mutant. Biochemical studies showed that this mutant was defective in GGTase I, but had normal levels of FTase and GGTase II activity *(8)*. The amino-acid change in the *cwg2-1* mutant was identified as a guanine to adenine substitution at position 202 of the cwg2 protein *(9)*. Deletion of the *cwg2*⁺ gene is lethal with the spores having deletion of the *cwg2*⁺ gene dividing two or three times before losing viability *(9)*.

The *S. pombe* gene encoding the α-subunit of GGTase I was identified by the yeast two-hybrid screen using cwg2 fused with the DNA binding domain of GAL4 as a bait *(9)*. This gene is termed *cwp1*⁺, which stands for cwg2 partner. The predicted molecular weight of the 294 amino acid cwp1 protein is 34.9 kDa. The *cwp1*⁺ gene product shares significant sequence similarity with the α-subunits of other FTase genes. There is a 33% identity and a 65% similarity between the *cwp1*⁺ gene product and the *S. cerevisiae RAM2* gene product; however, a slightly lower similarity score was obtained when the *cwp1*⁺ gene product was compared with the α-subunit of human FTase (27% identity and 60% similarity). Coexpression of *cwg2*⁺ and *cwp1*⁺ genes in *E. coli* results in the production of functional GGTase I.

The β-subunit of *S. pombe* FTase is encoded by a gene we termed *cpp1*⁺ for *cwp1*⁺ partner *(9)*. This gene encodes a protein of 382 amino-acid residues and exhibits 34% identity and 65% similarity with the *S. cerevisiae* Dpr1/Ram1 protein. Similar scores were observed when the *cpp1*⁺ gene product was compared with the human FTase β-subunit (33% identity and 64% similarity). In addition, amino-acid residues of *S. cerevisiae* Dpr1/Ram1, G149, S159, Y362, and Y366, which were found to be involved in the recognition of substrates (described previously), are conserved in the *cpp1*⁺ gene product. Coexpression of *cpp1*⁺ and *cwp1*⁺ in *E. coli* allows purification of the active FTase *(46)*.

3.2. S. pombe *as a Second Genetic System to Characterize FTase*

With the identification of the *S. pombe* genes encoding subunits of FTase, we can now exploit this second genetic system to characterize FTase. In addition, the *S. pombe* system offers an advantage over *S. cerevisiae* for the study of protein prenylation. It is possible to identify prenylated proteins by [^{3}H]mevalonic acid labeling. *S. pombe* cells can take up about 5% of the exogenously added mevalonic acid, whereas <0.5% is taken up by *S. cerevisiae* cells *(47)*. This makes *S. pombe* a unique system that offers both the advantages of the genetic manipulation of *S. cerevisiae* and the ease of [^{3}H]mevalonic acid labeling similar to mammalian cells. Therefore, characterization of FTase substrate proteins can be more easily carried out using the *S. pombe* system. It is also worth emphasizing that *S. pombe* cells are overall more similar to mammalian cells than are *S. cerevisiae*. Since recent results with FTase and GGTase I inhibitors point to the involvement of these enzymes in cell-cycle progression *(48)*, results obtained with *S. pombe* may be more relevant for understanding mammalian prenyltransferases.

4. DEVELOPMENT OF A MICROBIAL SCREEN AND IDENTIFICATION OF MANUMYCIN

4.1. Development of a Microbial Screen to Identify FTase Inhibitors

S. cerevisiae provided the ability to undertake a microbial screen to identify FTase inhibitors *(10–12)*. The development of this assay was a direct outcome of the characterization of the *dpr1/ram1* mutant. An allele of the *dpr1/ram1* mutation was obtained from the study of a heterotrimeric G-protein, which is involved in mating factor signal transduction in yeast *(10,24)*. Activation of this signal transduction pathway results in the arrest of cell growth. The βγ-subunits of the heterotrimeric G-protein activate the signal transduction when freed from the α-subunit. Disruption of the α-subunit gene, *GPA1*, is lethal because the βγ-subunits send a constitutive signal to arrest cell growth. However, it is possible to block this constitutive activation by inhibiting farnesylation of the γ-subunit, which is encoded by the *STE18* gene. Thus, compounds that act to inhibit farnesylation of Ste18 enable the *GPA1* disruptants to grow.

To carry out the microbial screen, we used a specially constructed yeast strain *(11,12)*, which contains a disruption of the *GPA1* gene on the chromosome and carries a plasmid containing the *GPA1* gene under the control of the *GAL1* promoter. The cells are grown in media containing galactose so that *GPA1* is expressed. When the cells are plated on glucose-containing media, *GPA1* is no longer expressed and the strain does not grow. A filter soaked in culture media containing microbial cells is placed on top of the plate. A zone of cell growth can be observed as a halo around filters that contain compounds inhibiting FTase. This provides a simple assay to screen a large number of compounds for their ability to act as FTase inhibitors. The size of the halo surrounding the filter provides an indication as to the strength and abundance of the compound.

Because this assay utilizes the suppression of growth arrest caused by the activation of the mating factor signal transduction pathway, inhibitors of this signal transduction pathway may also be picked up. For this reason, it is important to have a secondary assay to confirm that any positive results are owing to the direct inhibition of FTase. We developed a secondary assay that assesses the suppression of heat-shock sensitivity of yeast cells expressing $Ras2^{val19}$ *(12)*. A filter soaked in the compound solution is placed on top of a plate containing an overlay of yeast cells expressing $Ras2^{val19}$. Then, the plate is subjected to heat-shock treatment. If FTase inhibitors are present, farnesylation of $Ras2^{val19}$ is inhibited and the heat-shock sensitivity is suppressed, resulting again in a zone of growth around the filter. In addition, we also examine the effect of the positive compounds on the activity of purified FTase.

4.2. Identification of Manumycin

Manumycin produced by a *Streptomyces* strain was identified from an extensive screen of culture media from microbial cells *(11)*. Three closely related compounds—manumycin A, B, and C—are produced that contain a central cyclohexenone ring. One of the side chains resembles a farnesyl group, and the length of this side chain differs among the three manumycins. Manumycin A inhibits FTase with an IC_{50} value of 5 μ*M*, whereas the IC_{50} for GGTase I is 140 μ*M*. Kinetic analysis showed that manumycin acts as a competitive inhibitor of FTase with respect to one of its substrates, FPP and the K*i* value is 1.2 μ*M*.

The ability of manumycin to inhibit Ras activation was first shown by the suppression of heat shock sensitivity in yeast cells expressing $RAS2^{val19}$ *(12)*. Manumycin was also shown to be effective in blocking Ras activation in *Caenorhabditis elegans* cells *(49)*. In this organism, Ras activation results in multi-vulva phenotype. Manumycin suppressed this multi-vulva phenotype resulting from an activated *let-60 ras* mutation. However, manumycin did not suppress the multi-vulva phenotype resulting from mutations in the *lin-1* gene or the *lin-15* gene, which act downstream of *let-60 ras*. Gliotoxin, a different inhibitor of FTase, also exhibited similar effects on *C. elegans* multi-vulva phenotype.

The effects of manumycin on the growth of human tumor cells have also been investigated. Manumycin inhibited the growth of the human *HepG2* hepatoma cell line *(50)*. The manumycin treatment blocked farnesylation of Ras and MAP kinase activity but did not affect Rap1 geranylgeranylation or prenylation of 21–26 kDa proteins *(50)*. Furthermore, the growth of human colon carcinoma (*LoVo*) was significantly inhibited by manumycin and moderate inhibition was seen with hepatoma cells, *Mahlavu* and *PLC/PRF/5 (51)*. Manumycin also inhibited the growth of human pancreatic cancer cells (*SUIT-2, MIA PaCa-2, AsPC-1, BxPC-3*) in a dose-dependent manner *(52)*. The effect of manumycin on human pancreatic cancer cells was also assessed by examining the growth of a human pancreatic cancer cell line, *MIA PaCa-2* (with a point mutation in the *Ki-ras* gene), in nude mice *(53)*. The growth of inoculated tumors was significantly inhibited by treatment with manumycin. In another study, manumycin immediately and reversibly inhibited the growth of NIH3T3 cells expressing high levels of *N-Ras (54)*. The inhibition of these cells by manumycin did not modify the cell-death rate, suggesting that it acts as a cytostatic agent. The effect of manumycin on the growth of untransformed cells was assessed with one type of cells. In this study using NIH3T3 cells, it was shown that manumycin inhibited their growth in a reversible manner *(54)*.

5. NATURAL PRODUCT FTASE INHIBITORS

In addition to manumycin, a number of natural product farnesyltransferase inhibitors have been identified. Table 3 summarizes a variety of compounds obtained from diverse sources. Although some have been identified using the microbial screen, others have been identified by directly assaying for compounds that inhibit FTase activity.

Actinoplanic acid *(55)*, barcelonic acid A *(56)*, chaetomellic acids *(57)*, CP225,917 *(58)*, cylindrol A *(59)*, 10'-Desmethoxystreptonigrin *(60)*, fusidienol *(61)*, gliotoxin *(62)*, pepticinnamin E *(63)*, preussomerin G *(64)*, RPR113228 *(65)*, and SCH58540 *(66)* have been described in our previous review on FTase inhibitors *(1)*. For further details about these compounds, please refer to Sattler and Tamanoi *(1)*.

Andrastins A–D were isolated from the culture broth of *Penicillium sp. FO-3929*. These compounds share a common structure that can be classified as meroterpenoid fungal metabolites *(67)*. Differences in the side chains of the ring structure differentiate between the four compounds. Among the four, andrastin C has the highest FTase inhibition activity with IC_{50} of 13.3 μ*M*. Comparison of the inhibitory activity among andrastins as

Table 3
Natural Product Farnesyltransferase Inhibitors

Inhibitors	*FTase inhibition* IC_{50} *(μM)*	*Sources*	*References*
Actinoplanic acid	0.23	*Actinoplanes*	*(55)*
Andrastin A–D	13	*Penicillium*	*(67)*
Barceloneic acid A	40	*Fungus*	*(56)*
Chaetomellic acid A,B	0.1	*Chaetomella*	*(57)*
CP-225,917/CP263,114	6	*Fungus*	*(58)*
Cylindrol A	2.2	*Fungus*	*(59)*
10'-Desmethoxy-streptonigrin	21	*Streptomyces*	*(60)*
Fusidienol	0.3	*Fusidium griseum*	*(61)*
Gliotoxin	1.1	*Fungus*	*(62)*
Kurasoin A,B	59	*Paecilomyces*	*(68)*
Manumycin A	5	*Streptomyces*	*(11)*
Pepticinnamin E	0.1	*Streptomyces*	*(63)*
Preussomerin G	1.2	*Fungus*	*(64)*
RPR113228	2.1	*Fungus*	*(65)*
SCH58540	29	*Streptomyces*	*(66)*
Saquayamycin A–F	1.5	*Actinomycetes*	*(70)*

well as among structurally similar compounds such as citreohybridones, suggested that the inhibitory activity is greater when a methoxy or acetoxy residue is attached to the C-15 position rather than to the C-17 position of meroterpenoid ring. In addition, the FTase inhibition by andrastins appears to be reversible. Thus, this family of meroterpenoids provides a group of compounds useful for further probing the active site of FTase.

Kurasoins A and B were found from the culture broth of *Paecilomyces sp. FO-3684 (68)*. This fungus was isolated from a soil sample collected at Kurashiki City, Japan. Both compounds are acyloin compounds and have a 3-hydroxy-1-phenyl-2-butanone moiety in common. The IC_{50} values of kurasoins A and B against FTase were 59 and 58.7 μ*M*, respectively. These values are rather high and suggest that they may not be good FTase inhibitors. However, total synthesis of these compounds has recently been accomplished *(69)* and derivatives with higher potency may be obtained in the future.

Saquayamycin E and F were isolated from the *Actinomycetes* strain *MK290-AF1 (70)*. They belong to a family of compounds called saquayamycins, which include saquayamycin A,B, C, and D. These compounds as well as the related compound aquayamycin all exhibit the ability to inhibit FTase *(70)*, and their IC_{50} values are between 1 and 2 μ*M*. Kinetic analysis with Lineweaver-Burk plotting suggested that the Saquayamycins non-competitively inhibit the enzyme with respect to the Ras protein substrate.

In conclusion, a large number of natural products have been identified that exhibit FTase inhibitory activity. However, the potency is generally low with the IC_{50} value in the μ*M* range. Although these natural compounds are useful as lead compounds, extensive study to identify derivatives is necessary to obtain compounds with increased potency. Another caution concerning natural products is their possible toxicity towards mammalian cells. This is particularly true for antibiotics that have been shown to exhibit anti-bacterial or anti-fungal activities. Any toxicity associated with the compound needs to be critically evaluated to see whether the inhibitory activity against FTase is separated from the toxic activity before further tests can be conducted.

ACKNOWLEDGMENT

This work was supported by NIH grant CA41996. W. Yang is supported by Edwin D. Pauley Foundation. J. Urano is supported by USPHS National Research Service Award GM07185. N. Robinson is supported by Warsaw fellowship.

REFERENCES

1. Sattler I, Tamanoi F. Prenylation of Ras and inhibitors of prenyltransferases, in *Regulation of the RAS Signaling Network* (Maruta H, Burgess AW, eds.), Landes, Austin, TX, 1996; pp. 95–137.
2. Zhang FL, Casey PJ. Protein prenylation: molecular mechanisms and functional consequences. *Annu Rev Biochem* 1996; 65:241–269.
3. Del Villar K, Mitsuzawa H, Yang W, Sattler I, Tamanoi F. Amino acid substitutions that convert protein substrate specificity of farnesyltransferase to that of geranylgeranyltransferase type I. *J Biol Chem* 1997; 272:680–687.
4. Trueblood CE, Buyartchuk VL, Rine J. Substrate specificity determinants in the farnesyltransferase β-subunit. *Proc Natl Acad Sci USA* 1997; 94:10,774–10,779.
5. Schaffer WR, Rine J. Protein prenylation: genes, enzymes, targets, and functions. *Annu Rev Genet* 1992; 26:209–237.
6. Epand RM, Naider F, Becker JM. Lipid-mediated **a**-factor interactions with artificial membranes. *Methods Enzymol* 1995; 250:169–186.
7. Clarke S. Protein prenylation and methylation at carboxy-terminal cysteine residues. *Annu Rev Biochem* 1992; 61:355–386.
8. Diaz M, Sanchez Y, Bennett T, Sun CR, Godoy C, Tamanoi F, Duran A, Perez P. The *Schizosaccharomyces pombe cwg2*$^{+}$ gene codes for the β subunit of a geranylgeranyltransferase type I required for β-glucan synthesis. *EMBO J* 1993; 12:5245–5254.
9. Arellano M, Coll PM, Yang W, Duran A, Tamanoi F, Perez P. Characterization of the geranylgeranyl transferase type I from *Schizosaccharomyces pombe. Mol Micobiol* 1998; 29:1357–1367.
10. Finegold AA, Schafer WR, Rine J, Whiteway M, Tamanoi F. Common modifications of trimeric G proteins and ras protein: involvement of polyisoprenylation. *Science* 1990; 249:165–169.
11. Hara M, Akasaka K, Akinaga S, Okabe M, Nakano H, Gomez R, Wood D, Uh M, Tamanoi F. Identification of ras farnesyltransferase inhibitors by microbial screening. *Proc Natl Acad Sci USA* 1993; 90: 2281–2285.
12. Mitsuzawa H, Tamanoi F. *In vivo* assays for farnesyltransferase inhibitors with *Saccharomyces cerevisiae. Methods Enzymol* 1995; 250:43–51.
13. Judd SR, Tamanoi F. A genetic approach to the study of farnesylation. In *METHODS: Companion Methods Enzymol* 1990; 1:246–252.
14. Goodman LE, Judd SE, Farnsworth CC, Powers S, Gelb MH, Glomset JA, Tamanoi F. Mutants of *Saccharomyces cerevisiae* defective in the farnesylation of ras proteins. *Proc Natl Acad Sci USA* 1990; 87:9665–9669.
15. Goodman LE, Perou CM, Fujiyama A, Tamanoi F. Structure and expression of yeast *DPR1*, a gene essential for the processing and intracellular localization of ras proteins. *Yeast* 1988; 4:271–281.
16. He B, Chen P, Chen SY, Vancura KL, Michaelis S, Powers S. *RAM2*, an esential gene of yeast, and *RAM1* encode the two polypeptide components of the farnesyltransferase that prenylate **a**-factor and Ras proteins. *Proc Natl Acad Sci USA* 1991; 88:11,373–11,377.
17. Fujiyama A, Tamanoi F. Processing and fatty acylation of RAS1 and RAS2 proteins in *Saccharomyces cerevisiae. Proc Natl Acad Sci USA* 1986; 83:1266–1270.
18. Fujiyama A, Tamanoi F. RAS2 protein of *S cerevisiae* undergoes removal of methionine at N-terminus and removal of three amino acids at C-terminus. *J Biol Chem* 1990; 265:3362–3368.
19. Fujiyama A, Tsunasawa S, Tamanoi F, Sakiyama F. S-Farnesylation and methyl esterification of C-terminal domain of yeast RAS2 protein prior to fatty acid acylation. *J Biol Chem* 1991; 266:17,926–17,931.
20. Fujiyama A, Matsumoto K, Tamanoi F. A novel yeast mutant defective in the processing of ras proteins: assessment of the effect of the mutation on processing steps. *EMBO J* 1987; 6:223–228.
21. Powers S, Michaelis S, Broek D, et al. *RAM*, a gene of yeast required for a functional modification of RAS proteins and for production of mating pheromone **a**-factor. *Cell* 1986; 47:413–422.
22. Toda T, Uno I, Ishikawa T, Powers S, Kataoka T, Broek D, et al. In yeast, RAS proteins are controlling elements of adenylate cyclase. *Cell* 1985; 40:27–36.

23. Wilson KL, Herskowitz I. *STE16*, a new gene required for pheromone production by a cells of *Saccharomyces cerevisiae. Genetics* 1987; 115:441–449.
24. Nakayama N, Kaziro Y, Arai K, et al. Role of *STE* genes in the mating factor signaling pathway mediated by *GPA1* in *Saccharomyces cerevisiae. Mol Cell Biol* 1988; 8:3777–3783.
25. Gomez R, Goodman LE, Tripathy SK, O'Rourke E, Manne V, Tamanoi F. Purified yeast protein farnesyltransferase is structurally and functionally similar to its mammalian counterpart. *Biochem J* 1992; 289:25–31.
26. Finegold AA, Johnson DI, Farnsworth CC, Gelb MH, Judd SR, Glomset JA, Tamanoi F. Protein geranylgeranyltransferase of *Saccharomyces cerevisiae* is specific for Cys-Xaa-Xaa-Leu motif proteins and requires the *CDC43* gene product but not the *DPR1* gene product. *Proc Natl Acad Sci USA* 1991; 88: 4448–4452.
27. Mayer ML, Caplin B, Marshall MS. *CDC43* and *RAM2* encode the polypeptide subunits of a yeast type I protein geranylgeranyltransferase. *J Biol Chem* 1992; 267:20,589–20,593.
28. Ohya Y, Goebel M, Goodman LE, Petersen-Bjorn S, Friesen JD, Tamanoi F, Anraku Y. Yeast *CAL1* is structural and functional homologue to the *DPR1(RAM)* gene involved in ras processing. *J Biol Chem* 1991; 266:12,356–12,360.
29. Adams AE, Johnson DI, Longnecker RM, et al. *CDC42* and *CDC43*, two additional genes involved in budding and establishment of cell polarity in the yeast *Saccharomyces cerevisiae. J Cell Biol* 1990; 111: 131–142.
30. Drubin DG, Nelson WJ. Origins of cell polarity. *Cell* 1996; 84:335–344.
31. Trueblood CE, Ohya Y, Rine J. Genetic evidence for *in vivo* cross-specificity of the CaaX-box protein prenyltransferases farnesyltransferase and geranylgeranyltransferase-I in *Saccharomyces cerevisiae. Mol Cell Biol* 1993; 13:4260–4275.
32. Matsui Y, Toh-e A. Isolation and characterization of two novel ras superfamily genes in *Saccharomyces cerevisiae. Gene* 1992; 114:43–49.
33. Imai J, Toh-e A, Matsui Y. Genetic analysis of the *Saccharomyces cerevisiae RHO3* gene, encoding a Rho-type small GTPase, provides evidence for a role in bud formation. *Genetics* 1996; 142:359–369.
34. Robinson NGG, Guo L, Imai J, Toh EA, Matsui Y, Tamanoi F. Rho3 of *Saccharomyces cerevisiae*, which regulates the actin cytoskeleton and exocytosis, is a GTPase which interacts with Myo2 and Exo70. *Mol Cell Biol* 1999; 3580–3587.
35. Kurth DD, Farh L, Deschenes RJ. Functional consequence of mutating conserved residues of the yeast farnesyl-protein transferase β-subunit Ram1(Dpr1). *Biochemistry* 1997; 36:15,932–15,939.
36. Dolence JM, Cassidy PB, Mathis JR, Poulter CD. Yeast protein farnesyltransferase: steady-state kinetic studies of substrate binding. *Biochemistry* 1995; 34:16,687–16,694.
37. Dolence JM, Poulter CD. A mechanism for posttranslational modifications of proteins by yeast protein farnesyltransferase. *Proc Natl Acad Sci USA* 1995; 92:5008–5011.
38. Urano J, Tamanoi F. Reconstitution of yeast farnesyltransferase from individually purified subunits, in *Protein Lipidation Protocols* (Gelb M, ed.), Humana Press, Totowa, NJ, 1998; pp. 145–159.
39. Park HW, Boduluri SR, Moomaw JF, Casey PJ, Beese LS. Crystal structure of protein farnesyltransferase at 2.25 A resolution. *Science* 1997; 275:1800–1804.
40. Dunten P, Kammlott U, Crowther R, Weber D, Palermo R, Birktoft J. Protein farnesyltransferase: structure and implications for substrate binding. *Biochemistry* 1998; 37:7907–7912.
41. Mitsuzawa H, Esson K, Tamanoi F. Mutant farnesyltransferase β subunit of *Saccharomyces cerevisiae* that can substitute for geranylgeranyltransferase type I β subunit. *Proc Natl Acad Sci USA* 1995; 92: 1704–1708.
42. Dolence JM, Rozema DB, Poulter CD. Yeast protein farnesyltransferase. Site-directed mutagenesis of conserved residues in the β-subunit. *Biochemistry* 1997; 36:9246–9252.
43. Caplin BE, Ohya Y, Marshall MS. Amino acid residues that define both the isoprenoid and CAAX preferences of the *Saccharomyces cerevisiae* protein farnesyltransferase. *J Biol Chem* 1998; 273:9472–9479.
44. Park HW, Beese LS. Protein farnesyltransferase. *Curr Opin Str Biol* 1997; 7:873–880.
45. Arellano M, Duran A, Perez P. Rho1 GTPase activates the (1-3)β-D-glucan synthase and is involved in *Schizosaccharomyces pombe* morphogenesis. *EMBO J* 1996; 15:4584–4591.
46. Yang W, Urano J, Tamanoi F. Protein farnesylation is critical for maintaining normal cell morphology and canavanine resistance in *Schizosaccharomyces pombe. J Biol Chem* 2000; 275:429–438.
47. Ginnakouros T, Armstrong J, Magee AI. Protein prenylation in *Schizosaccharomyces pombe. FEBS Lett* 1992; 297:103–106.

48. Vogt A, Sun J, Qian Y Hamilton AD, Sebti SM. The geranylgeranyltransferase-I inhibitor GGTI-298 arrests human tumor cells in G_0/G_1 and induces $p21^{WAF1/CIP1/SDI1}$ in a p53-independent manner. *J Biol Chem* 1997; 272:27,224–27,229.
49. Hara M, Han M. Ras farnesyltransferase inhibitors suppress the phenotype resulting from an activated ras mutation in *Caenorhabditis elegans*. *Proc Natl Acad Sci USA* 1995; 92:3333–3337.
50. Nagase T, Kawata S, Tamura S, Matsuda Y, Inui Y, Yamasaki E, et al. Inhibition of cell growth of human hepatoma cell line (HepG2) by a farnesyl protein transferase inhibitor: a preferential suppression of ras farnesylation. *Intl J Cancer* 1996; 65:620–626.
51. Nagase T, Kawata S, Tamura S, Matsuda Y, Inui Y, Yamasaki E, et al. Manumycin and gliotoxin derivative KT7595 block Ras farnesylation and cell growth but do not disturb lamin farnesylation and localization in human tumor cells. *Br J Cancer* 1997; 76:1001–1010.
52. Kainuma O, Asano T, Hasegawa M, Kenmochi T, Nakagohri T, Tokoro Y, Isono K. Inhibition of growth and invasive activity of human pancreatic cancer cells by a farnesyltransferase inhibitor, manumycin. *Pancreas* 1997; 15:379–383.
53. Ito T, Kawata S, Tamura S, Igura T, Nagase T, Miyagawa JI, et al. Suppression of human pancreatic cancer growth in *BALB/c* nude mice by manumycin, a farnesyl:protein transferase inhibitor. *Jpn J Cancer Res* 1996; 87:113–116.
54. Servais P, Gulbis B, Fokan D, Galand P. Effects of the farnesyltransferase inhibitor UCF-1C/manumycin on growth and p21-ras post-translational processing in *NIH3T3* cells. *Intl J Cancer* 1998; 76:601–608.
55. Singh SB, Liesch JM, Lingham RB, et al. Actinoplanic acid A: a macrocyclic polycarboxylic acid which is a potent inhibitor of RAS farnesyl-protein transferase. *J Am Chem Soc* 1994; 116:11,606,11,607.
56. Jayasuriya H, Ball RG, Zinc DL, et al. Barcelonic acid A, a new farnesyl-protein transferase inhibitor from a phoma species. *J Nat Prod* 1995; 58:986–991.
57. Lingham RB, Silverman KC, Bills GF, et al. *Chaetomella acutiseta* produces chaetomellic acids A and B which are reversible inhibitors of farnesyl-protein transferase. *Appl Microbiol Biotechnol* 1993; 40:37–374.
58. Dabrah TT, Harwood J Jr, Huang LH, Jankovich ND, Kaneko T, Li J-C, et al. CP-225,917 and CP-263,114, novel Ras farnesylation inhibitors from an unidentified fungus. *J Antibiot* 1997; 50:1–7.
59. Singh SB, Zink DL, Bills GF, et al. Cylindrol A, a novel inhibitor of RAS farnesyl-prtoein transferase from *Cylindrocarpon lucidum*. *Tetrahedron Lett* 1995; 36:4935–4938.
60. Liu WC, Barbacid M, Bulgar M, et al. 10'-Desmethoxystreptonigrin, a novel analog of streptonigrin. *J Antibiot* 1992; 45:454–457.
61. Singh SB, Jones ET, Goetz MA, et al. Fusidienol: a novel inhibitor of RAS farnesyl-protein transferase from *Fusidium griseum*. *Tetrahedron Lett* 1994; 35:4693–4696.
62. Van der Pyl D, Inokoshi J, Shiomi K, et al. Inhibition of farnesyl-protein transferase by gliotoxin and acetylgliotoxin. *J Antibiot* 1992; 45:1802–1805.
63. Omura S, Van der Pyl D, Inokoshi J, et al. Pepticinnamins, new farnesy-protein transferase inhibitors produced by an actinomycete. *J Antibiot* 1993; 46:222–228.
64. Singh SB, Zink DL, Liesch JM, et al. Preussomerins and deoxypreussomerins: novel inhibitors of RAS farnesyl-protein transferase. *J Org Chem* 1994; 59:6296–6302.
65. Van der Pyl D, Cans P, Debernard JJ, et al. RPR113228, a novel farnesyl-protein transferase inhibitor produced by *Chrysoporium lobatum*. *J Antiobiot* 1995; 48:736,737.
66. Phife DW, Patton RW, Berrie R, et al. SCH58450:a novel farnesyl protein transferase inhibitor possessing a 6a,12a:7,12-diepoxybenz[a]anthracene ring system. *Tetrahydron Lett* 1995;36:6995–6998.
67. Omura S, Inokoshi J, Uchida R, Shiomi K, Masuma R, Kawakubo T, et al. Andrastins A-C, new protein farnesyltransferase inhibitors produced by *Penicillium sp FO-3929*. *J Antibiot* 1996; 49:414–417.
68. Uchida R, Shiomi K, Inokoshi J, Masuma R, Kawakubo T, Tanaka H, et al. Kurasoins A and B, new protein farnesyltransferase inhibitors produced by *Paecilomyces sp FO-3684*. *J Antibiot* 1996; 49: 932–934.
69. Sunazuka T, Hirose T, Ahi-Ming T, Uchda R, Shiomi K, Harigaya Y, Omura S. Synthesis and absolute structures of novel protein farnesyltransferase inhibitors, Kurasoins A and B. *J Antibiot* 1997; 50:453–455.
70. Sekizawa R, Inuma H, Naganawa H, Hamada M, Takeuchi T, Yamaizumi J, Umezawa K. Isolation of novel Saquayamycins as inhibitors of farnesyl-protein transferase. *J Antibiot* 1996; 49:487–490.

11 Effects of Farnesyltransferase Inhibitors on Cytoskeleton, Cell Transformation, and Tumorigenesis

The FTI-Rho Hypothesis

George C. Prendergast, PHD

CONTENTS

1. INTRODUCTION

Farnesyltransferase inhibitors (FTIs) are a novel class of nontoxic cancer therapeutics whose development was based on the discovery that oncogenic Ras must be post-translationally prenylated to function. Strikingly, FTIs can block or even reverse Ras-dependent tumor formation. However, several lines of work indicate that Ras prenylation does

From: *Farnesyltransferase Inhibitors in Cancer Therapy*
Edited by: S. M. Sebti and A. D. Hamilton © Humana Press Inc., Totowa, NJ

not have to be inhibited to achieve this. Therefore, the exact mechanism underlying FTI action has become a question of major interest.

Cell biological studies have shown that FTIs cause rapid and dramatic reversion of the Ras-transformed phenotype. In this chapter, we review findings that alterations in the cytoskeleton, proliferative capacity, and viability of transformed cells may be due at least in part to alteration of the function of farnesylated members of the Rho family of proteins. Prompted by these findings, we propose a new hypothesis for how FTIs may block or reverse tumor formation in vivo, through alteration of Rho-dependent pathways that control cell cycle and cell survival and that are subverted by Ras and other oncogenes in neoplastic cells. Through the new vantage offered by this model, we consider several key questions about FTI biology, including a) how FTIs can block transformed cell growth yet exert little effect on normal cells; b) why FTIs are cytotoxic in some tumor model systems but only cytostatic in others; c) how malignant cells derived from tumors susceptible to regression can still persist in FTI-treated animals; and d) how susceptible malignant cells may ultimately acquire drug resistance.

2. EFFECTS OF FTIs ON THE PROLIFERATION AND STRUCTURE OF NORMAL AND TRANSFORMED CELLS

Oncogenic Ras must be post-translationally farnesylated to transform cells. This discovery formed the foundation for developing inhibitors of the housekeeping enzyme farnesyltransferase (FTase) as a strategy to inhibit the growth of Ras-dependent tumors *(1)*. From a biological standpoint, this project has been very successful. Its success represents perhaps the first case in which basic research into the molecular biology of cancer has been rationally translated into a truly novel therapeutic modality.

Of the many structural classes of FTIs that have been developed *(2)*, most biological studies have employed CAAX peptidomimetic FTIs, which exhibit excellent potency, specificity, cell penetration, and little to no cell toxicity *(3,4)*. In cell culture, FTIs effectively inhibit Ras farnesylation and can selectively inhibit the anchorage-independent growth of Ras-transformed cells *(5–7)*. Associated with the loss of anchorage-independent potential is a reversion phenomenon in which cells flatten, enlarge, and acquire the morphological and growth regulatory characteristics of nontransformed parental cells *(8,9)*. Cells transformed by oncogenes such as *src*, which utilize Ras signaling, but not those such as *raf*, whose action is Ras-independent, are also inhibited by FTIs *(7,9)*. Notably, the anchorage-independent growth of many human tumor cell lines with multiple genetic alterations is also inhibited by FTIs *(10)*.

Consistent with their in vitro effects, FTIs block tumor formation in mouse xenograft models *(11–14)* and, even more dramatically, cause regression of tumors that arise in transgenic mice that harbor oncogenic *H-ras* or *N-ras* genes ("ras oncomice") *(15,16)*. Remarkably, given that Ras is required for normal cell growth and differentiation, in all in vitro and in vivo studies published to date FTIs have been found to be essentially nontoxic even at doses that completely block processing of H-Ras protein. In cell culture, concentrations sufficient to block the anchorage-independent growth of Ras-transformed cells have no cytotoxicity and at most only slight effects on proliferation of normal cells *(8,9)*. In mouse models, there is no apparent systemic toxicity at doses capable of blocking tumor growth *(12,15)*. Thus, FTIs apparently distinguish and target a unique aspect of transformed cell physiology.

3. FTIs DO NOT HAVE TO INHIBIT RAS FUNCTION TO SUPPRESS RAS TRANSFORMATION

Results from several studies have raised questions about the exact mechanism by which FTIs reverse cell transformation. Although it is clear that the FT inhibition is closely tied to the biological effects of FTIs, it is much less clear that inhibiting the farnesylation of Ras is important. First, the lack of cytostatic effects on normal cells suggests Ras function is not efficiently inhibited, because Ras is required for the proliferation of normal cells. Second, the kinetics of phenotypic reversion induced by FTI treatment of Ras-transformed cells are too rapid to be explained simply by loss of Ras function through inhibition of its farnesylation. Investigation of the reversion process induced by certain FTIs indicates that it is largely complete within 24 h of cell treatment *(9)*, even though Ras has a half-life of ~24 h *(17)*. Thus, cells can revert completely during a period in which steady-state levels of farnesylated oncogenic Ras are reduced only approx 50% *(9)*. Reversion does not appear to reflect the dominant inhibitory activity of soluble mutant Ras species (which are generated in drug-treated cells), because unfarnesylated Ras does not accumulate to significant steady-state levels *(9)*. This is also because only the Ras L61 allele, but not the Ras V12 mutant allele, which is used in all published experimental models, exhibits potent dominant inhibitory activity when soluble *(18)*. Additionally, FTIs can still inhibit the anchorage-independent growth of cells transformed with oncogenic Ras proteins engineered to function independently of farnesylation, owing to N-myristylation or geranylgeranylation *(19,20)* (G. C. Prendergast, unpublished observations). Similarly, the anchorage-independent growth of K-Ras-transformed cells remains partly susceptible to suppression by FTIs, even though the drugs do not inhibit K-Ras prenylation (owing to its ability to be geranylgeranylated by GGTase I when FTase activity is absent in cells) *(21–24)*. Lastly, the susceptibility of human tumor cell lines to FTIs does not correlate with their Ras status *(10)*. Thus, biological susceptibility can be separated to a significant degree from Ras inhibition. Taken together, these observations suggest that FTIs act by altering the farnesylation and therefore the activity of a non-Ras protein(s).

4. FTIs ALTER RHO FUNCTIONS: RHOB AND THE FTI-RHO HYPOTHESIS

The list of farnesylated proteins that constitute alternate targets of FTIs continues to grow *(25)*. At the current time, one class of appealing targets that have emerged are farnesylated Rho proteins, in particular RhoB, a member of the Rho/Rac family of small GTPases that regulate cytoskeletal actin, focal adhesion formation, cell adhesion signaling, and transcription (reviewed in refs. *26–28*). An initial clue that Rho alteration may be part of the drug mechanism was prompted by the observation that FTIs stimulate stress fiber formation and cell enlargement in normal cells *(9)*. RhoB has two features which make it a logical candidate target. First, unlike most Rho/Rac proteins, which are geranylgeranylated in cells, RhoB exists in two populations that are either farnesylated or geranylgeranylated *(29)*. Second, RhoB had been linked previously to cell growth regulation *(30,31)*. The hypothesis that the anti-transforming effects of FTIs were based at least in part on alteration of RhoB function *(9)* made several predictions corroborated in subsequent studies. First, FTIs specifically inhibited the farnesylation of RhoB in cells and this effect was correlated with loss of its cell growth-stimulating activity *(32)*. Second, consistent with its initial identification as a v-Src and epidermal growth factor (EGF)-responsive immediate early gene *(30)*, fully processed RhoB was demonstrated to be short-lived in cells, with a half-life of 2–4 h. This feature addressed the kinetic aspect of FTI biology mentioned earlier, because farnesylated RhoB would be rapidly depleted by drug treatment *(20)*. A further line of support for the status of RhoB as a drug target was that FTIs induced rapid relocalization in cells *(20)*. Third, a dominant inhibitory mutant of RhoB genetically mimicked the predicted effect of pharmacological inhibition, by blocking Ras transformation *(33)*. Thus, even though Ras transformation is associated with stress fiber dissolution *(34,35)*, certain Rho functions that may be affected by FTI treatment are apparently required for Ras transformation *(9)*. Lastly, Ras-transformed cells can be rendered drug resistant by ectopic expression of an *N*-myristylated RhoB species whose membrane localization is prenylation-independent *(20)*. The simplest interpretation of this result is that the biological effects of FTIs are mediated at least in part by altering RhoB prenylation patterns. Thus, although other farnesylated targets of FTIs should also be considered *(25,36,37)*, RhoB represents the first non-Ras target for which there is significant biochemical and biological evidence that altering its prenylation is a crucial step in the mechanism by which FTIs reverse the Ras-transformed phenotype.

5. CELL ADHESION STATUS: A RHO-REGULATED DETERMINANT OF FTI CYTOTOXICITY

To date, perhaps the most dramatic demonstration of the potential of FTIs as anticancer agents involves experiments using the transgenic *v-H-Ras* mouse model. In these animals, which harbor an oncogenic *v-H-ras* gene and therefore develop spontaneous carcinomas, FTI treatment leads to dramatic, nearly complete, tumor regression *(15)*. Although tumor eradication is certainly promising from a clinical standpoint, the basis of this regression was quite unclear, because FTIs are not cytotoxic against transformed cells in vitro at concentrations significantly beyond the minimal inhibitory concentration (MIC) required to inhibit Ras transformation *(9)*. Growth inhibitory effects of FTIs on transformed cells are revealed in soft agar culture, but cells remain viable, and it is clear that reverted cells can proliferate if anchorage is possible *(9)*. Oncogenic Ras promotes survival of epithelial cells *(38,39)*, which have not been investigated with regard to FTI response, but it is clear that FTIs are not cytotoxic but rather cytostatic to human carcinoma cell lines *(10)*. Thus, most in vitro experiments had illustrated cytostatic but not cytotoxic effects of FTIs, the latter of which were thought to be important for the rapid tumor regression observed in vivo.

Two complex and potentially problematic issues related to how FTIs cause tumor regression also are raised by findings from animal experiments. First, resistance to tumor regression was observed in some animals, a phenomenon that could be selected for by repeated cycles of drug exposure and withdrawal *(15)*. Second, even among tumors that appeared to regress completely, cessation of FTI treatment led to a rapid return of the tumor *(15)*, indicating that some malignant cells can persist even while the bulk of the tumor disappears. Tumor persistence, if also seen with the treatment of human cancer, would require continuous, long-term FTI treatment that could increase side effects and the development of resistance. Finally, although FTIs induce regression in oncomice models, in xenograph models FTIs appear to be only cytostatic *(11,12)*. Therefore, understanding the basis of tumor resistance and persistence, and determining the basis for cytotoxic vs cytostatic effects, may be important for maximizing the effectiveness of FTIs in the clinic.

The "FTI-Rho hypothesis" offers a new viewpoint and explanative power regarding these issues. Rho proteins have been implicated in focal adhesion formation and integrin signaling *(26)*, prompting experiments in which the effects of FTIs on Ras-transformed cells were compared under conditions where cell–cell or cell–matrix attachment were favored. Thus, cell attachment parameters have been shown to dictate the physiological response to FTIs *(40)*. Thus, Ras-transformed cells cultured in suspension, where cell–cell but not cell–substratum attachment is possible, respond to FTI treatment by undergoing apoptosis instead of reversion. Although this response might be predicted by studies on the ability of Ras to promote cell survival in the absence of substratum adhesion *(39)*, the cell-death mechanism appears to be based not on Ras but on a farnesylated Rho function consistent with farnesylated RhoB. First, the rapid kinetics and dose response are similar to those for reversion, inconsistent with a role for depletion of Ras, which is long-lived, but consistent with depletion of the short-lived RhoB protein. Second, as was the case with FTI-induced reversion, ectopic expression of myristylated RhoB blocked FTI-induced apoptosis. Two other features of cell death induced by FTIs were that it was p53-independent but inhibited by Bcl-X_L *(40)*, a member of the Bcl-2 family of apoptosis regulators that broadly influence cell-death responses. These findings demonstrated in principle that cell-attachment capabilities could dramatically influence the phenotypic response to FTIs and pointed to Rho-dependent integrin signaling pathways as a realm for understanding the drugs' antitransforming and antitumor properties. In future work, it will be important to determine whether FTIs may alter "inside-out" or "outside-in" signal transduction by integrins.

The identification of a link between FTI action and cell adhesion capacity suggests one way to explain nontoxic tumor regression and provides a starting point to address the question of drug resistance. First, the in vitro model suggests that lack of appropriate substratum attachment in vivo might facilitate FTI-induced apoptosis of tumor cells, thereby causing tumor regression. Tumor cells at privileged locations—perhaps in the periphery of the tumor, where normal anchorage cues exist and could be accessed by drug-treated cells—might allow such tumor cells to revert to a normal phenotype with regard to attachment properties, and therefore survive. Another explanation for persistence and resistance is suggested by the Bcl-X_L experiment. Genetic alterations that block the apoptotic response, such as Bcl-X_L overexpression, might arise in a percentage of tumor cells, thereby allowing them to survive drug treatment and regrow upon FTI removal. The recurrent tumor might then be resistant to tumor regression when FTI treatment is reimplemented. The ability of Bcl-X_L to defeat FTI-induced cell death raises serious clinical concerns about the ability of the drugs to cause regression of advanced invasive and metastatic cancers on their own. A significant proportion of cancers are known to involve Bcl-X_L and its related family member Bcl-2 *(41)*. However, even where these genes are not involved, one would anticipate an impediment to apoptosis in invasive cancers, since by their nature they have already evolved a reduced susceptibility to apoptosis elicited by loss of physiological adhesion. Consistent with this concept, FTIs do not cause regression of human tumors grown in xenograph models nor do they induce apoptosis of any human tumor cell lines that are deprived of adhesion by culturing on polyHEMA (P. Lebowitz and G. C. Prendergast, unpublished results). FTIs may therefore prove most effective to potentiate the action of other modalities, such as radiation or taxol *(42,43)*, or through exploiting their cytostatic properties as a means to extend remission after primary therapies by suppressing the proliferation of micrometastases.

6. FTI SUPPRESSION OF VEGF EXPRESSION IN TUMOR CELLS: A SECOND CYTOTOXIC MECHANISM LINKED TO RHO?

In addition to variant cell adhesion properties, another unique aspect of tumor cells is their dependence on angiogenesis. Because FTIs caused nontoxic tumor regression but did not completely eliminate the tumor *(15)*, two logical questions were 1) whether Ras might upregulate the expression of any angiogenic factors which are deregulated in palpable tumors and 2) whether FTIs might reverse such effects. In two rodent epithelial models, it has been demonstrated that oncogenic H-Ras can upregulate secretion of vascular endothelial growth factor (VEGF) *(44,45)*, a crucial angiogenic factor in cancer, and that FTI treatment can suppress this effect *(45)*. Although it is not yet clear that this mechanism operates in vivo, there are additional in vitro observations that would support such a role and that are consistent with a role for Rho inhibition. First, VEGF message levels that are elevated in H-Ras-transformed Rat1 cells are subject to suppression by FTI treatment, with kinetics consistent with RhoB inhibition/alteration (P. Lebowitz and G. C. Prendergast, unpublished results). Second, H-Ras-transformed cells, which are rendered FTI resistant by expression of myristylated RhoB *(20)*, exhibit resistance to FTI-induced suppression of VEGF message (G. C. Prendergast, unpublished results). Thus, in addition to possible effects on cell adhesion-dependent viability, FTIs might also mediate cell killing by indirectly inhibiting a pathway that leads to VEGF overexpression and therefore maintenance of tumor vasculature.

Recently, phosphatidylinositol 3-kinase (PI3'K) and the Akt kinase have been implicated in a transcriptional mechanism by which hypoxia activates VEGF expression in Ras-transformed cells *(46)*. Experiments using dominant inhibitory mutants suggest that Rho is not necessary for Ras to upregulate VEGF expression at the transcriptional level (W. Du and G. C. Prendergast, unpublished results). However, a role for Rho in the regulation of VEGF by FTIs in transformed cells cannot be ruled out, because Ras can also upregulate VEGF by a second mechanism that is independent of PI3'K *(47)*. In future work, it will be important to explore the linkage between the biological and genetic response to FTIs, and to determine whether their ability to suppress VEGF overexpression is a cause or effect of either tumor cell reversion and/or cytostatic effects induced by FTI treatment.

7. ELEVATION OF GERANYLGERANYLATED RHOB, A GAIN-OF-FUNCTION EFFECT OF FTI TREATMENT, IS SUFFICIENT TO MEDIATE CELL CYCLE INHIBITION BY FTIs

The simplest conception of how FTIs exert their biological effects is that they indirectly cause loss of function of farnesylated target proteins. However, because FTIs cause an increase in the levels of geranylgeranylated RhoB (RhoB-GG) as well as a loss of farnesylated RhoB (RhoB-F), it is possible that accumulation of RhoB-GG species with alteration localization may be part of the drug mechanism. In this manner, FTIs may act in part by a gain-of-function as well as loss-of-function effects, by increasing RhoB-GG levels. Consistent with different functions for RhoB-F and RhoB-GG, FTIs inhibit the growth promoting activity of RhoB in certain cell types, even though RhoB-GG persists at similar levels in drug-treated cells where RhoB-F has been depleted *(32)*. A precedent for the notion that differential prenylation alters function exists from studies on H-Ras, in which H-Ras-GG has been shown to inhibit and H-Ras-F to potentiate cell proliferation *(48)*. Supporting the likelihood that RhoB-GG is growth-inhibitory, ectopic expression of a solely geranylgeranylated form of RhoB in Ras-transformed cells causes morphological reversion, increased stress fiber formation, and loss of anchorage-independent growth potential *(49)*. These results raise the intriguing possibility that there are two parts to the FTI mechanism, one which is related to loss-of-function of farnesylated proteins such as RhoB-F, and a second related to gain-of-function mediated by the accumulation of RhoB-GG or other farnesylated proteins which can become geranylgeranylated in FTI-treated cells. Because increased levels of RhoB-GG are sufficient to inhibit the growth of Ras-transformed cells, RhoB-GG is a good candidate to mediate the cytostatic effects of FTIs seen in xenograph models and human tumor cell lines *(10–12,50)*. In future work, it will not only be important to identify other important non-Ras targets for inhibition by FTIs, but also to determine whether such targets have gain-of-function activities leading to growth inhibition or apoptosis when geranylgeranylated.

The realization that RhoB-GG levels are elevated following FTI treatment addresses a previous weakness of the FTI-Rho hypothesis. Given that Rho proteins cause stress fiber formation, alteration of Rho function in Ras-transformed cells by FTIs might have been expected to lead to the disappearance rather than the appearance of actin stress fibers. However, since FTIs cause an elevation in cells of RhoB-GG, which has a RhoA-like character. Therefore, one aspect of FTI treatment would be to increase RhoA-like function in cells and therefore to promote actin stress fiber formation, consistent with what is observed in both normal and transformed cells *(9)*. It is tempting to speculate that the loss of RhoB-F and the gain of RhoB-GG associated with stress fiber formation is compatible with anchorage-dependent but not with anchorage-independent growth. Although the reason for this difference is unclear, further investigation of RhoB functions in normal and transformed cells may shed light on the issue.

8. GENE EXPRESSION CHANGES MAY STABILIZE THE FTI-REVERTED PHENOTYPE

An intriguing feature of FTI biology is that drug treatment is required to initiate but not maintain phenotypic reversion of Ras-transformed cells. For example, after a single drug application, H-Ras-transformed cells revert to a normal morphology and proliferative capacity that will persist for up to 7–10 d *(9)*. Cells remain reverted even though FT activity and farnesylated H-Ras return to their initial steady-state levels within a few days. Experiments with the protein synthesis inhibitor cyclohexamide have shown that inhibition of new protein production significantly delays FTI-induced reversion (P. Lebowitz and G. C. Prendergast, unpublished observations). If gene induction is part of reversion, as this finding suggests, then long-lived products required for reversion may lengthen its persistence such that drug maintenance it is unnecessary. Identification of FTI-regulated genes may provide insights into this phenomenon, as well as offer a starting point for bottom-up investigation of target-to-gene regulatory pathways altered by FTI treatment. A role for Rho proteins in such pathways can be considered, based on the ability of RhoA and RhoB to activate serum response factor *(51,52)* (although in the case of RhoB this effect does not require protein prenylation *[52]*), the ability of RhoB to inhibit TGFβ-regulated genes *(53)*, and on the ability of RhoB to associate with and inhibit the activity of the zinc finger transcription factor DB1, perhaps by sequestration *(54)*. However, no matter how they are regulated, it is tempting to speculate that the relevant target genes participate somehow in integrin signaling. In future work, it will be important to determination which proteins are affected, and how they are altered, to gain insight into how the reverted phenotype is initiated as well as stabilized. Such work would also provide a starting point for a bottom-up approach to identify target-to-gene pathways inhibited by FTIs.

9. DIFFERENTIAL CELL ADHESION REQUIREMENTS AND ALTERNATE GERANYLGERANYLATION OF FTASE SUBSTRATES MIGHT UNDERLIE THE LACK OF CYTOSTATIC EFFECTS IN NORMAL CELLS

Given that FT is a housekeeping enzyme and that there are numerous farnesylated cellular proteins that are important to cell function (e.g., lamins *[55]*), it is puzzling that FT can be inhibited without detrimental effects to normal cells. Moreover, Ras is required for the proliferation of normal fibroblasts *(56)* and for the differentiation of certain cells, such as PC12 pheochromocytoma cells *(57)*, but FTIs neither inhibit the growth of normal fibroblasts nor block the ability of Ras to drive PC12 differentiation *(8,9,58)*. Rho-dependent cell adhesion signaling presents a realm to re-examine this question. Because normal cells have normal adhesion properties, one might expect them to be unaffected by drug treatment if FTIs wed proliferation to an appropriate cell-attachment response. Alternately, FTIs may target the action of aberrant or transformation-specific integrin activation events, which are acquired by transformed/tumor cells that can promote survival outside of a physiological adhesion context.

It is also possible that transformed cells are more sensitive than normal cells to FTIs simply because they rely more strongly on farnesyl-dependent functions in some general fashion. In normal cells, for example, protein geranylgeranylation appears to be crucial for cell-cycle progression and activation of certain growth factor receptors, whereas protein farnesylation is dispensable *(59,60)*. GGTase I crossprenylation of normally farnesylated proteins may provide a sort of shunt pathway that serves as a stopgap to cytostasis. K-Ras and RhoB are examples of proteins that are normally farnesylated in cells but that are completely geranylgeranylated following cell treatment with FTIs. Although the growth-related function of some FTase substrates is different when they are alternately geranylgeranylated—such is the case with normal H-Ras *(48)* and RhoB *(32)*—there are others, such as K-Ras (A. Cox, personal communication) where this does not seem to be the case. In the yeast *Saccharomyces cerevisiae*, growth inhibition caused by deletion of the FTase β-subunit can be suppressed by overexpression of the related GGTase I β-subunit *(61)*, implying that there are some farnesylated proteins that function normally with regard to proliferation when they are geranylgeranylated. Lastly, some geranylgeranylated proteins may be able to substitute functionally for loss of their farnesylated relatives. For example, the Ras-related protein TC12/R-Ras2 is normally geranylgeranylated and may be able to compensate for the loss of farnesylated Ras functions in normal cells *(62)*, which depend on these functions for proliferation *(56)*. Thus, the different cell-adhesion requirements for proliferation as well as the presence of crossprenylation shunt pathways may underlie the differential response of normal and transformed cells to FTI treatment.

10. CONCLUSION

Results from clinical trials of FTIs now underway may provide further hints as to how these agents act. If they prove to be sufficiently safe and effective for treatment of human cancer, further cell biological studies will be critical, not only to define the mechanism of action, but also to provide the basis to combat the non-MDR-based drug resistance that has been observed to arise in vitro and in animal models *(10,15,63)*. Even if FTIs prove clinically problematic, from an experimental standpoint they still offer an excellent probe of cell processes that are specific to or extensively required by transformed and malignant cells. Further analysis of farnesylated Rho family and other non-Ras proteins affected by FTIs may help define these transformation-specific mechanisms, the definition of which truly represents a "holy grail" of research in cancer biology.

REFERENCES

1. Gibbs J, Oliff A, Kohl NE. Farnesyltransferase inhibitors: Ras research yields a potential cancer therapeutic. *Cell* 1994; 77:175–178.
2. Leonard DM. Ras farnesyltransferase: a new therapeutic target. *J Med Chem* 1997; 40:2971–2990.
3. Sebti SM, Hamilton AD. Inhibition of Ras prenylation: a novel approach to cancer chemotherapy. *Pharmacol Ther* 1997; 74:103–114.
4. Gibbs JB, Oliff A. The potential of farnesyltransferase inhibitors as cancer chemotherapeutics. *Ann Rev Pharm Tox* 1997; 37:143–166.
5. Garcia AM, Rowell C, Ackermann K, Kowalczyk JJ, Lewis MD. Peptidomimetic inhibitors of Ras farnesylation and function in whole cells. *J Biol Chem* 1993; 268:18,415–18,418.
6. James GL, Goldstein JL, Brown MS, Rawson TE, Somers TC, McDowell RS, et al. Benzodiazepine peptidomimetics: potent inhibitors of Ras farnesylation in animal cells. *Science* 1993; 260:1937–1942.
7. Kohl NE, Mosser SD, deSolms SJ, Giuliani EA, Pompliano DL, Graham SL, et al. Selective inhibition of *ras*-dependent transformation by a farnesyltransferase inhibitor. *Science* 1993; 260:1934–1937.
8. James GL, Brown MS, Cobb MH, Goldstein JL. Benzodiazepine peptidomimetic BZA-5B interrupts the MAP kinase activation pathway in H-Ras-transformed Rat-1 cells, but not in untransformed cells. *J Biol Chem* 1994; 269:27,705–27,714.
9. Prendergast GC, Davide JP, deSolms SJ, Giuliani E, Graham S, Gibbs JB, et al. Farnesyltransferase inhibition causes morphological reversion of *ras*-transformed cells by a complex mechanism that involves regulation of the actin cytoskeleton. *Mol Cell Biol* 1994; 14:4193–4202.
10. Sepp-Lorenzino L, Ma Z, Rands E, Kohl NE, Gibbs JB, Oliff A, Rosen N. A peptidomimetic inhibitor of farnesyl:protein transferase blocks the anchorage-dependent and -independent growth of human tumor cell lines. *Cancer Res* 1995; 55:5302–5309.
11. Kohl NE, Redner F, Mosser S, Guiliani EA, deSolms SJ, Conner MW, et al. Protein farnesyltransferase inhibitors block the growth of ras-dependent tumors in nude mice. *Proc Natl Acad Sci USA* 1994; 91: 9141–9145.
12. Nagasu T, Yoshimatsu K, Rowell C, Lewis MD, Garcia AM. Inhibition of human tumor xenograft growth by treatment with the farnesyltransferase inhibitor B956. *Cancer Res* 1995; 55:5310–5314.
13. Sun J, Qian Y, Hamilton AD, Sebti SM. Ras CAAX peptidomimetic FTI 276 selectively blocks tumor growth in nude mice of a human lung carcinoma with K-Ras mutation and p53 deletion. *Cancer Res* 1995; 55:4243–4247.
14. Sun J, Qian Y, Hamilton AD, Sebti SM. Both farnesyltransferase and geranylgeranyltransferase I inhibitors are required for inhibition of oncogenic K-Ras prenylation but each alone is sufficient to suppress human tumor growth in nude mouse xenografts. *Oncogene* 1998; 16:1467–1473.
15. Kohl NE, et al. Inhibition of farnesyltransferase induces regression of mammary and salivary carcinomas in *ras* transgenic mice. *Nat Med* 1995; 1:792–797.
16. Mangues R, Corral T, Kohl NE, Symmans WF, Lu S, Malumbres M, et al. Antitumor effect of a farnesyl protein transferase inhibitor in mammary and lymphoid tumors overexpressing N-ras in transgenic mice. *Cancer Res* 1998; 58:1253–1259.

17. Shih TY, Stokes PE, Smythers GW, Dhar R, Oroszlan S. Characterization of the phosphorylation sites and the surrounding amino acid sequences of the p21 transforming proteins coded for by the Harvey and Kirsten strains of murine sarcoma viruses. *J Biol Chem* 1982; 257:11,767–11,773.
18. Gibbs JB, Schaber MD, Schofield TL, Scolnick EM, Sigal IS. *Xenopus* oocyte germinal-vesicle breakdown induced by [Val^{12}]Ras is inhibited by a cytosol-localized Ras mutant. *Proc Natl Acad Sci USA* 1989; 86:6630–6634.
19. Cox AD, Garcia AM, Westwick JK, Kowalczyk JJ, Lewis MD, Brenner DA, Der CJ. The CAAX peptidomimetic compound B581 specifically blocks farnesylated, but not geranylgeranylated or myristylated, oncogenic ras signaling and transformation. *J Biol Chem* 1994; 269:1–4.
20. Lebowitz PF, Davide JP, Prendergast GC. Evidence that farnesyl transferase inhibitors suppress Ras transformation by interfering with Rho activity. *Mol Cell Biol* 1995; 15:6613–6622.
21. James GL, Goldstein JL, Brown MS. Polylysine and CVIM sequences of K-RasB dictate specificity of prenylation and confer resistance to benzodiazepine peptidomimetic in vitro. *J Biol Chem* 1995; 270: 6221–6226.
22. Lerner EC, Zhang TT, Knowles DB, Qian YM, Hamilton AD, Sebti SM. Inhibition of the prenylation of K-Ras, but not H- or N-Ras, is highly resistant to CAAX peptidomimetics and requires both a farnesyltransferase and a geranylgeranyltransferase-I inhibitor in human tumor cell lines. *Oncogene* 1997; 15:1283–1288.
23. Rowell CA, Kowalczyk JJ, Lewis MD, Garcia AM. Direct demonstration of geranylgeranylation and farnesylation of Ki-Ras in vivo. *J Biol Chem* 1997; 272:14,093–14,097.
24. Whyte DB, Kirschmeier P, Hockenberry TN, Nunez-Olivia I, James L, et al. K- and N-ras geranylgeranylated in cells treated with farnesyl protein transferase inhibitors. *J Biol Chem* 1997; 272:14,459–14,464.
25. Cox AD, Der CJ. Farnesyltransferase inhibitors and cancer treatment: targeting simply Ras? *Biochim Biophys Acta* 1997; 1333:F51–F71.
26. Symons M. Rho family GTPases: the cytoskeleton and beyond. *Trends Biochem Sci* 1996; 21:178–181.
27. Tapon N, Hall A. Rho, Rac and Cdc42 GTPases regulate the organization of the actin cytoskeleton. *Curr Opin Cell Biol* 1997; 9:86–92.
28. Van Aelst L, D'Souza-Schorey C. Rho GTPases and signaling networks. *Genes Dev* 1997; 11:2295–2322.
29. Adamson P, Marshall CJ, Hall A, Tilbrook PA. Post-translational modification of p21rho proteins. *J Biol Chem* 1992; 267:20,033–20,038.
30. Jahner D, Hunter T. The stimulation of quiescent rat fibroblasts by v-*src* and v-*fps* oncogenic protein-tyrosine kinases leads to the induction of a subset of immediate early genes. *Oncogene* 1991; 6:1259–1268.
31. Jahner D, Hunter T. The *ras*-related gene *rhoB* is an immediate-early gene inducible by v-Fps, epidermal growth factor, and platelet-derived growth factor in rat fibroblasts. *Mol Cell Biol* 1991; 11:3682–3690.
32. Lebowitz P, Casey PJ, Prendergast GC, Thissen J. Farnesyltransferase inhibitors alter the prenylation and growth-stimulating function of RhoB. *J Biol Chem* 1997; 272:15,591–15,594.
33. Prendergast GC, Khosravi-Far R, Solski P, Kurzawa H, Lebowitz P, Der CJ. Critical role of Rho in cell transformation by oncogenic Ras. *Oncogene* 1995; 10:2289–2296.
34. Bar-Sagi D, Feramisco JR. Induction of membrane ruffling and fluid-phase pinocytosis in quiescent fibroblasts by *ras* proteins. *Science* 1986; 233:1061–1066.
35. Ridley AJ, Hall A. The small GTP-binding protein rho regulates the assembly of focal adhesions and actin stress fibers in response to growth factors. *Cell* 1992; 70:389–399.
36. Foster R, Hu KQ, Lu Y, Nolan KM, Thissen J, Settleman J. Identification of a novel human Rho protein with unusual properties: GTPase deficiency and in vivo farnesylation. *Mol Cell Biol* 1996; 16:2689–2699.
37. Murphy D, Saffrich R, Grummt M, Gournier H, Rybin V, Rubino M, et al. Endosome dynamics regulated by a Rho protein. *Nature* 1996; 384:427–432.
38. Lin H-J, Eviner V, Prendergast GC, White E. Activated H-*ras* rescues E1A-induced apoptosis and cooperates with E1A to overcome p53-dependent growth arrest. *Mol Cell Biol* 1995; 15:4536–4544.
39. Rak J, Mitsuhashi Y, Erdos V, Huang S-N, Filmus J, Kerbel RS. Massive programmed cell death in intestinal epithelial cells induced by three-dimensional growth conditions: suppression by mutant c-H-ras oncogene expression. *J Cell Biol* 1995; 131:1587–1598.
40. Lebowitz PF, Sakamuro D, Prendergast GC. Farnesyltransferase inhibitors induce apoptosis in Ras-transformed cells denied substratum attachment. *Cancer Res* 1997; 57:708–713.
41. Krajewska M, Moss SF, Krajewski S, Song K, Holt PR, Reed JC. Elevated expression of Bcl-x and reduced Bak in primary colorectal adenocarcinomas. *Cancer Res* 1996; 56:2422–2427.
42. Bernhard EJ, Kao G, Cox AD, Sebti SM, Hamilton AD, Muschel RJ, McKenna WG. The farnesyltransferase inhibitor FTI-277 radiosensitizes H-ras-transformed rat embryo fibroblasts. *Cancer Res* 1996; 56:1727–1730.

43. Moasser MM, Sepp-Lorenzino L, Kohl NE, Oliff A, Balog A, Su DS, et al. Farnesyl transferase inhibitors cause enhanced mitotic sensitivity to taxol and epothilones. *Proc Natl Acad Sci USA* 1998; 95: 1369–1374.
44. Larcher F, Robles AK, Duran H, Murillas R, Quintanilla M, Cano A, et al. Upregulation of vascular endothelial growth factor/vascular permeability factor in mouse skin carcinogenesis correlates with malignant progression state and activate H-ras expression levels. *Cancer Res* 1996; 56:5391–5396.
45. Rak J, Mitsuhashi Y, Bayko L, Filmus J, Shirasawa T, Sasazuki T, Kerbel RS. Mutant ras oncogenes upregulate VEGF/vPF expression: implications for induction and inhibition of tumor angiogenesis. *Cancer Res* 1995; 55:4575–4580.
46. Mazure NM, Chen EY, Laderoute KR, Giaccia AJ. Induction of vascular endothelial growth factor by hypoxia is modulated by a phosphatidylinositol 3-kinase/Akt signaling pathway in Ha-ras-transformed cells through a hypoxia inducible factor-1 transcriptional element. *Blood* 1997; 90:3322–3331.
47. Arbiser JL, Moses MA, Fernandez CA, Ghiso N, Cao Y, Klauber N, et al. Oncogenic H-ras stimulates tumor angiogenesis by two distinct pathways. *Proc Natl Acad Sci USA* 1997; 94:861–866.
48. Cox AD, Hisaka MM, Buss JE, Der CJ. Specific isoprenoid modification is required for function of normal, but not oncogenic, Ras function. *Mol Cell Biol* 1992; 12:2606–2615.
49. Du W, Lebowitz P, Prendergast GC. Cell growth inhibition by farnesyltransferase inhibitors is mediated by gain of geranylgeranylated RhoB. *Mol Cell Biol* 1999; 19:1831–1840.
50. Miquel K, Pradines A, Sun J, Qian Y, Hamilton AD, Sebti SM, Favre G. GGTI-298 induces G0-G1 block and apoptosis whereas FTI-277 causes G2-M enrichment in A549 cells. *Cancer Res* 1997; 57: 1846–1850.
51. Hill CS, Wynne J, Treisman R. The Rho family GTPases RhoA, Rac1, and CDC42Hs regulate transcriptional activation by SRF. *Cell* 1995; 81:1159–1170.
52. Lebowitz P, Du W, Prendergast GC. Prenylation of RhoB is required for its cell transforming functions but not its ability to activate SRE-dependent transcription. *J Biol Chem* 1997; 272:16,093–16,096.
53. Engel ME, Datta PK, Moses HL. RhoB is stabilized by transforming growth factor β and antagonizes transcriptional activation. *J Biol Chem* 1998; 273:9921–9926.
54. Lebowitz P, Prendergast GC. Functional interaction between RhoB and the transcription factor DB1. *Cell Adhes Comm* 1998; 6:279–288.
55. Farnsworth CC, Wolda SL, Gelb MH, Glomset JA. Human lamin B contains a farnesylated cysteine residue. *J Biol Chem* 1989; 264:20,422–20,429.
56. Mulcahy LS, Smith MR, Stacey DW. Requirement for ras proto-oncogene function during serum-stimulated growth of NIH3T3 cells. *Nature* 1985; 313:241–243.
57. Hagag N, Halegoua S, Viola M. Inhibition of growth factor-induced differentiation of PC12 cells by microinjection of antibody to ras p21. *Nature* 1986; 319:680–682.
58. Dalton MB, Fantle KS, Bechtold HA, DeMaio L, Evans RM, Krystosek A, Sinensky M. The farnesyl protein transferase inhibitor BZA-5B blocks farnesylation of nuclear lamins and p21ras but does not affect their function or localization. *Cancer Res* 1995; 55:3295–3304.
59. McGuire TF, Qian Y, Vogt A, Hamilton AD, Sebti SM. Platelet-derived growth factor receptor tyrosine phosphorylation requires protein geranylgeranylation but not farnesylation. *J Biol Chem* 1996; 271: 27,402–27,407.
60. Vogt A, Qian Y, McGuire TF, Hamilton AD, Sebti SM. Protein geranylgeranylation, not farnesylation, is required for the G1 to S phase transition in mouse fibroblasts. *Oncogene* 1996; 13:1991–1999;
61. Trueblood CE, Ohya Y, Rine J. Genetic evidence for in vivo cross-specificity of the CaaX-box protein prenyltransferases farnesyltransferase and geranylgeranyltransferase-I in *Saccharomyces cerevisiae*. *Mol Cell Biol* 1993; 13:4260–4275.
62. Graham SM, Vojtek AB, Huff SY, Cox AD, Clark GJ, Cooper JA, Der CJ. TC21 causes transformation by Raf-independent signaling pathways. *Mol Cell Biol* 1996; 16:6132–6140.
63. Prendergast GC, Davide JP, Lebowitz PF, Wechsler-Reya R, Kohl NE. Resistance of a variant ras-transformed cell line to phenotypic reversion by farnesyl transferase inhibitors. *Cancer Res* 1996; 56: 2626–2632.

12 Prenyltransferase Inhibitors as Radiosensitizers

Eric J. Bernhard, PHD, *Ruth J. Muschel,* MD, PHD,
Elizabeth Cohen-Jonathan, MD, PHD,
Gilles Favre, PHD, *Andrew D. Hamilton,* PHD,
Saïd M. Sebti, PHD, *and W. Gillies McKenna,* MD, PHD

CONTENTS

1. INTRODUCTION

Radiation therapy is frequently used in the treatment of a number of different tumors. However, the effectiveness of radiotherapy is limited by the ability of normal tissues adjacent to tumors to tolerate radiation in the doses required to kill or sterilize tumor cells. This limitation is compounded by the presence in tumors of radiation-resistant cells that may arise as a result of environmental factors, such as hypoxic regions in tumors, the expression of growth factors that can reduce radiation sensitivity, or tumor cell intrinsic radiation resistance that may be imparted through the activation of certain oncogenes. Ras oncogenes in particular may contribute to radiation resistance, because they have been shown to increase radiation resistance in many experimental systems, and are mutated in an estimated 30% of all human tumors. Basic fibroblast growth factor (bFGF) has also been implicated in increased radiation resistance and is over-expressed in certain tumors, particularly glioblastomas.

From: *Farnesyltransferase Inhibitors in Cancer Therapy*
Edited by: S. M. Sebti and A. D. Hamilton © Humana Press Inc., Totowa, NJ

The discovery of prenyltransferase inhibitors has provided a potential means of targeting tumor cells with increased radiation resistance owing either to activation of oncogenes or to expression of certain growth factors. By inhibiting the function of prenylated proteins such as Ras that are involved in signal transduction, the use of these compounds may significantly improve the results obtained after radiation therapy for tumors in which growth factor or oncogene signaling contribute to radiation resistance. This chapter focuses on the evidence for radiation resistance mediated by Ras oncogenes and the growth factor bFGF, and describes abrogation of this radiation resistance using prenyltransferase inhibitors. The possible mechanisms mediating the effects of prenyltransferase inhibitors on radiation resistance will also be discussed.

1.2. Evidence for Ras-Mediated Radiation Resistance

One factor known to increase tumor cell resistance to radiation is the presence of activated oncogenes. In vitro studies have demonstrated that the expression of *Ras* oncogenes can increase radioresistance in NIH 3T3 cells *(1–4)*, rat embryo fibroblasts (REF) *(5,6)*, and rhabdomyosarcoma cells *(7)*, as well as human osteosarcoma (HOS) cells *(8,9)* and mammary carcinoma cells *(10)*, although increased radioresistance was not seen in all cell types after *Ras* transfection *(11,12)*. In human fibroblast lines, unlike rodent lines, transfection with H-Ras alone resulted in neither transformation nor radioresistance; however when SV40 T was subsequently introduced into ras expressing cells, a larger increase in radiation resistance was observed in cells expressing H-ras than in cells expressing only the T antigen *(13)*. The human HaCaT keratinocyte line also showed little change in radiosensitivity at higher doses (D_0) after transfection with H-Ras; however, low-dose radioresistance was modestly increased in 5 of 6 clones *(14)*. Taken together, these studies support a role for Ras activation in contributing to radiation resistance in both rodent and human cells, and led us to explore further the role of Ras in radioresistance and the possibility of reducing radiation resistance by targeting Ras for inactivation.

1.3. The Ras Pathway and Radiation Resistance

Multiple signal transduction pathways converge on Ras, many of these originating with growth factor receptor binding to ligand. Several growth factors that signal through Ras pathways including KGF (keratinocyte growth factor) and bFGF have been shown to have a radioprotective effect *(15–19)*. Radiation itself has also been shown to activate growth factor receptors that signal to Ras *(20,21)*. The activation of tyrosine kinase receptors, such as the receptor for bFGF, results in recruitment of Grb2 and SOS (guanine nucleotide exchange factor). SOS then promotes release of guanosine diphosphate (GDP) and binding of guanosine triphosphate (GTP) by Ras, resulting in Ras activation and signaling to multiple downstream pathways *(22)*. The Raf-MAP kinase pathway is one such pathway. Activated Ras directly interacts with Raf-1, localizing it to the plasma membrane, an obligate step in Raf phosphorylation *(23)*. Kasid *(24)* has shown that Raf activation via Ras is stimulated by ionizing radiation. Activated Raf-1 stimulates the MAP Kinase (ERK) pathway, which controls the transcription of immediate early genes, such as fos. Raf also binds cdc25, a protein phosphatase involved in progression of the cell cycle through distinct checkpoints *(25)*. Truncation of the regulatory sequences of Raf can create a constitutively active form of the protein and Kasid and others *(3)* have shown that an activated form of Raf will cause radioresistance when transfected into 3T3 cells.

Kasid *(26)* has also shown that treatment of radioresistant human cell lines with antisense Raf can increase radiosensitivity. Other oncogenes involved in Ras signaling through the Raf-MAP Kinase pathway are also known to confer radiation resistance in some systems. These include, but are not limited to, *mos (3,27)*, *ets*, and *sis (3,27)*. Thus, both Ras and bFGF appear to signal through a common pathway that lies upstream of MAP kinase and that has been implicated in radiation resistance. However, other possible pathways for radiation resistance exist.

The rac-rho pathway is also activated by Ras. Both rac and rho are GTPases related to Ras. Rho B is modified by prenylation, and this modification is necessary for some, but not all of its activities *(28)*. Rac controls the assembly of actin filaments at the plasma membrane resulting morphologically in the formation of membrane ruffling whereas rho regulates organization of the actin cytoskeleton and stress fiber formation *(29)*. It was recently noted that the rac-rho pathway is involved in promoting transcription, and that both the Raf-MAP Kinase and rac-rho pathways are required for entry into S phase *(30)*. Specifically, it was noted that the rho pathway activated Jun Kinase (JNK), an enzyme that phosphorylates and activates jun, an immediate early gene that complexes with fos to form the transcription factor AP-1. ERK and JNK are now known to be activated by different downstream pathways of Ras *(31,32)*. Thus, it is possible that Ras-mediated radiation resistance is signaled either through the Raf-MAP kinase pathway, or through the rac-rho pathway. Alternatively, because both the Raf-MAP Kinase and rac-rho pathways are required for entry into S phase, a third possibility is that Ras induced radiation-resistance also requires the integration of signaling from both of these pathways. One advantage of the prenyltransferase inhibitors in targeting radiation resistant cells is that whichever path downstream of Ras is responsible for increased radiation resistance, blocking Ras activity should lead to a decrease in signaling through that pathway in cells expressing activated Ras, and thus reduce radiation resistance.

1.4. Inhibition of Ras Activity and Its Effect on Radiation Resistance

An initial approach to abrogating the effect of Ras on radiation resistance was carried out by Miller et al. *(8)* using the drug lovastatin. Lovastatin, a fungal compound, inhibits HMG-CoA reductase and mevalonate biosynthesis *(33)*. The mevalonate pathway leads to the synthesis of the polyisoprenoid intermediates necessary to the farnesylation or geranylgeranylation of such proteins as Ras. Ras activation requires not only the presence of the activating mutation resulting in constitutive GTP binding but also insertion into the inner cell membrane, which occurs as a result of protein prenylation *(34)*. Thus, inhibitors of the rate limiting step of that pathway, the conversion of HMG-CoA into mevalonate by HMG-CoA reductase will inhibit isoprenylation of Ras and other proteins. However, inhibitors of HMG-CoA reductase will also inhibit synthesis of other molecules, such as cholesterol, that are synthesized from isoprenoids. Miller et al. *(8)* demonstrated that H-Ras when introduced into HOS cells induced increased radioresistance and that treatment of the cells with lovastatin resulted in reversion of survival to that of the parental cells. In this system, the degree of radioresistance was correlated with the levels of isoprenylated Ras. Thus, Miller suggested that lovastatin could reverse the effect of Ras on radioresistance. The drug lovastatin is, however, highly non-specific in its effects on Ras and interferes with numerous biochemical pathways involved in cholesterol, steroid, and lipid metabolism within the cell, thus introducing the possibility of many confounding variables.

It is now possible to take a much more specific approach to the same problem using inhibitors of prenylation that are not only more specific for Ras, but do not interfere with other aspects of cholesterol and lipid metabolism *(35)*. Prenylation of Ras is carried out by farnesyl- or geranylgeranyltransferase enzymes, which recognize four amino-acid sequences (known as CAAX boxes) at the carboxyl terminal end of Ras. CAAX box sequences have been used as the basis for the design of several prenyltransferase inhibitors with specificity for either the farnesyl- or the geranylgeranyltransferase enzymes, as discussed elsewhere in this book. We have used these inhibitors to treat cells expressing activated Ras oncogenes prior to irradiation and shown that this results in decreased radiation survival. In these cells, the presumptive target of inhibition is the prenylation of activated Ras. However, as we will discuss below, the possibility exists that other pathways implicated in radiation resistance may also be blocked by prenyltransferase inhibitors.

2. STUDIES OF PRENYLTRANSFERASE EFFECTS ON RAS-TRANSFORMED REF RADIOSENSITIVITY

2.1. Farnesyltransferase (FTase) Inhibitors Effectively Inhibit Prenylation of H-Ras

The role of the H-*Ras* oncogene in radiation resistance was initially studied using early passage REF transformed with either H-*Ras* alone, v-, or c-*myc* alone, or v-*myc* together with H-*Ras*. Using this model, it was established that transfection with H-*Ras* resulted in increased radioresistance, whereas transfection with v- or c-*myc* alone did not *(5,36)* (Fig. 1). The finding that activated H-*Ras* oncogene expression in REF cells caused an increase in radiation resistance led to the hypothesis that inhibiting H-Ras activity in these cells should reduce their radiation resistance if Ras activation was the necessary factor for the observed increase in radioresistance.

To test this hypothesis, REF cells were treated with the farnesyltransferase (FTase) inhibitor FTI-277. The inhibition of H-Ras farnesylation was monitored by Western-blot analysis of Ras protein migration on sodium dodecyl sulfate (SDS)-polyacrylamide gels. Prenylated forms of Ras have been shown to migrate more rapidly than unprenylated Ras on SDS-polyacrylamide gels *(37,38)*. Treatment of H-ras^{V12} oncogene transformed REF cells (3.7, 4R, and 5R) with various doses of FTI-277 (2.5–10 μM) for 24 h resulted in the majority of the H-Ras being unprocessed (Fig. 2). However, cells expressing wild-type c-H-Ras were less susceptible to the effects of this inhibitor. Treatment of untransformed REF resulted in detectable H-Ras in the slower moving, unprocessed form, although the majority of the H-Ras remained in the farnesylated form in spite of the treatment with FTI-277 at doses up to 30 μM. MR4 cells (REF immortalized by v-myc) appear to express very low levels of H-Ras (we could only detect Ras protein in these cells using a pan-Ras antibody) and little change in the migration of Ras was seen with treatment up to 10 μM FTI-277 after 24 h.

The targeting of prenyltransferases to a particular Ras protein is in large part dictated by the CAAX recognition sequence found at the carboxyl terminal portion of Ras and other prenylated proteins *(39)*. The cysteine within the carboxyl terminal end of H-Ras, CVLS, is the target for prenylation by FTase. In contrast, CAAX sequences terminating in leucine have a greatly reduced affinity for farnesyltransferase and are geranylgeranylated by geranylgeranyltransferase I (GGTase I) during processing *(40)*. A chimeric H-ras^{V12} with CVLL as the CAAX motif is fully transforming in NIH 3T3 cells and Rat-1

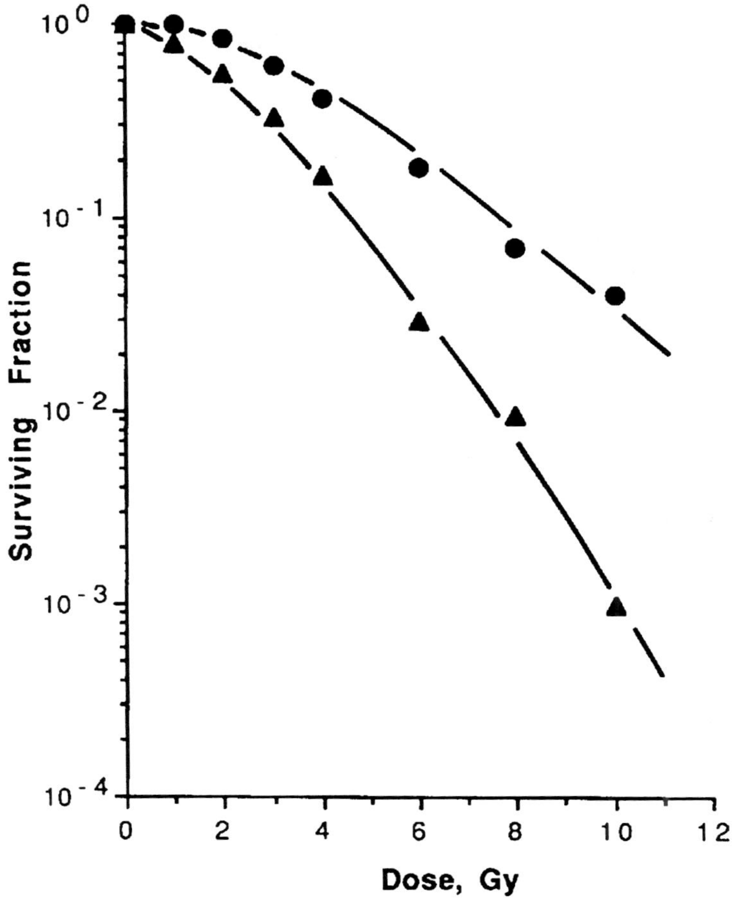

Fig. 1. Survival of REF cells transformed with H-Ras plus myc or immortalized with myc alone. Log-phase cultures of rat embryo fibroblasts transformed with H-ras^{V12} plus v-myc (●), or immortalized with c-myc (▲) were irradiated and plated for clonogenic survival. Cultures were stained and scored for the formation of colonies after 2 wk. Surviving fraction, plotted on the Y-axis is defined as: Number of colonies formed / (Number of cells plated) • (Plating efficiency).

cells. This altered H-Ras also transformed primary REF in co-transfection with v-myc, but at a lower efficiency than the farnesylated form *(41)*. Compared to 3.7 or 5R, cells transformed with H-ras^{V12} CVLL (REF-GG) adhere poorly to tissue culture dishes and do not form discrete colonies. These cells serve as useful controls in our experiments since the H-ras^{V12} CVLL should be impervious to the effects of FTI-277, but should be sensitive to GGTase I inhibition of geranylgeranyltransferase. When REF-GG were treated with up to 10 μ*M* FTI-277, no change in mobility of H-Ras-CVLL was observed (Fig. 2), thus demonstrating the specificity of the FTI-277 inhibitor for FTase over geranylgeranyltransferase. In contrast, the geranylgeranyltransferase inhibitor GGTI-298 *(42,43)* demonstrated effective inhibition of the geranylgeranylation of the chimeric H-ras^{V12} with the CVLL recognition sequence for geranylgeranyltransferase at inhibitor doses as low as 4 μ*M*, whereas H-ras^{V12} expressed by the 5R cell line that has the FTase recognition sequence CVLS, and is thus farnesylated, was only partially inhibited at a dose of 32 μ*M* (Fig. 3).

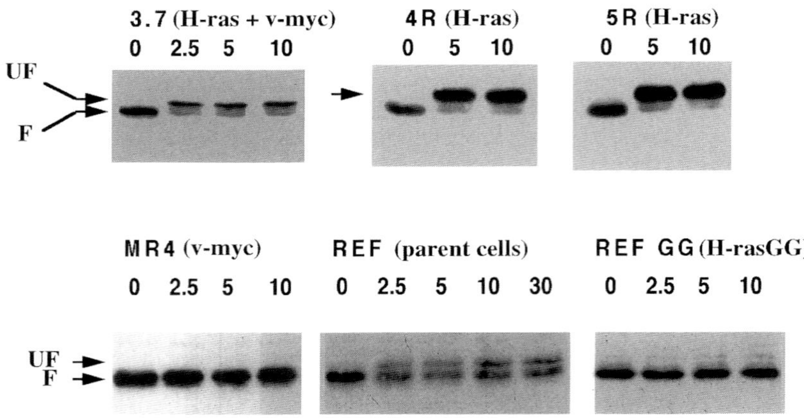

Fig. 2. Effects on Ras farnesylation of 24-h treatment with FTI-277. Cells in log-phase culture were treated with the indicated dose of FTI-277 (μ*M*). Cell lysates for the indicated cell types were obtained after 24 h and analyzed by Western blotting. H-Ras specific antibody was used for all blots except MR4 where pan Ras-specific antibody results are shown owing to very low levels of H-Ras expression. Farnesylated (F) and unfarnesylated (UF) Ras are indicated.

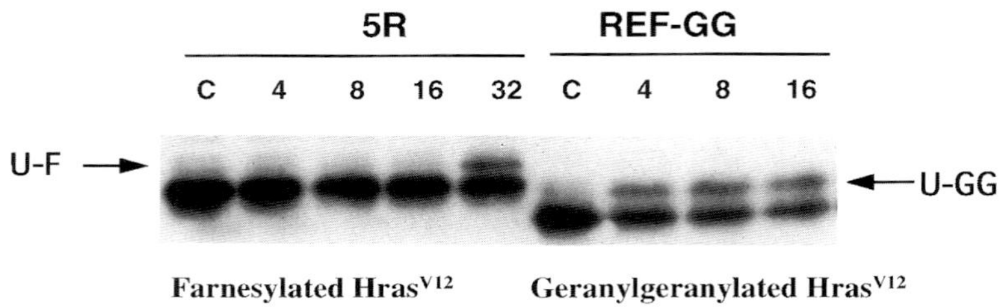

Fig. 3. Effects on Ras geranylgeranylation of 24-h treatment with GGTI-298. Cells in log phase culture were treated with the indicated dose of GGTI-298 (μ*M*) and Ras status monitored as above. 5R cells are transformed with H-ras^{V12}. REF-GG are transformed with an H-ras^{V12} chimeric molecule that is prenylated by geranylgeranyltransferase. Unfarnesylated Ras expressed in 5R cells (U-F) and ungeranylgeranylated Ras expressed in REF-GG cells (U-GG) after treatment with GGTI-298 are indicated.

2.2. Effect of FTI-277 and GGTI-298 Treatment on Apoptosis After Irradiation

We have previously shown that exposure of REF cells immortalized by v-myc to ionizing radiation results in high levels of apoptosis, whereas cells transformed by H-Ras plus v-myc have substantially lower levels of apoptosis. This demonstrated that transformed cells expressing activated H-Ras, which were more resistant to radiation killing, were also more resistant to the induction of apoptosis by radiation than cells that lacked H-Ras expression. It further implied that loss of H-Ras activity in these cells would lead to increased radiation-induced apoptosis and decreased clonogenic survival. As a first test of this prediction, the effect of inhibiting H-Ras farnesylation on apoptosis after irradiation was examined. Cells were irradiated with 10 Gy and concurrently treated with various concentrations of FTI-277. The extent of apoptosis induced by the irradiation of 3.7 cells, which express activated H-Ras and v-myc, was greatly enhanced by treatment with FTI-277 (Fig. 4). The maximum effect was seen at 5 μ*M* with a significant increase also seen at 2.5 μ*M*. Thus, at doses of FTI that inhibit H-ras^{V12} farnesylation, the level

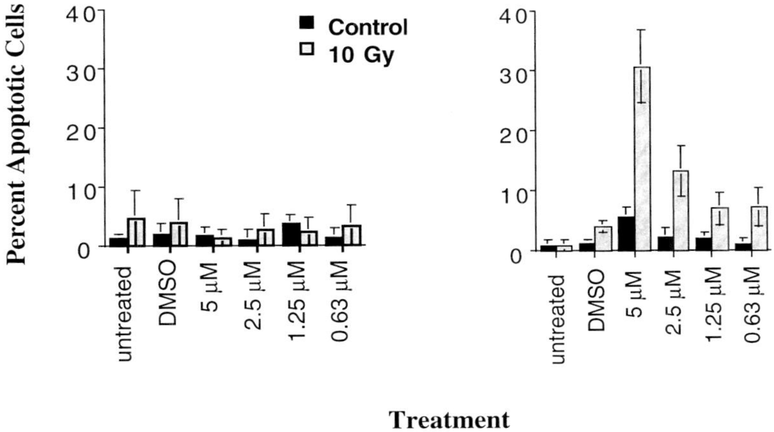

Fig. 4. Apoptosis after irradiation and FTI-277 treatment. Early passage rat embryo fibroblast cells (REF) or myc + Ras transformed REF cells (3.7) were treated with the indicated concentration of FTI-277 at the time of irradiation with 10 Gy. Apoptosis was quantitated 24 h later by scoring for changes in nuclear morphology after staining with propidium iodide.

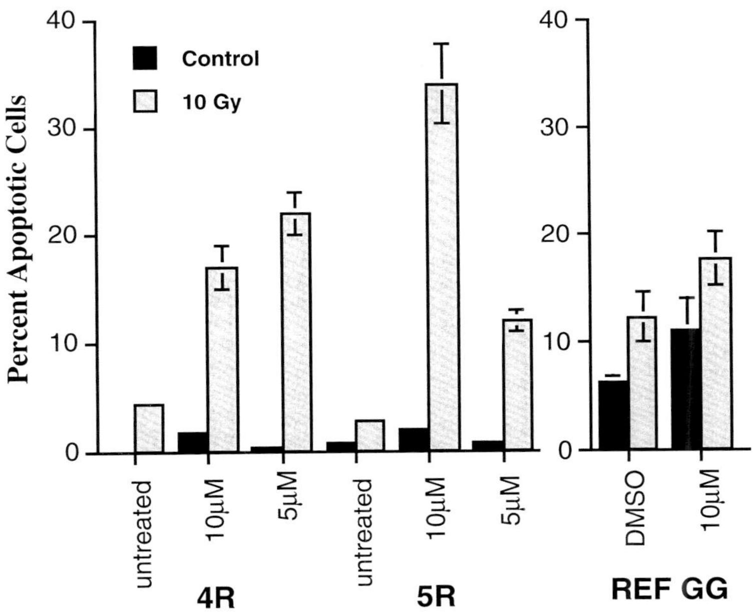

Fig. 5. FTI-277-mediated enhancement in radiation-induced apoptosis is specific for cells expressing farnesylated H-Ras. Cells were treated with FTI-277 at the time of irradiation and assessed 24 h later for apoptosis, as shown in Fig. 4. 4R and 5R cells are transformed with H-ras^{V12}. REF-GG are transformed with an H-ras^{V12} chimeric molecule that is prenylated by geranylgeranyltransferase.

of apoptosis after irradiation is increased. This increase is specific for cells expressing activated H-Ras, as FTI-277 treatment of REF cells, which are untransformed, caused no increase in apoptosis after irradiation.

The ability of FTI-277 to augment irradiation-induced apoptosis in cells transformed by the H-Ras oncogene was confirmed in 4R and 5R cells (Fig. 5). These are two independent clones of REF cells transformed by H-ras^{V12} alone. These results demonstrate

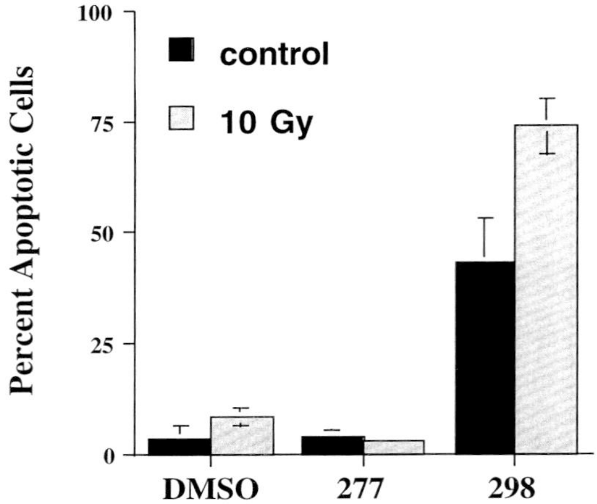

Fig. 6. Apoptosis in REF cells expressing geranylgeranylated Ras is only induced by a geranylgeranyltransferase inhibitor. REF-GG cells were irradiated in the presence of 5 μ*M* FTI-277 or 8 μ*M* GGTI-298 and scored for the presence of apoptotic cells 24 h after irradiation, as shown in Fig. 4.

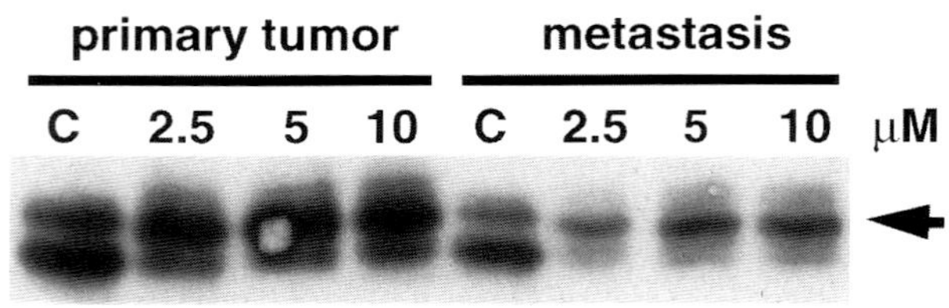

Fig. 7. Inhibition of H-Ras farnesylation in H-Ras transformed murine prostate cells by FTI-277. Mouse prostate tumor cells transformed by retroviral transduction with the H-ras^{V12} + myc oncogenes were treated with the indicated doses (μ*M*) of FTI-277. After 24 h samples were harvested for Western-blot analysis with anti-H-Ras antibody. The upper band (arrow) corresponds to unfarnesylated H-Ras. C, controls.

that the presence of an activated H-Ras alone is sufficient to cause increased radiation-induced apoptosis after FTI-277 treatment.

A further control for the specificity of apoptosis induction after irradiation was obtained using the REF-GG cell line. Because the H-ras^{V12} expressed by these cells was not affected by FTI-277 treatment, the level of apoptosis after irradiation should not be increased after FTI-277 treatment. These cells had a relatively high baseline level of apoptosis of about 6% (Fig. 5). This was increased by irradiation to 12%. Treatment of these cells with FTI-277 slightly increased the baseline level of apoptosis, but had no significant effect on enhancing the extent of apoptosis after irradiation. In contrast, treatment of REF-GG cells with the geranylgeranyltransferase inhibitor GGTI-298 significantly increased both the basal level of apoptosis in these cells and the apoptosis observed after irradiation (Fig. 6). Thus, the increase in apoptosis seen after irradiation and prenyltransferase inhibitor treatment appears to correlate with inhibition of oncogenic H-Ras prenylation.

We have extended the observations obtained in the REF model system to mouse prostate tumor cells derived by transduction of H-Ras and v-myc into mouse embryo urogenital sinus cells *(44,45)*. Treatment of either primary tumor cells or a metastatic clone of this tumor line showed a dose-dependent reduction of the farnesylated form and the accumulation of the more slowly migrating, unprocessed form of H-Ras (Fig. 7). Thus, FTI-277 is an effective inhibitor of Ras farnesylation in transformed prostatic epithelial cells.

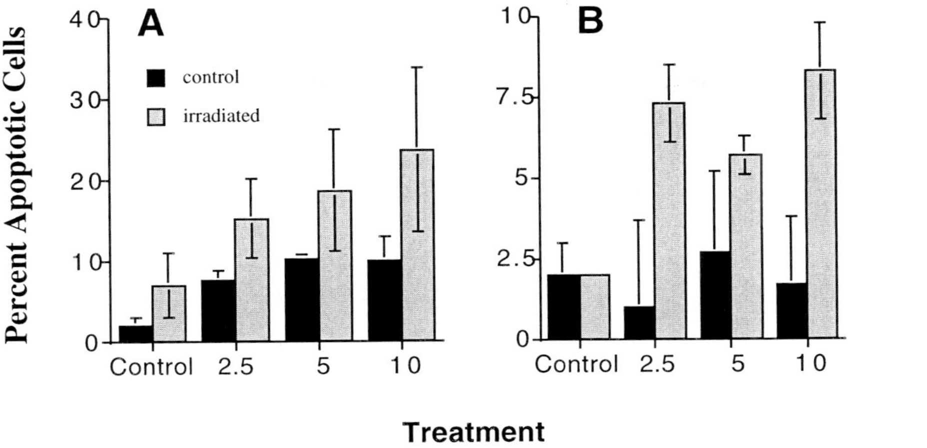

Fig. 8. Increased radiation-induced apoptosis in Ras transformed mouse prostate tumor cells. Mouse prostate tumor cells transformed by retroviral transduction with the H-ras^{V12} + myc oncogenes were treated with the indicated concentrations of FTI-277 (μ*M*) and irradiated 24 h later with 10 Gy. Apoptosis was quantitated 24 h after irradiation by scoring for changes in nuclear morphology visualized by epifluorescent microscopy after staining of cells with propidium iodide. (**A**) Prostate tumor cells cultured from a primary tumor. (**B**) Prostate tumor cells cultured from an isolated metastasis derived from the tumor shown in (**A**).

When these cells were examined for radiation-induced apoptosis after treatment with FTI-277 at the doses shown to inhibit Ras farnesylation, a significant increase in apoptosis was seen (Fig. 8). Thus, inhibition of farnesylation in prostate cells as well as fibroblasts resulted in increased radiation-induced apoptosis.

2.3. Decreased Radiation Survival of Ras-Transformed Cells After FTI-277 Treatment

We have previously shown that H-Ras-transformed REF are significantly more radioresistant than REF and that REF immortalized by myc are not altered in radiation survival compared to the parental REF. Therefore, inhibition of H-Ras activity using a farnesylation inhibitor might be expected to reduce radiation resistance in cells with oncogenic H-Ras. Cells were irradiated with the indicated doses of ionizing radiation and treated with FTI-277 for 24 h after irradiation (Fig. 9). The survival curves for 3.7 and 5R showed them to be more resistant to radiation than MR4 or REF, with MR4 being slightly more resistant than REF. After treatment with FTI-277, the radioresistance of 3.7 and 5R were reduced to a level of survival similar to that seen in MR4, the myc immortalized REF or of REF themselves. The shoulders of the survival curves were reduced as were the overall slopes. Exposure of the cell line MR4 or REF to FTI-277 had no effect on radiation survival. These results indicate that FTI-277 can act as a specific radiosensitizer of cells expressing an activated H-Ras oncogene, but that the inhibitor has no effect on other cells.

Radiosensitization of murine prostate tumor cells by FTI-277 treatment was also observed. Survival after 2 Gy irradiation of H-Ras plus v-myc transformed mouse prostate tumor cells was reduced from 0.85 to 0.36 when cells were treated with 5 μ*M* FTI-277 prior to irradiation. This demonstrates that radiosensitization can be obtained not only in

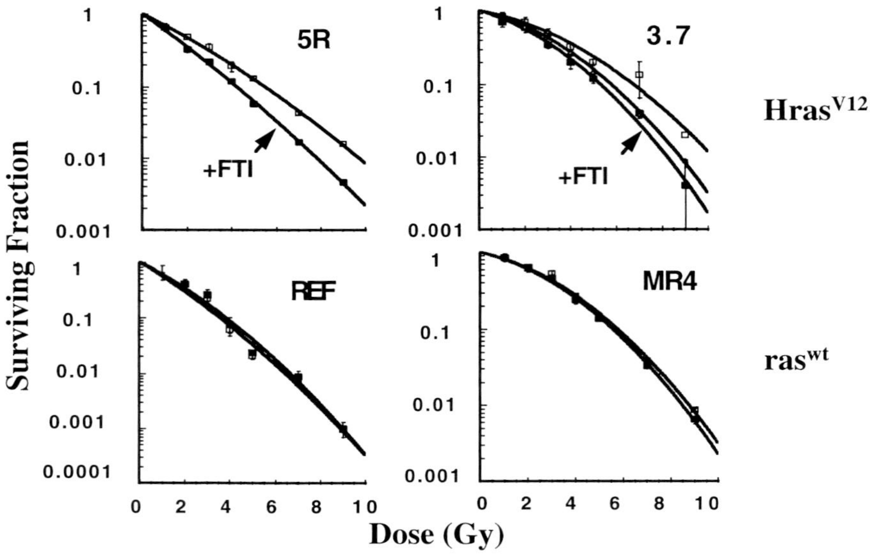

Fig. 9. Clonogenic survival after treatment with farnesyltransferase inhibitor. Immediately prior to irradiation FTI-277 was added at concentrations of 2.5 μ*M* (3.7 cells) or 5 μ*M* (5R, MR4, and REF). The inhibitor was diluted out 24 h later resulting in a final concentration of 1 μ*M* (3.7 cells) or 2 μ*M* (5R, MR4, and REF). Plating efficiency for MR4 and 5R were unaffected by FTI-277 and were 100% and 32–38%, respectively. FTI-277 reduced plating efficiency in 3.7 and REF by 50% from values of 75% and 5%, respectively. Thus the results seen here cannot be accounted for by toxicity of the drug. The data points shown represent the mean of at least three dishes. The open symbols (□, △) indicate the results from untreated cells and the closed symbols (■) indicate cells treated with FTI-277. In the panel 3.7, the open triangles (△) show the results from untreated MR4 cells.

sarcomas, which are of mesenchymal origin, such as the fibroblast derived 3.7 and 5R tumor cells, but in tumors of endothelial origin such as prostate tumors. Thus, in two rodent cell models, one endothelial and the other of mesenchymal origin, we have demonstrated radiosensitization after exposure of Ras-transformed cells to the FTase inhibitor FTI-277. These results led to further studies attempting to address the question of the role of Ras in the radiosensitivity of human cells expressing activated Ras alleles.

3. RADIOSENSITIZATION OF HUMAN TUMORS EXPRESSING ACTIVATED RAS ONCOGENES

In order to extend our findings to human cells, we examined a panel of human tumor lines with naturally occurring Ras mutations. The T24 bladder carcinoma line and the HS578T breast cancer cell line expressing H-ras^{V12} were used for studies on the effects of H-Ras activation and its inhibition. Twenty-four hours after adding increasing amounts of FTI-277, the percentage of unfarnesylated Ras was determined (Fig. 10). Increasing inhibition of farnesylation was seen with 95% inhibition achieved in T24 cells at 10 μ*M* FTI-277. Significant inhibition of H-Ras farnesylation was also seen in the HS578T breast line expressing H-ras^{V12} and three cell lines expressing wild-type Ras at 5 μ*M*.

Once the conditions for inhibiting H-Ras prenylation had been established, we examined the effects of inhibiting H-Ras farnesylation on the radiation survival. Two methods were used to determine survival: 1) standard clonogenic assays over a range of radiation doses, and 2) limiting dilution analysis of cell clonogenicity after 2 Gy irradiation. We

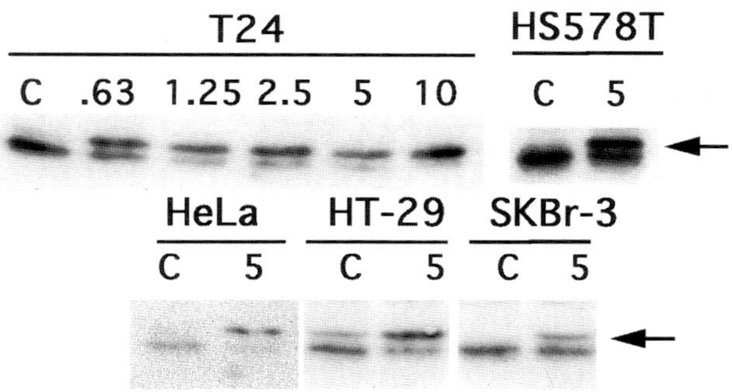

Fig. 10. Inhibition of H-Ras farnesylation in human tumor cells by FTI-277. Cells with mutated (T24 and HS578T) or wild-type H-Ras (HeLa, HT-29, and SKBr-3) were treated with the indicated concentration (μ*M*) of FTI-277 for 24 h or with an equal volume of 10 m*M* dithiothreitol (DTT) in DMSO carrier (C, control). Cell lysates were then obtained for Western-blot analysis. Blots were probed with monoclonal antibody (MAb) to H-Ras. Arrows indicate migration of unfarnesylated H-Ras.

used both assays because in some cell lines pretreatment with FTI-277 inhibited cell attachment after plating of cells. In limiting dilution analysis, the effect of inhibitor treatment was determined by comparing the frequency of colonies arising in microtiter wells inoculated with varying cell numbers in the range calculated to yield an average of one cell per well. In contrast to clonogenic assays carried out by plating cells in culture dishes and scoring for colony formation, the limiting dilution analysis of cell clonogenicity is not influenced by loose attachment of cells immediately after plating, because individual microtiter wells are scored simply as positive or negative for clonogens, not for colony number *(46–48)*. Thus, loose adherence at plating and secondary colony formation could not influence this measurement. The data are presented as the natural log of negative wells vs the number of cells plated. The effect of treatment on clonogenic survival was determined by obtaining the linear regression for the data from each treatment group and comparing the slopes of the resulting lines. Presented in this way, the steeper the slope of the linear regression, the greater the clonogenic survival.

Both methods of determining radiation survival were used to determine the effects of treating T24 cells with 5 μ*M* FTI-277 for 24 h prior to irradiation and 24 h after irradiation. Five μ*M* FTI-277 was the dose of inhibitor chosen as inhibition of farnesylation was seen in all cells at this dose, and the treatment was not significantly cytotoxic (not shown). The clonogenic survival of T24 cells after irradiation from 1 to 4 Gy in the presence of inhibitor was reduced at all radiation doses (Fig. 11A). The surviving fraction after 2 Gy (SF_2) measured by clonogenic survival was reduced from 0.68 to 0.45 by treatment with FTI-277. By extrapolating from the limiting dilution analysis (Fig. 11B), the SF_2 was reduced from 0.86 to 0.5. Thus, both methods detected an equivalent reduction in surviving fraction of T24 cells after FTI-277 treatment.

The effect of inhibiting farnesylation was also tested on HS578T, a breast tumor cell line with an H-Ras^{V12} mutation. FTI-277 reduced the survival of HS578T after 2 Gy from 0.79 to 0.63 when radiation survival was adjusted for the cytotoxicity of the inhibitor treatment.

In order to ensure that the radiosensitization seen after FTI-277 treatment was a property common to FTase inhibitors, and not a unique characteristic of FTI-277, we determined the effects of treating T24 cells with another FTase inhibitor, L744,832 (Merck Pharmaceuticals). This inhibitor was independently designed and synthesized and has

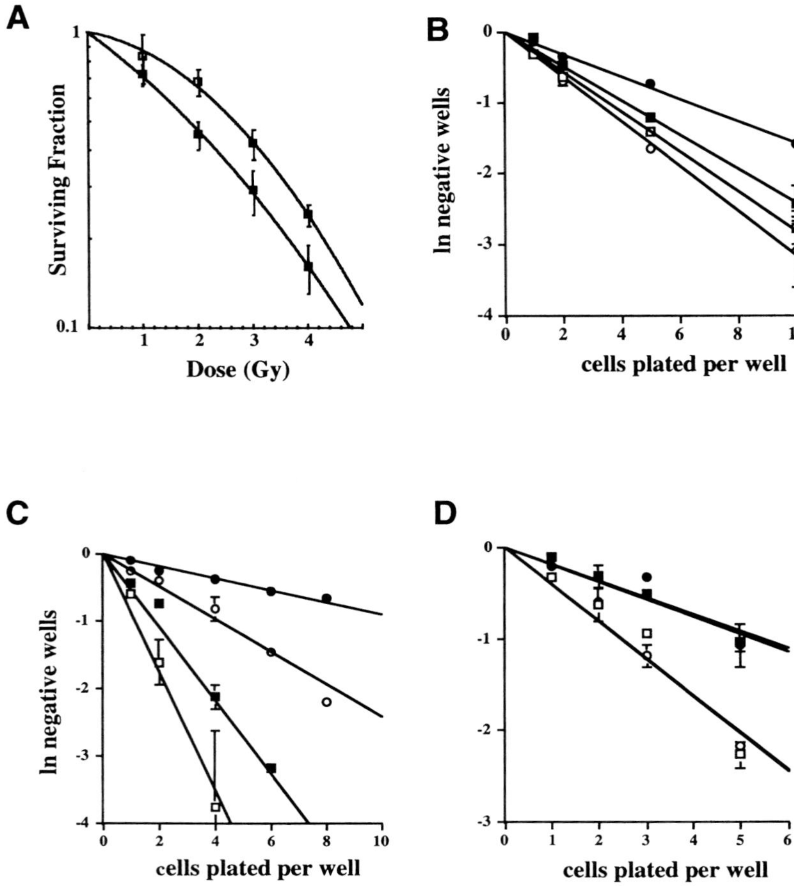

Fig. 11. Radiosensitization of human tumor cells expressing activated H-Ras. (**A**) T24 cells were pretreated with 5 μ*M*FTI-277 (■) or with an equal volume of 10 m*M* DTT in DMSO carrier (□) prior to plating for clonogenic survival determination. Dishes were then irradiated with 1–4 Gy as indicated and cultured an additional 24 h. The culture medium was then removed, and cultures were then refed with drug-free medium and allowed to grow for an additional 8 d (controls) or 2 wk (FTI-277 treated) prior to staining for colony formation. The points shown are the mean and standard deviations obtained from at least three plates. The plating efficiency of unirradiated cultures treated with inhibitor was 0.77; the plating efficiency of control cells was 0.85.

T24 cells (**B**,**C**) or SKBr-3 cells (**D**) were plated at the indicated cell number per well in 96-well microtiter plates after 24 h pretreatment with inhibitor (○,●) or with an equal concentration of DTT/DMSO carrier (□,■). Dishes were then irradiated with 2 Gy (solid symbols) and cultured an additional 24 h. All cultures were then re-fed with drug-free medium to obtain a 10-fold dilution of the inhibitor and allowed to grow for 3 wk. Inhibitor treatment was 5 μ*M* FTI-277 in (**B**) and (**D**), and 5 μ*M* L744,832 in (**C**). The points shown are the mean and standard deviations obtained for duplicate plates. Linear regression in all instances had an r^2 value above 0.95.

significant structural and pharmacological differences from FTI-277. L744,832 also effectively blocked H-Ras farnesylation within 24 h of treatment (Fig. 12). L744,832 by itself reduced colony formation by 73%. It also caused a significant reduction in clonogenic survival after 2 Gy irradiation in T24 cells (Fig. 11C). The surviving fraction of irradiated cells treated with this inhibitor was nearly half that of cells irradiated without inhibitor treatment (0.37 vs 0.62) when corrected for decreased clonogenicity. The cumu-

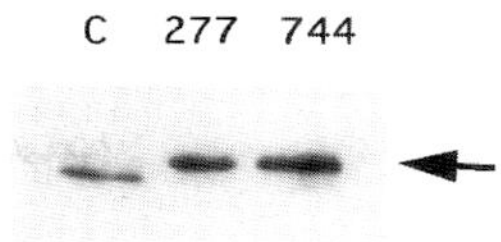

Fig. 12. Inhibition of H-Ras farnesylation by L744,832. T24 cells were treated with 5 μ*M* FTI-277 (277), 2 μ*M* L744,832 (744), or DTT/DMSO carrier (C, Control) for 24 h prior to lysis for protein analysis by Western blotting. Blots were probed with MAb to H-Ras. Arrow indicates migration of unfarnesylated H-Ras.

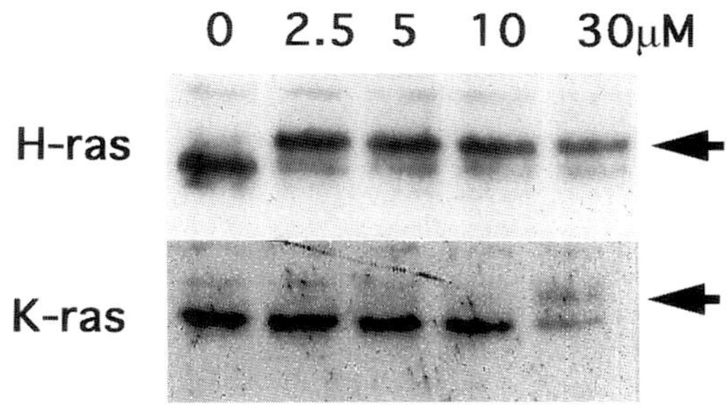

Fig. 13. Inhibition of K-Ras prenylation by FTI-277. SW480 cells were treated with the indicated concentrations of FTI-277 (μ*M*) for 48 h. Cell lysates were analyzed by Western blotting with either an H-Ras MAb (top) or a K-Ras MAb (bottom). Arrow indicated unfarnesylated Ras bands.

lative effect was a reduction in clonogenic survival in cells receiving both inhibitor and radiation to 10% of untreated control cells.

To assess whether FTI-277 radiosensitization was specific to cells with activated H-Ras, we examined survival of cells expressing wild-type Ras after FTI-277 treatment. The HT-29 colon carcinoma and SKBr-3 breast cancer cell lines were assessed for survival after 2 Gy irradiation in the presence of 5 μ*M* FTI-277, a dose of inhibitor that was documented to inhibit wild-type H-Ras prenylation in these cells (Fig. 10). The limiting dilution analysis is shown in Fig. 11D for SKBr-3. The SF_2 obtained in this analysis was 0.47 for controls and 0.45 in FTI-277-treated cells, showing that inhibition of farnesylation had no effect on radiation survival in these cells. FTI-277 treatment alone showed some toxicity to HT-29 cells reducing clonogenicity by 35%; however, no significant change in the SF_2 of HT-29 cells was seen after correcting for the toxicity of the drug ($SF_2 = 0.78$ in controls and 0.82 in treated cells). Thus, FTI-277 treatment of cells under conditions that inhibited wild-type H-Ras prenylation did not increase radiation-induced cell death in cells that do not have activated H-*Ras* oncogenes. These findings are consistent with a correlation between radiation sensitization by FTI-277 and the presence of activated H-Ras oncogenes.

Because the effect of FTI-277 treatment is largely specific for H-Ras over K-Ras, we next asked whether K-Ras prenylation could be blocked by FTI-277. As shown in Fig. 13, the SW480 colon carcinoma cell line expressing H-Ras and activated K-Ras showed altered migration of H-Ras with as little as 2.5 μ*M* FTI-277 treatment, whereas altered migration of K-Ras became evident only at 30 μ*M* FTI-277. At this dose, FTI-277 inhibits both farnesylation and geranylgeranylation *(38)*. Thus, although FTI-277 specifically inhibits farnesylation of H-Ras and K-Ras remains prenylated at doses of FTI-277 below 30 μ*M*, at 30 μ*M* some inhibition of K-Ras prenylation was seen. We then used the dose of 30 μ*M* FTI-277 to determine whether SW480 cells could be radiosensitized. The results

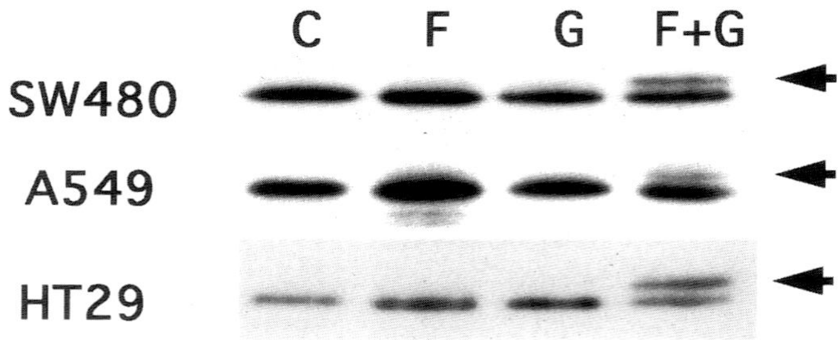

Fig. 14. Inhibition of K-Ras prenylation by combined FTI + GGTI treatment. Log phase cultures of human colon (SW480, HT29) or lung (A549) carcinomas were treated with 5 μ*M* FTI-277, 8 μ*M* GGTI-298, or with both 5 μ*M* FTI-277 plus 8 μ*M* GGTI-298 for 48 h after which cell lysates were obtained and analyzed by Western blotting with MAbs to K-Ras. Control (C) cultures were treated with carrier alone.

of clonogenic survival assays demonstrated the possibility of radiosensitizing human tumor cells expressing activated K-Ras using FTI-277 alone, although the degree of radiosensitization was modest (not shown).

Because the dose of FTI-277 required to inhibit K-Ras prenylation was a dose where inhibition of GGTase I would be expected, we investigated the possibility that combined FTI plus GGTI treatment could be used as a more effective means of inhibiting K-Ras prenylation and increasing radiation sensitivity. Our findings demonstrate that combining FTI-277 with GGTI-298 results in increased inhibition of K-Ras prenylation in cells with either mutant (SW480 and A549) or wild-type K-Ras (HT-29) (Fig. 14).

Because combined prenyltransferase inhibitor treatment was effective in inhibiting K-Ras prenylation, this combination of inhibitors was used in clonogenic survival assays to test the effect of this treatment on radiation survival in cells with either mutant or wild-type K-Ras (Fig. 15). The radiation survival of the SW480 cell line expressing mutant K-Ras was reduced by the treatment with combined FTI-277 + GGTI-298. In contrast, the radiation survival of the HT-29 cell line expressing wild-type Ras was not decreased. HeLa cell radiation survival was similarly unaffected by inhibitor treatment.

This observation was extended to the A549 lung carcinoma cell line expressing activating mutations in K-Ras. As shown in Fig. 16, A549 cells also demonstrated significant radiosensitization at 2 Gy after FTI + GGTI treatment. The SF_2 derived from the limiting dilution analysis showed a reduction from 0.53 for controls to 0.15 for inhibitor-treated cells.

These findings demonstrate that combined treatment with FTI + GGTI acts synergistically to inhibit prenylation of K-Ras and that this treatment is also effective in radiosensitizing human tumor cells expressing an activated *K-Ras* oncogene product. In contrast to the increase in radiation-induced cell death caused by prenyltransferase inhibitors in cells with activated Ras, the radiation survival of tumor cell lines that do not express activated Ras was not altered after prenyltransferase inhibitor treatment. The HT29 colon carcinoma (Fig. 14), the SKBr-3 breast carcinoma cells and HeLa cervical carcinoma cells that express wild-type Ras demonstrated no significant decrease in radiation survival after treatment with FTI-277 plus GGTI-298 at doses shown to inhibit K-Ras prenylation (Fig. 14). The inhibitors themselves did, however, reduce clonogenicity in these cells to a variable extent that showed no correlation with Ras status. Thus, radiosensitization again correlated with the expression of oncogenic Ras, whereas inhibition of clonogenicity did not.

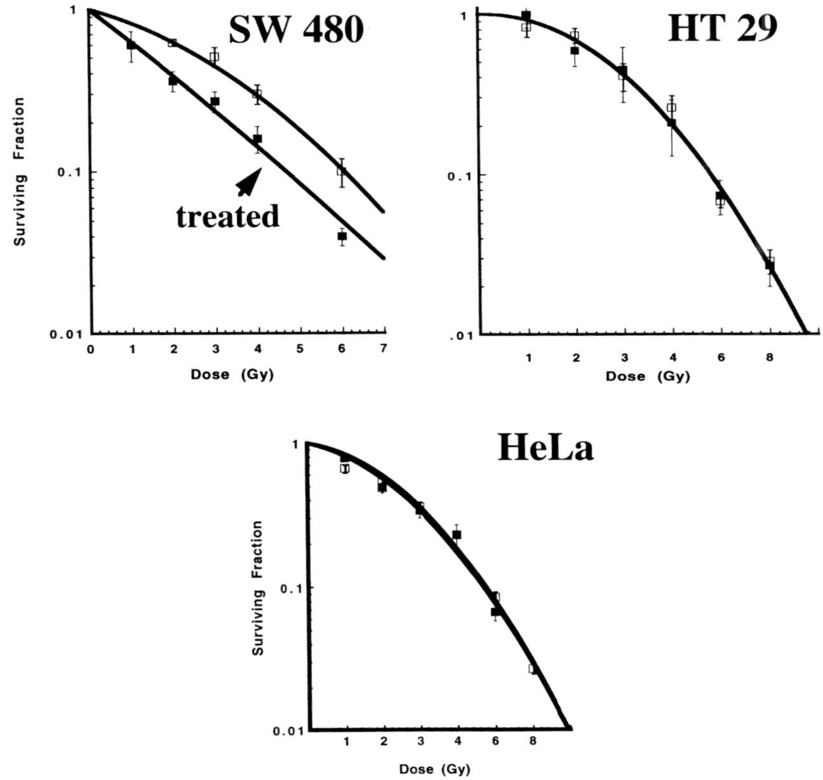

Fig. 15. Radiation survival of cells with mutant but not wild-type K-Ras is reduced after inhibition of K-Ras prenylation by FTI + GGTI. SW480, HT-29 and HeLa cells were treated for 24 h with 5 μ*M* FTI-277 + 8 μ*M* GGTI-298 before irradiation for clonogenic survival determination. Inhibitor treatment was maintained for 24 h after irradiation at which time medium was replaced with inhibitor, free medium. Control cells were treated in the same way as inhibitor-treated cells, but with an equal amount of drug-free diluent. (□), control cells; (■), FTI + GGTI-treated cells.

These findings are in agreement with the results obtained in the REF studies where radiosensitization was only seen in cells with activated Ras, and support the hypothesis that radiosensitization obtained after prenyltransferase treatment may result from inhibition of Ras directly. Two potential alternative explanations for these results presented here are that some other pathway is required in addition to signaling by activated Ras for radiation resistance, and that a prenylated protein in such a pathway is the true target of inhibition. Another possibility is that a component of one of the Ras signaling pathways is the critical target of the prenyltransferase inhibitors where radiation resistance is concerned. A potential candidate could be the rho B protein, which is also prenylated. In the latter case, inhibition of signaling at rho B would block the signal imparting radiation resistance to cells, even in the presence of Ras activation.

Although the degree of radiosensitization after prenyltransferase treatment of cells with activated Ras was not large, fractionated doses such as those used in clinical radiotherapy can cause even small increases in cell killing to appreciably improve outcome. This is because clinical radiotherapy involves the delivery of small daily doses of radiation over many weeks of treatment. This has the effect of amplifying small differences in radiosensitivity to the power of the number of treatments delivered (typically 30 or

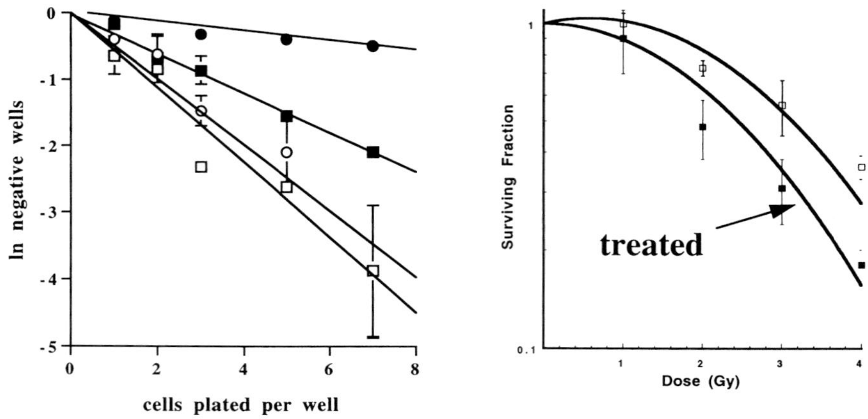

Fig. 16. Radiation survival of A549 lung carcinoma cells after inhibition of K-Ras prenylation by FTI + GGTI. (**A**) A549 cells were treated with 5 μ*M* FTI-277 + 8 μ*M* GGTI-298 for 24 h prior to plating in medium containing inhibitor or carrier in 96-well microtiter plates. Plates were then irradiated with 2 Gy. Inhibitors were diluted 1:10 by addition of medium 24 h after irradiation and the cultures scored for presence of colonies after two weeks. (□), controls; (■), 2 Gy irradiated; (○), inhibitor-treated; (●), inhibitor-treated and 2 Gy-irradiated. (**B**) Cells were treated for 24 h with 5 μ*M* FTI-277 + 8 μ*M* GGTI-298 before irradiation for clonogenic survival determination. Inhibitor treatment was maintained for 24 h after irradiation, at which time medium was replaced with inhibitor-free medium. Control cells were treated with an equal amount of drug-free diluent. (□), control cells; (■), FTI + GGTI-treated cells.

more treatments are delivered when treatment is delivered with curative intent *[49–51]*). Thus, even small differences in radiosensitivity may have a very large impact on clinical outcome in cancer treatment *(50,51)*. It is also important to note that this has in fact been demonstrated to be true in a prospective clinical trial. In the study of cervical cancer by West et al. *(52)*, a difference in median SF_2 between 0.38 and 0.54 was significant for survival at the $p = 0.01$ level. This difference is less than the difference in SF_2 we have seen using FTI-277 in human cells expressing activated Ras. For this reason we believe that the degree of radiosensitization obtained in our studies may translate into a significant increase in radiosensitivity in a clinical setting.

4. bFGF AND RADIATION RESISTANCE

4.1. Evidence of a Role for Growth Factors in Radiation Resistance

Several growth factors have been found to be radioprotective, especially when given before irradiation. Among these are the cytokines IL1, granulocyte colony stimulating factor (G-CSF), and some members of the FGF family. Irradiation can also cause the release of cytokines. For example, irradiation of mononuclear phagocytes can cause release of platelet-derived growth factor (PDGF), tumor necrosis factor-alpha (TNF-α) and insulin-like growth factor-1 (IGF-I) *(53)*, which may, in addition to altering cell survival, play a significant role in the pathogenesis of irradiation-induced pulmonary fibrosis *(53)*. Thus, the role of growth factors at the cellular and organism level is complex, and the net effect of growth factor release by host and tumor cells on the radiosensitivity of either host or tumor will depend on the site of release, and the cells involved in the response to these factors. The work described here focuses on the effect of bFGF in the radiation resistance of transformed cells.

bFGF expression has been associated with advanced stage or poor prognosis in a number of solid tumors, including pancreatic and renal malignancies *(54–56)*, and is frequently expressed by glioblastoma cells *(57–60)*. bFGF has been shown in several studies to be a radioprotector agent both for hematopoietic tissues and endothelial cells. In vitro, exogenous bFGF has been implicated in protection of bovine endothelial cells (BAEC) from the lethal effects of ionizing radiation via an autocrine loop *(61)*. This radioprotective effect is not owing to preferential repair of DNA breaks but rather to an inhibition of interphase apoptosis *(16)* involving protein kinase C *(62)*. Langley et al. *(63)* reported that bFGF has a radioprotective effect in microvessels cells, and that either bFGF withdrawal or ionizing radiation induce apoptosis in confluent monolayers of capillary endothelial cells, and that radiation apoptosis was decreased but not abolished in the presence of bFGF.

Studies in vivo have also shown that bFGF is radioprotective when administered before total body irradiation. This effect was attributed to myeloprotection, and did not appear to affect the radiation response of tumors *(64–66)*. Thus, protection appeared to be specific for normal tissue in these studies. A radioprotective effect was also reported by Fuks et al. *(16)*, who showed that bFGF prevented lethal radiation-induced pneumonitis in C3H/HeJ mice. The authors suggested that this was owing to bFGF protection of pulmonary endothelial cells from radiation-induced apoptosis. Another study by Tee and Travis *(67)*, however, failed to detect protection by bFGF from death owing to classical radiation pneumonitis in two different strains of mice. They also observed that the incidence of apoptotic bodies did not exceed 1%, were scattered throughout the lung, and were not located selectively in endothelial cells *(67)*. Thus, the exact role, if any, of bFGF in radiation-induced pneumonitis is debatable. Another member of the FGF family, FGF4, when transfected in adrenal cortical carcinoma cells, caused enhanced cell survival to ionizing radiation, an effect that was correlated with a pronounced increase in the duration of G2 arrest *(68)*. Khan et al. *(18)* have also shown that keratinocytes growth factor (KGF), a member of the FGF7 family administered intravenously before total body irradiation increased the survival of irradiated murine intestinal crypt cells in the duodenum, jejunum, and ileum. These studies have shown that the cellular response to radiation can include the production of cytokines, and that certain of these can, in some cases provide a measure of protection to irradiated cells. This conclusion led to examination of the possibility that inhibiting signaling from these cytokine pathways could diminish the protective effect of cytokine production.

4.2. bFGF Signaling Pathways

The biological activity of bFGF is mediated through a family of high-affinity plasma membrane receptors (reviewed in ref. *69*). Ligand binding to the high-affinity receptors activates the receptor by triggering autophosphorylation of its tyrosine kinase domain (reviewed in ref. *70*). This activation first induces the binding of phosphorylated receptors to at least three proteins: phospholipase C γ, pp60src, and Shc *(71–73)*. Signaling from the FGF receptor is also linked to the Ras pathway through a newly reported FRS2 protein that is tyrosine phosphorylated in response to FGF stimulation and acts to recruit Grb2/Sos leading to Ras activation *(74)*.

4.3. Dissection of the Role of the Different bFGF Isoforms

Several forms of bFGF have been described in most bFGF producing cell types *(75)*. They result from an alternative initiation of translation of one single mRNA: at an AUG codon (18-kDa form) or at three different upstream CUG (24-, 21.5- and 21-kDa forms) *(76)*. In vitro, the relative amount of the four bFGF forms varies according to the cell type: the higher molecular weight CUG-initiated forms are mainly detected in transformed cell lines, whereas normal cell types mostly overproduce the 18-kDa form *(77)*. Different roles have been described for the low and high molecular weight forms. Although constitutive expression of the 18-kDa form leads to an in vitro transformed phenotype, constitutive expression of 21-, 21.5-, and 24-kDa forms is immortalizing *(78)* after transfection with the respective cDNA. CUG initiated forms contain an amino-terminal sequence responsible for nuclear localization of these proteins *(79)*, whereas the 18-kDa form is mainly cytoplasmic *(75,79)*. Recent results have demonstrated that the 18-kDa form may be externalized and acts via specific cell surface receptors, the higher molecular weight CUG initiated forms on the other hand modulate cell proliferation by an intracellular mechanism independent of membrane receptors *(80)*.

4.4. Induction of Radiation Resistance by the 24-kDa Form of bFGF

In order to examine the relative contributions of the CUG- or AUG-initiated forms in the radioprotective effect attributed to bFGF, cDNA encoding each form were transfected in HeLa cells. HeLa cells were transfected with retroviral vector PINA encoding the 24-kDa form (HeLa 3A cells), or the 18-kDa bFGF form (HeLa 5A cells), or the vector alone (HeLa PINA cells). Figure 17 shows the bFGF protein expression profile of the selected clones. Four bands were detected in HeLa and HeLa PINA cells corresponding to the 24-, 21.5-, 21-, and 18-kDa bFGF isoforms. In contrast, HeLa 3A cells only expressed the 24-kDa form. Overexpression of the 24-kDa form was associated with downregulation of the expression of the other parental forms as already described in bovine aortic cells *(78)*. HeLa 5A cells mainly produced the 18-kDa form of bFGF with apparent downregulation of the endogenous high molecular weight forms.

The radiation survival of these cell lines was then compared in order to assess the contribution of the 18- and 24-kDa forms to bFGF-mediated radioresistance. As shown in Fig. 18, a significant increase in radioresistance was seen in the 24-kDa bFGF transfected cell line compared to wild-type or HeLa PINA cells. In contrast, HeLa 5A cells did not display any significant increase in clonogenic survival compared to the HeLa PINA.

Because exogenous bFGF was reported to inhibit apoptosis in BAEC *(61,62)*, we compared the extent of apoptosis induction 24 h after 5 and 10 Gy irradiation between radioresistant HeLa 3A and HeLa wild-type. No apoptosis was detected in either HeLa wild-type or in HeLa 3A (not shown), suggesting that apoptosis was not involved in irradiation-induced death in these cells.

These findings demonstrate that both high and low molecular weight bFGF isoforms exhibit a distinct capacity to modify cellular radiosensitivity. However, in the HeLa cell sytem used for analysis of bFGF radiation resistance after transfection with different bFGF isoforms, only those cells that were engineered to produce the 24-kDa isoform exhibited a significant increase in clonogenic survival after irradiation *(81)*. The mechanism through which endogenous expression of the 24-kDa bFGF isoform mediates radiation resistance is not known, but recent work has demonstrated that in neurons, an

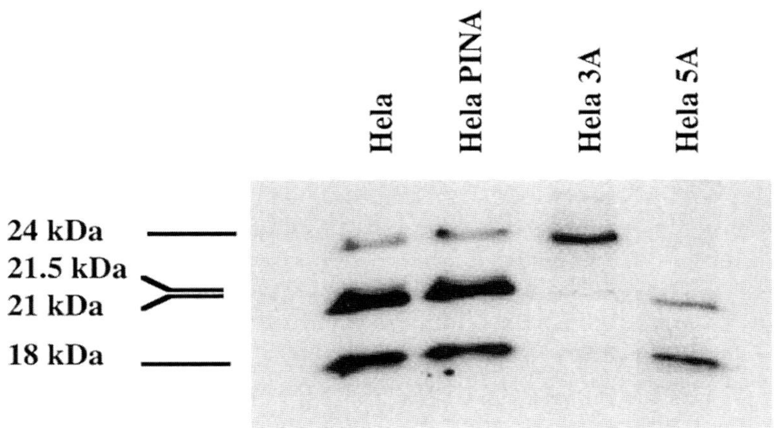

Fig. 17. bFGF protein expression from nonirradiated HeLa wild-type, HeLa PINA, HeLa 3A, and HeLa 5A cell lines. 40 μg of extracted proteins from each cell line were loaded on 12.5% SDS-PAGE and Western blotting was performed. The migration of the different bFGF isoforms is indicated by molecular weight.

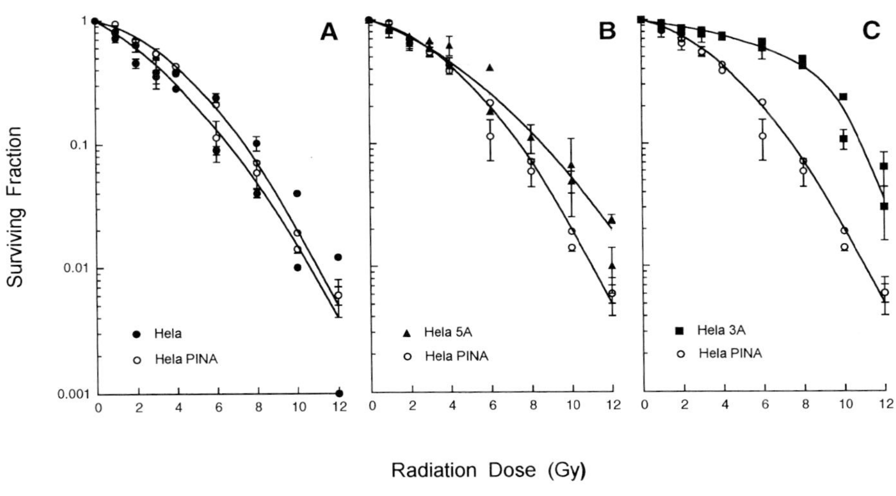

Fig. 18. Radiation survival curves of HeLa, HeLa PINA, HeLa 3A, HeLa5A. Cells were exposed to various ionizing radiation doses and the radiosensitivity of the different cell lines calculated using the clonogenic survival assay. A semi-logarithmic plot of data from the different cell lines is shown. Radiation survival curves were obtained using the quadratic linear model. (**A**) (●), HeLa and (○), HeLa PINA cells; (**B**) (○), HeLa PINA and (■), HeLa 3A; (**C**) (○), HeLa PINA and (▲), HeLa 5A. Each bar represents the mean ± SD of three different experiments for one selected clone of each transfected cell line. (SD representing less than 1% of variation do not appear on the graph).

intracellular bFGF receptor is present in the nuclear membrane and that this FGF^R binds acidic FGF and with a lower affinity than bFGF (reviewed in ref. *82*). The 18-kDa form is secreted and binds to membrane tyrosine kinase receptors leading to the activation of serial intracellular messengers including Ras and Raf, but no interaction between the 24-kDa endogenous pathway and these oncogenes has as yet been described. It cannot yet be excluded that Ras or a small G protein in the Ras family transduces a signal originating with the 24-kDa bFGF isoform. Alternatively, the pathway of radiation resistance induced by the 24-kDa bFGF isoform may be altogether distinct from that observed in Ras transformed cells, and define a new mechanism of radiation resistance.

4.5. Radiosensitization of 24-kDa bFGF Expressing Cells by FTI-277

As discussed previously, FTI was shown to radiosensitize human cells expressing activated Ras. Given the possibility that bFGF radiation resistance was also mediated by a prenylated protein, we analyzed the effect of the specific inhibitors FTI-277 on the radiosensitivity of the radioresistant HeLa cells expressing the 24-kDa bFGF isoform. We first compared the effect of inhibiting protein farnesylation on the ability of radioresistant 24-kDa bFGF transfected HeLa cells (HeLa 3A) and control cells (HeLa PINA) to survive irradiation. Cells were treated with the specific FTase inhibitor, FTI-277 (20 μ*M*) for 48 h prior to irradiation. At these concentrations, no inhibition of cellular proliferation was detected. After 48 h of treatment with FTI-277, Ha-Ras processing was inhibited as evidenced by the appearance of the slowly migrating unprocessed form as documented by Western-blot analysis (Fig. 19A), whereas no effect was seen on the processing of the geranylgeranylated Rap1A protein (Fig. 19B).

We next determined the effect of FTase inhibition on radiation survival of HeLa 3A and HeLa PINA cell lines in clonogenic assays (Fig. 20). FTI-277 did not affect the sensitivity of HeLa PINA to radiation treatment. In contrast, the radioresistance of HeLa 3A was dramatically reduced in the presence of 20 μ*M* FTI-277 prior to irradiation. Thus, FTI-277 increases the sensitivity of HeLa 3A but not HeLa PINA cells to ionizing radiation.

To be certain that the radiosensitization of cells expressing the 24-kDa bFGF isoform by FTI-277 was not owing to a change in the distribution of the different isoforms, particularly to a switch from the 24-kDa to the 18-kDa isoform, cells were treated as described previously with FTI-277 and the cellular content of the various bFGF isoforms was analyzed for HeLa PINA and HeLa 3A. FTI-277 treatment did not alter bFGF distribution in HeLa 3A and in HeLa PINA (not shown). Thus, the radiosensitizer effect of FTI-277 on HeLa 3A could not be attributed to changes in bFGF isoforms expression.

We next asked whether the radiosensitizer effect of FTI-277 was caused by induction of apoptosis as we had shown in Ha-Ras transfected REF (*see* ref. *41* and Fig. 5). No radiation-induced apoptosis was detected after FTI-277 treatment of either HeLa 3A or HeLa PINA as assessed at 4, 24, and 48 h after irradiation. However, morphological examination of irradiated HeLa PINA and HeLa3A cells treated with FTI-277 prior to irradiation revealed a significant increase in the percentage of giant cells in the HeLa3A cells, whereas no effect was noticed for radiosensitive HeLa PINA cells (not shown). This finding is consistent with FTI-277 radiosensitization of HeLa cells expressing the 24-kDa isoform through increased postmitotic cell death rather than by induction of apoptosis.

5. CONCLUSIONS

We have presented studies that demonstrate the role of *ras* oncogene activation in contributing to radiation resistance in both rodent and human cells. We have also shown that the 24-kDa isoform of bFGF contributes to the radiation survival of HeLa cells overexpressing this protein. These studies demonstrate two mechanisms whereby tumor cells can aquire increased resistance to radiotherapy. It is apparent that both of these mechanisms involve the generation of intracellular signals, but it is not clear that the signals emanating from activated Ras, and those initiated by the 24-kDa isoform of bFGF, are

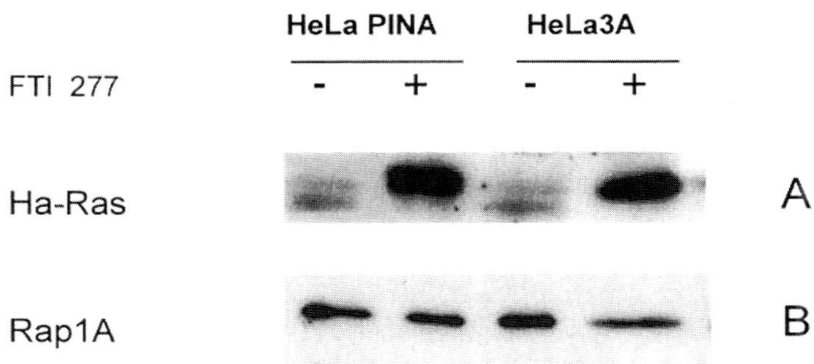

Fig. 19. FTI-277 inhibition of H-Ras and Rap1A processing. Cells were treated for 48 h with FTI-277 (20 μ*M*) (**A**,**B**). Cellular proteins were analyzed by Western blotting using anti Ha-Ras antibody (**A**) and anti-Rap 1A antibody (**B**).

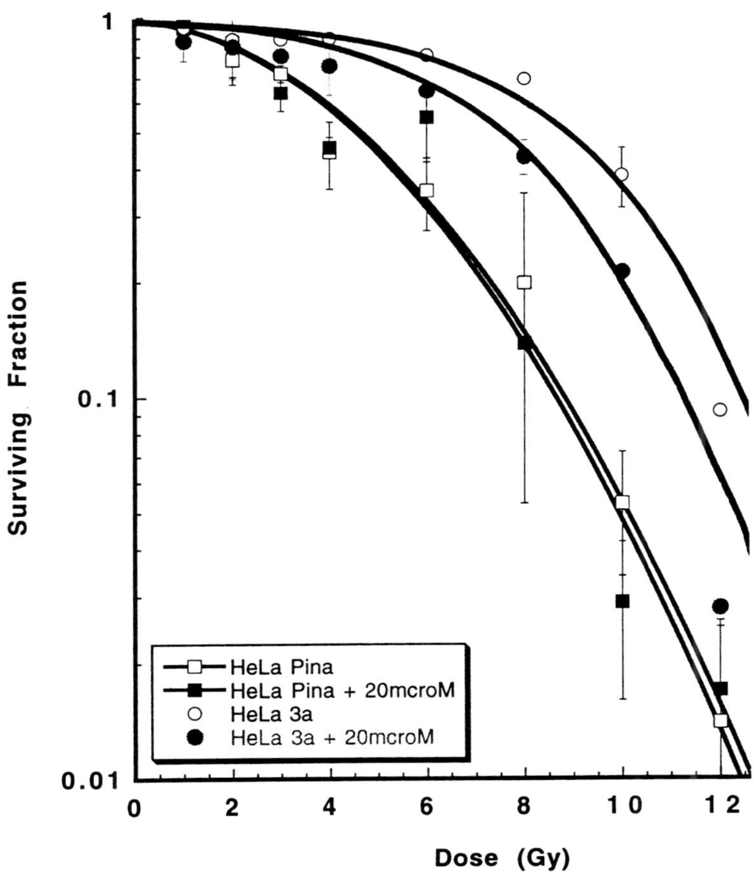

Fig. 20. FTI-277 only radiosensitizes HeLa 3A. (□,■), HeLa PINA, or (○,●), HeLa 3A cells were treated for 48 h with FTI-277 (20 μ*M*) (closed symbols) or with vehicle (open symbols), and were exposed to various ionizing radiation doses. Clonogenic assays were scored 1 wk later. Data represent the means of three independent experiments (SD representing <1% of variation do not appear on the graph).

in any way linked. Continued study of the signal transduction pathways mediating radiation resistance in both systems will further define the relationship or the separation of these paths. Both pathways, however, are subject to inhibition by prenyltransferase inhibitors. In the case of Ras activation, the data suggest that the target of the inhibitors is the activated Ras itself, although as discussed earlier, other proteins in the Ras pathways are targets of prenyltransferase inhibition, and may prove to be essential contributors to the signaling required for radiation resistance. In the case of bFGF, the target of the FTase inhibitors is not known. Because the major contributor to the radiation resistance observed here was the 24-kDa isoform, and this isoform is localized to the nucleus, it is unlikely that the signal from this protein involves Ras. The signal responsible for radioresistance generated by overexpression of the 24-kDa bFGF isoform may, however, feed into a pathway that is downstream of Ras, or require an intact Ras signaling pathway for its effect.

It is apparent from these studies that prenyltransferase inhibitors are promising as candidates for radiosensitization of tumor cells. The potential for therapeutic application comes from the specificity of this approach for certain tumor cells, and the apparent lack of radiosensitization of cells that do not express either activated Ras or the 24-kDa bFGF protein. The specificity of this approach to radiosensitization may permit more efficient sterilization of tumor cells by radiation at doses that are tolerable to the surrounding tissues.

6. ACKNOWLEDGMENTS

The authors wish to thank the following for their contribution to this work: Yimin Qian, Christine Toulas, Vincent Bakanauskas, JunMin Wu, and Lorretta Knight. The work presented was supported by NIH grants: CA73820 (E.J.B.), CA75138 (R.J.M., W.G.M., E.J.B.), CA67771 (S.M.S., A.D.H.), and a grant from the US Department of Defense PC970493 (E.J.B.). We also thank the editors at the American Association for Cancer Research for permission to use material previously published in *Cancer Research (36,41,81,83).*

REFERENCES

1. FitzGerald TJ, Daugherty C, Kase K, Rothstein LA, McKenna M, Greenberger JS. Activated human N-Ras oncogene enhances x-irradiation repair of mammalian cells in vitro less effectively at low dose rate. *Am J Clin Oncol* 1985; 8:517–522.
2. Sklar MD. Increased resistance to cis-diamminedichloroplatinum(II) in NIH 3T3 cells transformed by Ras oncogenes. *Cancer Res* 1988; 48:793–797.
3. Pirollo K, Tong Y, Villegas Z, Chen Y, Chang E. Oncogene-transformed NIH 3T3 cells display radiation resistance levels indicative of a signal transduction pathway leading to the radiation-resistant phenotype. *Radiat Res* 1993; 135:234–243.
4. Samid D, Miller AC, Rimoldi D, Gafner J, Clark EP. Increased radiation resistance in transformed and nontransformed cells with elevated Ras proto-oncogene expression. *Radiat Res* 1991; 126:244–250.
5. McKenna WG, Weiss MA, Bakanauskas VJ, Sandler H, Kelsten M, Biaglow J, et al. The role of the Hras oncogene in radiation resistance and metastasis. *Intl J Rad Onc Biol Phys* 1990; 18:849–860.
6. Ling CC, Endlich B. Radioresistance induced by oncogenic transformation. *Radiat Res* 1989; 120: 267–279.
7. Hermens A, Bentvelzen P. Influence of the H-Ras oncogene on radiation responses of a rat rhabdomyosarcoma cell line. *Cancer Res* 1992; 52:3073–3082.
8. Miller AC, Kariko K, Myers CE, Clark EP, Samid D. Increased radioresistance of EJ*ras*-transformed human osteosarcoma cells and its modulation by lovastatin, an inhibitor of $p21^{ras}$ isoprenylation. *Intl J Cancer* 1993; 53:302–307.
9. Miller AC, Gafner J, Clark EP, Samid D. Differences in radiation-induced micronuclei yields of human cells: influence of Ras gene expression and protein localization. *Intl J Radiat Biol* 1993; 64:547–554.

10. Bruyneel EA, Storme GA, Schallier DC, Van den Berge DL, Hilgard P, Mareel MM. Evidence for abrogation of oncogene-induced radioresistance of mammary cancer cells by hexadecylphosphocholine in vitro. *Eur J Cancer* 1993; 29A:1958–1963.
11. Harris JF, Chambers AF, Tam ASK, Some *Ras*-transformed cells have increased radiosensitivity and decreased repair of sublethal readiation damage. *Somat Cell Mol Genet* 1990; 16:39–48.
12. Alapetite C, Baroche C, Remvikos Y, Goubin G, Moustacchi E. Studies on the influence of the presence of an activated *Ras* oncogene on the *in vitro* radiosensitivity of human mammary epithelial cells. *Intl J Radiat Biol* 1991; 59:385–396.
13. Su L-N, Little JB. Transformation and radiosensitivity of human diploid skin fibroblasts transfected with activated *RAS* oncogene and SV40 T-antigen. *Intl J Radiat Biol* 1992; 62:201–210.
14. Mendonca MS, Boukamp P, Stanbridge EJ, Redpath JL. The radiosensitivity of human keratinocytes: influence of activated c-H-Ras oncogene expression and tumorigenicity. *Intl J Radiat Biol* 1991; 59: 1195–1206.
15. Farrell CL, Bready JV, Rex KL, Chen JN, DiPalma CR, Whitcomb KL, et al. Keratinocyte growth factor protects mice from chemotherapy and radiation-induced gastrointestinal injury and mortality. *Cancer Res* 1998; 58:933–939.
16. Fuks Z, Persaud RS, Alfieri A, McLoughlin M, Ehleiter D, Schwartz JL, et al. Basic fibroblast growth factor protects endothelial cells against radiation-induced programmed cell death in vitro and in vivo. *Cancer Res* 1994; 54:2582–2590.
17. Haimovitz-Friedman A, Vlodavsky I, Chaudhuri A, Witte L, Fuks Z. Autocrine effects of fibroblast growth factor in repair of radiation damage in endothelial cells. *Cancer Res* 1991; 51:2552–2558.
18. Khan WB, Shui C, Ning S, Knox SJ. Enhancement of murine intestinal stem cell survival after irradiation by keratinocyte growth factor. *Radiat Res* 1997; 148:248–253.
19. Yi ES, Williams ST, Lee H, Malicki DM, Chin EM, Yin S, et al. Keratinocyte growth factor ameliorates radiation- and bleomycin-induced lung injury and mortality. *Am J Pathol* 1996; 149:1963–1970.
20. Schmidt-Ullrich RK, Mikkelsen RB, Dent P, Todd DG, Valerie K, Kavanagh BD, et al. Radiation-induced proliferation of the human A431 squamous carcinoma cells is dependent on EGFR tyrosine phosphorylation. *Oncogene* 1997; 15:1191–1197.
21. Kavanagh BD, Dent P, Schmidt-Ullrich RK, Chen P, Mikkelsen RB. Calcium-dependent stimulation of mitogen-activated protein kinase activity in A431 cells by low doses of ionizing radiation. *Radiat Res* 1998; 149:579–587.
22. Vojtek AB, Der CJ. Increasing complexity of the Ras signaling pathway. *J Biol Chem* 1998; 273:19,925–19,928.
23. Marais R, Light Y, Paterson HF, Marshall CJ. Ras recruits Raf-1 to the plasma membrane for activation by tyrosine phosphorylation. *EMBO J* 1995; 14:3136–3145.
24. Kasid U, Suy S, Dent P, Ray S, Whiteside TL, Sturgill TW. Activation of Raf by ionizing radiation. *Nature* 1996; 382:813–816.
25. Galaktionov K, Jessus C, Beach D. Raf1 interaction with cdc25 phosphatase ties mitogenic signal transduction to cell cycle activation. *Genes Devel* 1995; 9:1046–1058.
26. Kasid U, Pfeifer A, Brennan T, Beckett M, Weichselbaum RR, Dritschilo A, Mark GE. Effect of antisense c-Raf-1 on tumorigenicity and radiation sensitivity of a human squamous carcinoma. *Science* 1989; 243:1354–1356.
27. Suzuki K, Watanabe M, Miyoshi J. Differences in effects of oncogenes on resistance of gamma rays, ultraviolet light, and heat shock. *Radiat Res* 1992; 129:157–162.
28. Lebowitz PF, Du W, Prendergast GC. Prenylation of RhoB is required for its cell transforming function but not its ability to activate serum response element-dependent transcription. *J Biol Chem* 1997; 272: 16,093–16,095.
29. Olson MF, Ashworth A, Hall A. An essential role for rho, rac and cdc42 GTPases in cell cycle progression through G1. *Science* 1995; 269:1270–1272.
30. Joneson T, White MA, Wigler MH, Bar-Sagi D. Stimulation of membrane ruffling and MAP kinase activation by distinct effectors of RAS. *Science* 1996; 271:810–812.
31. Minden A, Lin A, McMahon M, Lange-Carter C, Derijard B, Davis R, et al. Differential activation of ERK and JNK mitogen-activated protein kinases by Raf-1 and MEKK. *Science* 1994; 266:1719–1723.
32. Lim L, Manser E, Leung T, Hall C. Regulation of phosphorylation pathways by p21 GTPases. The p21 Ras-related Rho subfamily and its role in phosphorylation signalling pathways. *Eur J Biochem* 1996; 242:171–185.
33. Alberts AW. Biochemistry and biology of lovastatin. *Am J Cardiol* 1988; 62:10J–15J.

34. Kato K, Cox AD, Hisaka MM, Graham SM, Buss JE, Der CJ. Isoprenoid addition to Ras protein is the critical modification for its membrane association and transforming activity. *Proc Natl Acad Sci USA* 1992; 89:6403–6407.
35. Sebti S, Hamilton A. Inhibition of Ras prenylation: a novel approach to cancer chemotherapy. *Pharmacol Ther* 1997; 74:103–114.
36. McKenna WG, Weiss MC, Endlich B, Ling CC, Bakanauskas VJ, Kelsten ML, Muschel RJ. Synergistic effect of the v-myc oncogene with Hras on radioresistance. *Cancer Res* 1990; 50:97–102.
37. Gutierrez L, Magee AI, Marshall CJ, Hancock JF. Post-translational processing of p21^{ras} is two-step and involves carboxyl-methylation and carboxy-terminal proteolysis. *EMBO J* 1989; 8:1093–1098.
38. Lerner E, Qian Y, Hamilton A, Sebti S. Disruption of oncogenic K-ras4B processing and signaling by a potent geranylgeranyltransferase I inhibitor. *J Biol Chem* 1995; 270:26,770–26,773.
39. Reiss Y, Stradley SJ, Gierasch LM, Brown MS, Goldstein JL. Sequence requirements for peptide recognition by rat brain p21^{ras} protein farnesyltransferase. *Proc Natl Acad Sci USA* 1991; 88:732–736.
40. Cox AD, Hisaka MM, Buss JE, Der CJ. Specific isoprenoid modification is required for function of normal, but not oncogenic, Ras protein. *Mol Cell Biol* 1992; 12:2606–2615.
41. Bernhard EJ, Kao G, Cox AD, Sebti SM, Hamilton AD, Muschel RJ, McKenna WG. The farnesyltransferase inhibitor FTI-277 radiosensitizes H-Ras-transformed rat embryo fibroblasts. *Cancer Res* 1996; 56:1727–1730.
42. McGuire TF, Qian Y, Vogt A, Hamilton AD, Sebti SM. Platelet-derived growth factor receptor tyrosine phosphorylation requires protein geranylgeranylation but not farnesylation. *J Biol Chem* 1996; 271: 27,402–27,407.
43. Vogt A, Qian Y, McGuire TF, Hamilton AD, Sebti SM. Protein geranylgeranylation, not farnesylation, is required for the G1 to S phase transition in mouse fibroblasts. *Oncogene* 1996; 13:1991–1999.
44. Thompson TC, Southgate J, Kitchener G, Land H. Multistage carcinogenesis induced by Ras and myc oncogenes in a reconstituted organ. *Cell* 1989; 56:917–930.
45. Thompson TC, Truong LD, Timme TL, Kadmon D, McCune BK, Flanders KC, et al. Transgenic models for the study of prostate cancer. *Cancer* 1993; 71:S1165–S1171.
46. Lefkovits I. Limiting dilution analysis. *Immunol Methods* 1979; 355–370.
47. Thilly WG, DeLuca JG, Furth EE, Hoppe HI, Kaden DA, Krolewski JJ, et al. Gene-locus mutation analysis in diploid human lymphoblast lines, in *Chemical Mutagens* Vol. 6 (Serres FJD, Hollaender A, eds.), Plenum, New York, 1980, pp. 331–364.
48. Grenman R, Burk D, Virolainen E, Buick RN, Church J, Schwartz DR. Clonogenic cell assay for anchorage-dependent squamous carcinoma cell lines using limiting dilution. *Intl J Cancer* 1989; 44:131–136.
49. Fertil B, Malaise EP. Inherent cellular radiosensitivity as a basic concept for human tumor radiotherapy. *Intl J Rad Onc Biol Phys* 1981; 7:621–629.
50. Steel GG. Cellular senitivity to low dose-rate irradiation focuses the problem of tumour radioresistance. *Radiother Oncol* 1991; 20:71–83.
51. Thames HD, Schultheiss TE, Hendry JH, Tucker SL, Dubray BM, Brock WA. Can modest escalations of dose be detected as increased tumor control? *Int J Rad Onc* 1992; 22:241–246.
52. West CM, Davidson SE, Burt PA, Hunter RD. The intrinsic radiosensitivity of cervical carcinoma: correlations with clinical data. *Intl J Rad Onc* 1995; 31:841–846.
53. Thornton S, Walsh B, Bennett S, Robbins J, Foulcher E, Morgan G, et al. Both in vitro and in vivo irradiation are associated with induction of macrophage-derived fibroblast growth factors. *Clin Exp Immunol* 1996; 103:67–73.
54. Morrison RS, Yamaguchi F, Saya H, Bruner JM, Yahanda AM, Donehower LA, Berger M. Basic fibroblast growth factor and fibroblast growth factor receptor I are implicated in the growth of human astrocytomas. *J Neuro-Oncol* 1994; 18:207–216.
55. Yamanaka Y, Friess H, Buchler M, Beger HG, Uchida E, Onda M, et al. Overexpression of acidic and basic fibroblast growth factors in human pancreatic cancer correlates with advanced tumor stage. *Cancer Res* 1993; 53:5289–5296.
56. Nanus DM, Schmitz-Drager BJ, Motzer RJ, Lee AC, Vlamis V, Cordon-Cardo C, et al. Expression of basic fibroblast growth factor in primary human renal tumors: correlation with poor survival. *J Natl Cancer Inst* 1993; 85:1597–1599.
57. Takahashi JSH, Yasuda Y, Ito N, Ohta M, Jaye M, Fukumoto MO, et al. Gene expression of fibroblast growth factor receptors in the tissues of human gliomas and meningiomas. *Biochem Biophys Res Commun* 1991; 177:1–7.
58. Jensen RL. Growth factor-mediated angiogenesis in the malignant progression of glial tumors: a review. *Surgical Neurol* 1998; 49:189–195.

59. Joy A, Moffett J, Neary K, Mordechai E, Stachowiak EK, Coons S, et al. Nuclear accumulation of FGF-2 is associated with proliferation of human astrocytes and glioma cells. *Oncogene* 1997; 14:171–183.
60. Chinot O. [Biological profiles of malignant gliomas]. *Pathologie Biologie* 1995; 43:224–232.
61. Haimovitz-Friedman A, Vlodavsky I, Chaudhuri A, Witte L, Fuks Z. Autocrine effects of fibroblast growth factor in repair of radiation damage in endothelial cells. *Cancer Res* 1991; 51:2552–2558.
62. Haimovitz-Friedman A, Balaban N, McLoughlin M, Ehleiter D, Michaeli J, Vlodavsky I, Fuks Z. Protein kinase C mediates basic fibroblast growth factor protection of endothelial cells against radiation-induced apoptosis. *Cancer Res* 1994; 54:2591–2597.
63. Langley R, Bump E, Quartuccio S, Medeiros D, Braunhut S. Radiation-induced apoptosis in microvascular endothelial cells. *Br J Cancer* 1997; 75:666–672.
64. Ding I, Huang K, Wang X, Greig JR, Miller RW, Okunieff P. Radioprotection of hematopoietic tissue by fibroblast growth factors in fractionated radiation experiments. *Acta Oncologica* 1997; 36:337–340.
65. Ding I, Huang K, Snyder ML, Cook J, Zhang L, Wersto N, Okunieff P. Tumor growth and tumor radiosensitivity in mice given myeloprotective doses of fibroblast growth factors. *J Natl Cancer Inst* 1996; 88:1399–1404.
66. Okunieff P, Abraham EH, Moini M, Snyder ML, Gloe TR, Capogrossi MC, Ding I. Basic fibroblast growth factor radioprotects bone marrow and not RIF1 tumor. *Acta Oncol* 1995; 34:435–438.
67. Tee PG, Travis EL. Basic fibroblast growth factor does not protect against classical radiation pneumonitis in two strains of mice. *Cancer Res* 1995; 55:298–302.
68. Jung M, Kern FG, Jorgensen TJ, McLeskey SW, Blair OC, Dritschilo A. Fibroblast growth factor-4 enhanced G2 arrest and cell survival following ionizing radiation. *Cancer Res* 1994; 54:5194–5197.
69. Jaye M, Schlessinger J, Dionne C. Fibroblast growth factor receptor tyrosine kinases: molecular analysis and signal transduction. *Biochim Biophys Acta* 1992; 1135:185–199.
70. Ullrich A, Schlessinger J. Signal transduction by receptors with tyrosine kinase activity. *Cell* 1990; 61: 203–212.
71. Zhan XPCHX, Friesel R, Maciag T. Association of fibroblast growth factor receptor-1 with c-src correlates with association between c-src and cortactin. *J Biol Chem* 1994; 269:20,221–20,224.
72. Vainikka S, Joukov VWS, Bergman M, Pelicci PG, Alitalo K. Signal transduction by fibroblast growth factor receptor-4 (FGF4). *J Biol Chem* 1994; 269:18,320–18,326.
73. Shi EKM, Xu J, Wang F, Hou J, McKeehan WL. Control of basic fibroblast growth factor receptor kinase signal transduction by heterodimerization of combinatorial splice variants. *Mol Cell Biol* 1993; 13:3907–3918.
74. Kouhara H, Hadari YR, Spivak-Kroizman T, Schilling J, Bar-Sagi D, Lax I, Schlessinger J. A lipid-anchored Grb2-binding protein that links FGF receptor activation to the Ras/MAPK signaling pathway. *Cell* 1997; 89:693–702.
75. Renko N, Quarto N, Morimoto T, Rifkin D. Nuclear and cytoplasmic localization of different basic fibroblast growth factor species. *J Cell Physiol* 1990; 144:108–114.
76. Prats H, Kaghad M, Prats AC, Klagsbrun M, Lelias JM, Liauzun P, Chalon P, et al. High molecular mass forms of basic fibroblast growth factor are initiated by alternative CUG. *Proc Natl Acad Sci USA* 1989; 86:1836–1840.
77. Vagner STC, Galy B, Audigier S, Gensac MC, Amalric F, Bayard F, et al. Translation of CUG- but not AUG-initiated forms of human fibroblast growth factor 2 is activated in transformed and stressed cells. *J Cell Biol* 1996; 135:1391–1402.
78. Couderc B, Prats H, Bayard F, Amalric F. Potential oncogenic effects of basic fibroblast growth factor requires cooperation between CUG and AUG-initiated forms. *Cell Regul* 1991; 2:708–718.
79. Bugler BAF, Prats H. Alternative initiation of translation determines cytoplasmic or nuclear localization of basic fibroblast growth factor. *Mol Cell Biol* 1991; 11:543–547.
80. Bikfalvi A, Klein S, Pintucci G, Quarto NMP, Rifkin DB. Differential modulation of cell phenotype by different molecular weight forms of basic fibroblast growth factor: possible intracellular signaling by the high molecular forms. *J Cell Biol* 1995; 129:233–243.
81. Cohen-Jonathan E, Toulas C, Monteil S, Couderc B, Maret A, Bard JJ, et al. Radioresistance induced by the high molecular forms of the basic fibroblast growth factor is associated with an increased G2 delay and a hyperphosphorylation of p34CDC2 in HeLa cells. *Cancer Res* 1997; 57:1364–1370.
82. Stachowiak M, Moffett JMP, Tucholski J, Stachowiak E. Growth factor regulation of cell growth and proliferation in the nervous system. A new intracrine nuclear mechanism. *Mol Neurobiol* 1997; 15:257–283.
83. Bernhard EJ, McKenna WG, Hamilton AD, Sebti SM, Qian Y, Wu J, Muschel RJ. Inhibiting Ras prenylation increases the radiosensitivity of human tumor cell lines with activating mutations of *Ras* oncogenes. *Cancer Res* 1998; 58:1754–1761.

13 Farnesyltransferase and Geranylgeranyltransferase I Inhibitors as Novel Agents for Cancer and Cardiovascular Diseases

Saïd M. Sebti, PHD *and Andrew D. Hamilton,* PHD

CONTENTS

From: *Farnesyltransferase Inhibitors in Cancer Therapy*
Edited by: S. M. Sebti and A. D. Hamilton © Humana Press Inc., Totowa, NJ

1. INTRODUCTION

1.1. Ras and Oncogenic Transformation

Cancer is believed to result from an accumulation of genetic alterations that cause loss of function of tumor suppressor genes and gain of function of oncogenes. Among the most thoroughly studied oncogenes are those that encode Ras proteins. Ras proteins are guanosine triphosphate/guanosine diphosphate (GTP/GDP) binding guanosine triphosphate phosphatases (GTPases) that play a critical role in a variety of signal transduction pathways in the cell *(1–7)*. The three mammalian ras genes encode four highly homologous plasma membrane-bound G-proteins (H-, N-, KA- and KB-Ras) that cycle between their GTP (active)- and GDP (inactive)- forms to switch on and off signals from the cell surface to the nucleus *(4–8)*. For example, binding of platelet-derived growth factor (PDGF) to its receptor tyrosine kinase results in autophosphorylation, which creates phosphotyrosines that serve as binding sites for several key signaling molecules such as phospholipase C-γ1 and phosphatidyl inositol-3 kinase (PI-3 Kinase) *(9–15)*. Most important for Ras activation is the recruitment of the growth factor receptor binding protein 2 (GRB-2), which is complexed to a ras GTP/GDP exchanger, mammalian son of seven less-1 (m-SOS-1), *(6,7)*. The exchanger m-SOS-1 binds GDP Ras and catalyzes the exchange of GDP for GTP, which results in Ras activation *(6,7)*. GTP-bound Ras can then activate several effectors that result in triggering a variety of signaling pathways, such as the PI-3 kinase/AKT-2 survival pathway and the mitogen-activated protein (MAP) kinase cascade. In the latter, GTP-Ras recruits a ser/thr kinase, c-raf- 1, to the plasma membrane where it gets activated and in turn activates a series of MAP kinases *(5,8)*. These Ras-dependent signals are turned off by hydrolysis of the bound GTP, which returns Ras to its GDP-bound state. The GTPase activity is intrinsic to the Ras protein but requires the GTPase activating protein, Ras-GAP *(4)*. Thus, Ras is activated by guanine nucleotide exchange factors, such as m-SOS-1, and turned off by GTPase activating proteins such as GAP. Mutations in the Ras sequence (i.e., amino acids 12, 13, and 61) that lock Ras in its GTP-bound form result in a growth factor-independent, constitutively activated signal that leads to uncontrolled growth *(16–19)*. Mutated Ras no longer needs exchange factors for activation and GAP does not turn it off. This Ras-dependent uncontrolled growth is believed to be directly implicated in a large number of human cancers, because 30% of human tumors express ras oncogenes with these transforming mutations *(1,2)*.

Because Ras mutations are common in human cancers, several approaches over the last two decades have been attempted to reverse the aberrant Ras function. Among these were efforts to reverse the low GTPase activity of mutated Ras. This approach failed, most probably because of the small structural differences between wild-type and mutated Ras that could not be exploited pharmacologically. Antisense and ribozyme approaches that target selectively Ras mutated sequences so far have not yielded a therapeutic agent. A third approach exploits the discovery that the Ras protein requires farnesylation for its localization to the plasma membrane and for its malignant transforming activity *(20–28)*. Farnesylation is a lipid post-translational modification that increases the hydrophobicity of Ras allowing it to associate with the plasma membrane, where it is needed for recruiting Ras effectors, such as raf-1. Thus, preventing Ras from going to its plasma membrane microenvironment by blocking its farnesylation short-circuits oncogenic growth signal to the nucleus.

1.2. Farnesyltransferase and Geranylgeranyltransferase I

The first step in Ras post-translational modifications that leads to membrane association is farnesylation *(29)*. This lipid modification is catalyzed by farnesyltransferase (FTase), which transfers farnesyl from farnesylpyrophosphate (FPP), a cholesterol biosynthesis intermediate *(30)*, to the cysteine of the carboxyl terminal CAAX tetrapeptide of Ras (C = Cys; A = any amino acid; and X = Met or Ser) *(29)*. Farnesylation is followed by peptidase cleavage of AAX and carboxymethylation of the resulting farnesylated cysteine *(22,23)*. H-, N-, and Ka-Ras proteins are also palmitoylated on cysteines upstream of the farnesylated cysteine and these additional lipid modifications further stabilize the interaction of Ras with the plasma membrane *(22,23,26)*. Kb-Ras on the other hand, is not palmitoylated, but contains a polylysine stretch upstream of the farnesylated cysteine that further stabilizes the interaction of Kb-Ras with the plasma membrane *(26)*. Farnesylation is a prerequisite for all subsequent Ras post-translational modifications. Furthermore, farnesylation was shown to be absolutely required for Ras malignant transformation *(25,31)*.

In addition to the Ras proteins, FTase farnesylates several other cellular proteins *(29,32,33)*. The protein substrates of FTase all share a common feature: a CAAX at their carboxyl terminal, where X is most often a methionine, serine, cysteine, alanine, or glutamine *(29,32,33)*. Proteins terminating with CAAX boxes, where X is leucine or isoleucine, are modified with the 20-carbon cholesterol biosynthesis intermediate geranylgeranyl pyrophosphate (GGPP) *(27,34,35)*. The enzyme responsible for this post-translational modification is geranylgeranyltransferase I (GGTase I). FTase and GGTase I are α/β heterodimeric zinc metalloproteins that share the α-subunit *(36,37)*. The β-subunit of FTase has been shown to bind the protein substrate, as well as the prenyl substrate FPP. Among the substrates for FTase are Ras proteins, lamin B and transducin *(27)*. GGTase I substrates are more numerous and include most γ-subunits of heterotrimeric G-proteins and a large number of small G-proteins such as those in the Rho family *(27)*. A member of this family, Rho B, was shown to be geranylgeranylated and farnesylated *(38–40)*. Although H-Ras is exclusively farnesylated, recent in vitro and in vivo data suggested that Kb-Ras and N-Ras become geranylgeranylated when FTase is inhibited *(41–45)*. For more detail about the biochemistry of FTase and GGTase I, please refer to Chapter 2.

1.3. Inhibitors of FTase and GGTase I

The discovery that Ras is lipid-modified by a farnesyl group and that farnesylation is required for Ras transforming activity has led to an intense search for inhibitors of FTase (reviewed in refs. *46–48*). The ultimate goal of this search was to identify pharmacological agents that suppress selectively oncogenic Ras function and hence block cancerous growth. To this end, two approaches have been taken. One is based on rational design, using as targets the two substrates of FTase, the CAAX tetrapeptide and/or the isoprenoid FPP. The other approach uses targeted random screens from either natural products or chemical libraries.

Because CAAX tetrapeptides are farnesylated by FTase as efficiently as the corresponding full-length protein and also are potent (10–200 n*M*) competitive inhibitors of FTase *(29)*, several groups from academic and pharmaceutical labs targeted the CAAX tetrapeptide as a novel anticancer drug development strategy (Table 1). The major efforts of these groups focused on improving stability of the tetrapeptides towards proteolytic

Table 1
Chemical Structures of Various FTIs

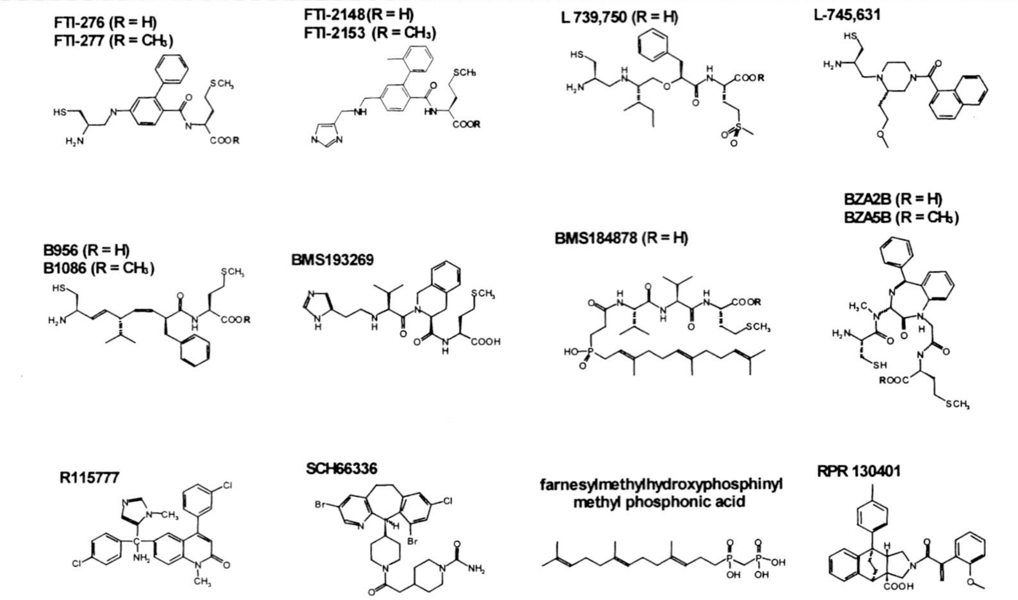

degradation and increasing their cellular uptake *(46–48)*. A range of approaches were applied to the CAAX peptide backbone to prevent proteolysis. Table 1 shows the structures of a variety of agents that were identified over the last seven years. Table 1 also shows examples of FPP mimics as well as compounds that were identified from chemical libraries. Many of these have been described thoroughly in other chapters throughout this book. In this chapter, we describe our efforts in developing CAAX peptidomimetics as novel agents for cancer and cardiovascular diseases. Because of space limitation, we describe only those efforts relevant to this book topic. We have, however, cited most of our prenylation work in the references *(41,42,49–72)*.

2. DESIGN OF CAAX PEPTIDOMIMETICS AS INHIBITORS OF FTASE AND GGTASE I

Because CAAM tetrapeptides are potent inhibitors of FTase (IC_{50}s ranging from 20 n*M* for CVFM to 200 n*M* for CVIM), we have used the CAAM scaffold as a target to design peptidomimetics that are cell-permeable and that are less susceptible to protease degradation (reviewed in ref. *46*). Our initial strategy was to replace the central aliphatic dipeptide "AA" by a moiety that is hydrophobic in nature and that lacks peptidic features *(41, 42,60,62)*. FTI-276 is a third-generation CAAM peptidomimetic where reduced cysteine was linked to methionine by an aromatic spacer, 2-phenyl-4 amino benzoic acid *(49)* (Fig. 1). Using a similar strategy we have also made several GGTase I inhibitors. GGTI-297 is a CAAL peptidomimetic where reduced cysteine is linked to leucine by 2-naphthyl-4 aminobenzoic acid *(52,55)* (Fig. 1). To eliminate more peptidic features from these molecules, we replaced cysteine by several moieties and found imidazoles to yield highly potent and selective inhibitors *(72)*. For example, FTI-2148 and GGTI-2154 (Fig. 1) are

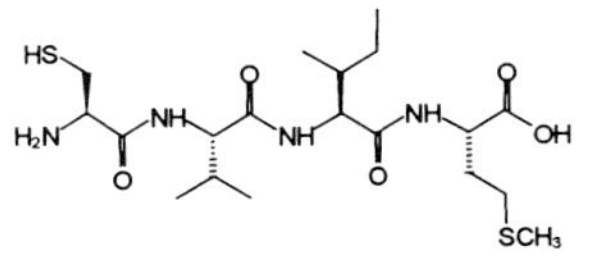

C - V - I - M

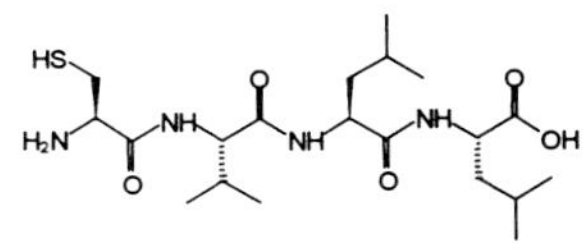

C - V - L - L

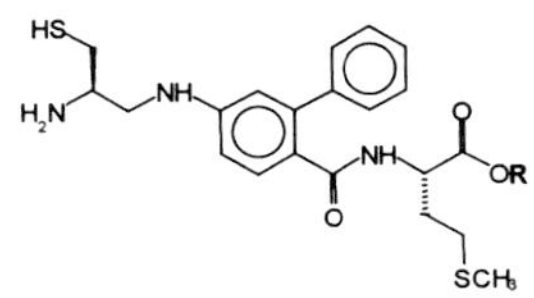

FTI-276 R = H
FTI-277 R = CH_3

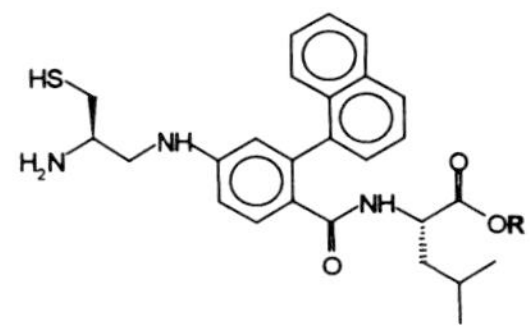

GGTI-297 R = H
GGTI-298 R = CH_3

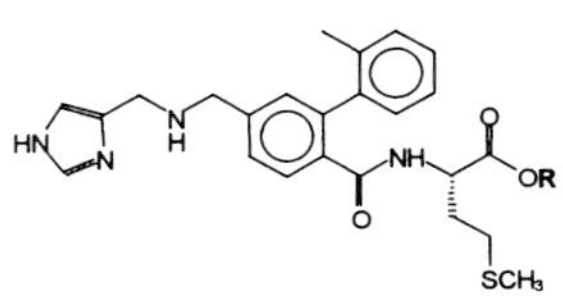

FTI-2148 R = H
FTI-2153 R = CH_3

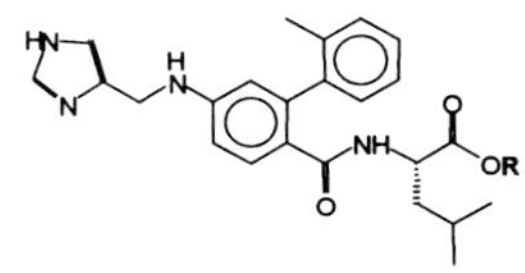

GGTI-2154 R = H
GGTI-2166 R = CH_3

Fig. 1. Design of CAAX peptidomimetics as FTIs and GGTIs.

Table 2
Inhibition of IC_{50}s for FTIs and GGTIs

	***In vitro* Activity**		**Selectivity**	**Whole Cell Processing (methylesters)**		**Selectivity**
Compound/ methylester	FTase (nM)	GGTase I (nM)	Fold	Ras (μM)	Rap1 (μM)	Fold
FTI-276/FTI-277	0.5	50	100	0.3	>30	>100
FTI-2148/FTI-2153	1.4	1700	1200	0.03	>30	>1000
GGTI-297/GGTI-298	203	55	4	>15	5	>3
GGTI-2154/GGTI-2166	5600	21	266	>30	0.3	>100

highly potent and selective noncysteine-containing FTase and GGTase I inhibitors (*see* Table 2). For more details about the design of these compounds and for extensive structure activity relationship (SAR) studies, please refer to Chapter 4.

3. SELECTIVE INHIBITION OF FARNESYLATION AND GERANYLGERANYLATION IN VITRO AND IN WHOLE CELLS

The potency of the FTIs and GGTIs that we have made (Fig. 1) to inhibit FTase and GGTase I was determined as described previously *(49,52,55)*. Table 2 shows the IC_{50} values of the inhibitors. FTI-276 (IC_{50} = 0.5 n*M*) is three times more potent than FTI-2148 (IC_{50} = 1.4 n*M*) at inhibiting FTase. Table 2 shows that GGTI-2154 (IC_{50} = 21 n*M*) is 2.5 times more potent than GGTI-297 (IC_{50} = 55 n*M*) at inhibiting GGTase I. Table 2 also demonstrates that both FTIs are highly selective for FTase over GGTase I, whereas only GGTI-2154 but not GGTI-297 is highly selective for GGTase I over FTase.

We next evaluated the ability of the FTIs and GGTIs to inhibit protein farnesylation and geranylgeranylation in whole cells by determining their ability to inhibit H-Ras and Rap1A processing in NIH3T3 cells, as described previously *(49,52,55)*. Here we used FTI-277, FTI-2153, GGTI-298 and GGTI-2166 which are the methyl ester prodrugs of FTI-276, FTI-2148, GGTI-297 and GGTI-2154, respectively. We and others have shown that the methyl ester prodrugs facilitate the inhibitor's cellular uptake and hence increase whole cell inhibitory activity *(49,50,73–75)*. Table 2 shows that the methyl esters FTI-277 and FTI-2153 are able to inhibit H-Ras processing with IC_{50}s of 300 and 30 n*M*, respectively. This inhibition is highly selective for farnesylation since the IC_{50} of the processing of Rap1A was higher than 30 μ*M*. Furthermore, GGTIs also inhibited Rap1A processing selectively. Table 2 shows that GGTI-298 and GGTI-2166 inhibited Rap1A processing with IC_{50}s of 5 and 0.3 μ*M*, respectively. The selectivity for GGTI-298 and GGTI-2166 were over 3- and 100-fold, respectively (Table 2).

4. FTIs ANTAGONIZE ONCOGENIC H-RAS SIGNALING

We next determined the ability of FTIs to inhibit oncogenic H-Ras signaling in NIH3T3 cells transformed by GTP-locked H-Ras, as described previously *(49,52)*. NIH3T3 cells were treated with inhibitors and the lysates immunoblotted with MAPK antibodies. FTI-277 and FTI-2153 suppressed oncogenic H-Ras activation of MAPK at 3 and 0.03 μ*M*, respectively *(72)*. FTI-277 inhibited MAPK activation by oncogenic farnesylated H-Ras (CVLS) and not by oncogenic geranylgeranylated H-Ras (CVLL), (where CVLS was mutated to CVLL), further confirming the selectivity of FTI-277 for FTase over GGTase I *(49)*. Furthermore, FTI-277 suppressed oncogenic H-Ras, but not oncogenic Raf, activation of MAPK indicating the specificity of FTI-277 for its intended target *(49)*.

5. BOTH FTIs AND GGTIs ARE REQUIRED FOR INHIBITION OF ONCOGENIC K-RAS PRENYLATION BUT EACH ALONE IS SUFFICIENT TO SUPPRESS HUMAN TUMOR GROWTH IN SOFT AGAR AND NUDE MOUSE XENOGRAFTS

Our earlier studies demonstrated that in NIH3T3 cells transformed with oncogenic H-Ras and K_B-Ras, inhibition of K_B-Ras prenylation and MAPK activation required much higher concentrations of FTI-277 than inhibition of H-Ras prenylation and MAPK activation *(41,49)*. Furthermore, K_B-Ras was shown to be a substrate for GGTase I and becomes geranylgeranylated when FTase is inhibited (*43-45*; *see also* Chapter 7). Our results with three human tumor cell lines give further support for this. Figure 2 shows that in a lung (Calu-1) and two pancreatic (Panc-1 and Colo-357) carcinoma cell lines, inhibition of the prenylation of K-Ras is highly resistant to FTI-277 and requires cotreatment with FTI-277 and GGTI-298 *(42,54)*. Another lung carcinoma cell line (A-549) was resistant to the cotreatment (Fig. 2).

The results from the K-Ras prenylation studies summarized previously suggest that inhibition of the growth of human tumors that contain mutated K-Ras may require cotreatment with both FTIs and GGTIs. This is not the case, however. We found that FTI-277 alone is a highly potent antitumor agent. For example, Fig. 3A shows that the growth of the human pancreatic cell line Colo-357 in soft agar was inhibited at concentrations of FTI-277 as low as 25 μ*M* and was blocked at 50 μ*M* *(42)*. Yet, at these concentrations K-Ras prenylation was not inhibited (Fig. 3B) *(42)*. Furthermore, the ability of FTI-276

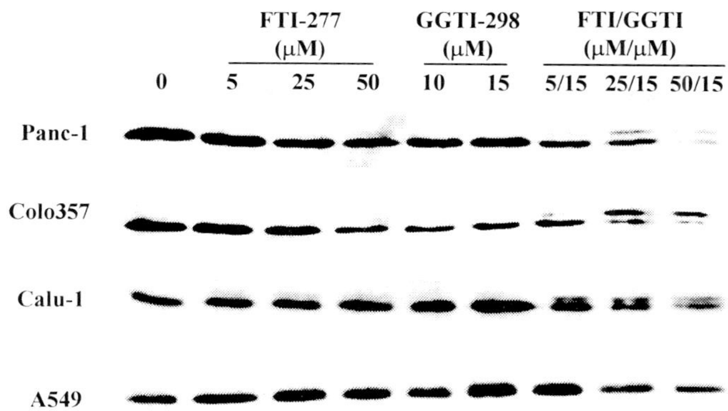

Fig. 2. Inhibition of K-Ras prenylation requires co-treatment with both FTI and GGTI.

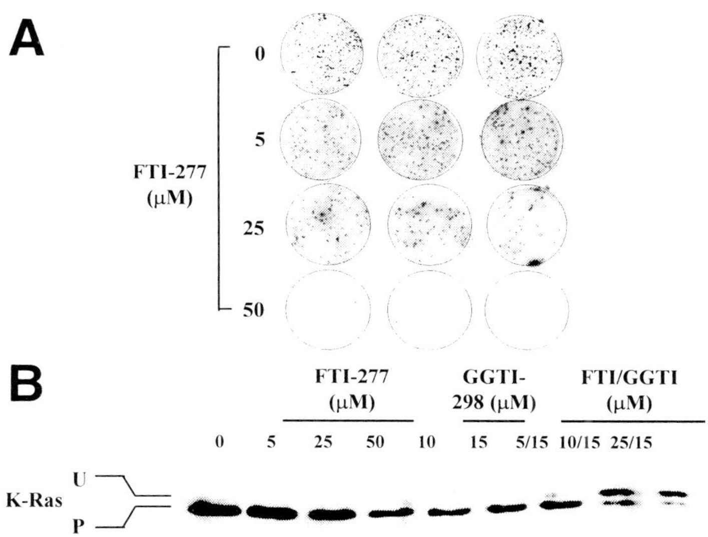

Fig. 3. Inhibition of K-Ras prenylation and soft agar growth of the human pancreataic cell line Colo357.

alone to inhibit human tumor growth in whole animals was confirmed by showing that daily injections of 50 mg/kg of FTI-276 resulted in 80% growth inhibition of A-549 cells grown subcutaneously (s.c.) under the skin of nude mice (Fig. 4) *(50,54)*. Importantly, FTI-276 was also able to suppress tumor growth in nude mice of another human lung carcinoma (Calu-1) with multiple genetic alternations, such as Ras mutation, p53 deletion, and a silenced CDK inhibitor, p16 *(50)*. This antitumor activity was not restricted to FTI-276, in that the highly selective nonthiol FTase inhibitor, FTI-2148, was also efficacious at inhibiting tumor growth (Fig. 4) *(72)*. Thus, despite their inability to inhibit K-Ras prenylation, FTIs are highly potent antitumor agents. These experiments and others in the field suggest that farnesylated targets in addition to K-Ras may play an important role in transformation. However, K-Ras may still be the important target for FTIs mechanism of antitumor activity.

This would be the case if in FTI-treated human tumors, the newly geranylgeranylated K-Ras *(43,44)* can no longer drive transformation or may even antagonize transformation. So far the only experiments that show that geranylgeranylated K_B-Ras can drive trans-

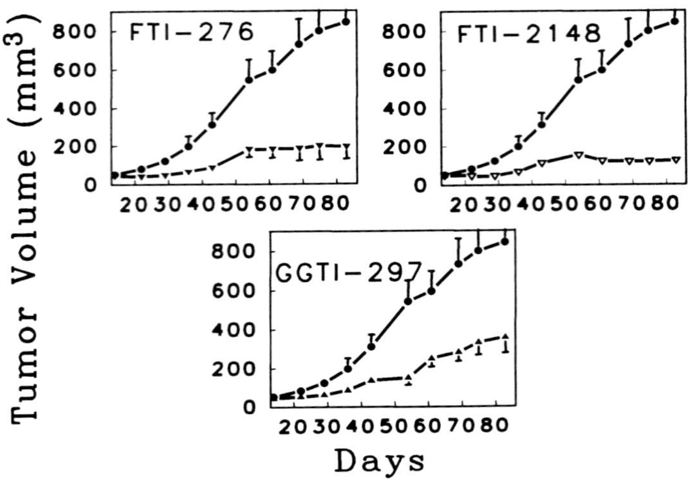

Fig. 4. Antitumor efficacy in A-549 human lung.

formation are those where this form of K_B-Ras is overexpressed in NIH-3T3 cells, but not expressed at its physiological levels as is the case in human tumors treated with FTIs.

It is also very important to note that we have found that GGTIs alone also inhibited human tumor growth in nude mice but not as effectively as FTIs. For example, GGTI-297 inhibited the growth of A-549 tumors by 60% (Fig. 4) *(54)*. The nonthiol derivative GGTI-2154 (Table 1) was also effective at inhibiting human tumor growth in vivo *(72)*. The fact that GGTase I inhibitors have antitumor activity of their own also suggests that some substrates for GGTase I are important for malignant transformation *(54,72)*. This is consistent with reports that demonstrated that two substrates for GGTase I (RhoA and Racl) are transforming in their GTP-locked form *(76,77)*. Another attractive target is RhoB, which was shown to be both farnesylated and geranylgeranylated *(38–40,78)*. However, in order for RhoB to be a target, one would need to hypothesize that farnesylated RhoB is required for transformation, and/or geranylgeranylated RhoB antagonizes transformation (*see* Chapter 11 for more details). Our recent data argues against RhoB being a target for FTIs in human cancer cells since both farnesylated and geranylgeranylated RhoB were found to inhibit soft agar growth, foci formation, and growth in nude mice *(86)*.

The ability of our FTIs to suppress human tumor growth is not limited to the aforementioned s.c. xenograft model. Recently, in collaboration with Dr. Ian Pollack (University of Pittsburgh), we have demonstrated that FTI-276 was also highly effective as an antitumor agent against a human malignant glioma implanted intracranially in nude mice. In this model, all animals receiving vehicle died by d 32, whereas all but one of those receiving FTI-276 survived until d 50 when the experiment was terminated Fig. 5 *(79)*. Furthermore, the antitumor activity of FTI-276 in immunocompetent mice was also very impressive. Indeed, in collaboration with Dr. Ming You (Medical College of Ohio), we have recently shown that FTI-276 was very effective at decreasing the number of lung tumors formed per mouse (Fig. 6) as well as the size of tumors (Fig. 7) in a carcinogen (NNK)-induced lung tumorigenesis model using immunocompetent A/J mice *(80)*.

In addition to its impressive antitumor efficacy in animal models, FTI-276 is not toxic. For example, animals treated for over 2 mo with FTI-276 (50 mpk/d) (Fig. 4) showed no weight loss or overall toxicity *(50,54,72)*. Further evidence of lack of toxicity came from work involving the effects of our FTIs and GGTIs on activated Ras1-induced eye

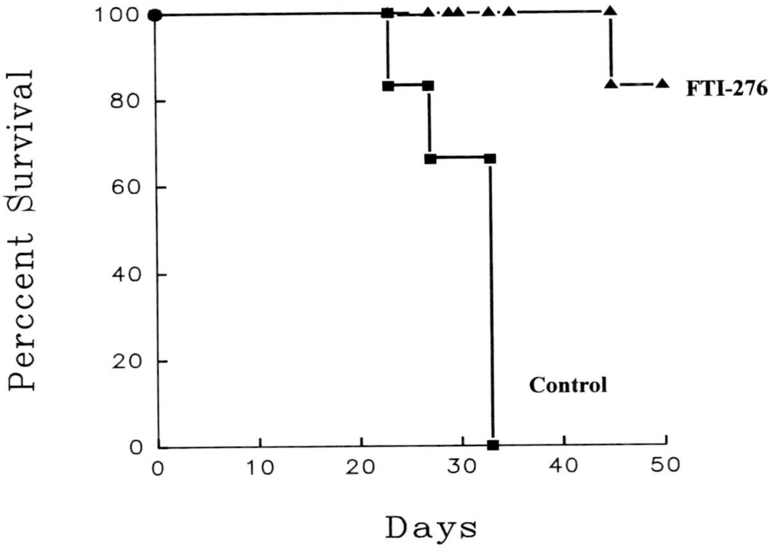

Fig. 5. FTI-276 increases survival time of nude mice intracranially implanted with U-87 human malignant glioma. Data suggests that FTI-276 crosses the blood–brain barrier.

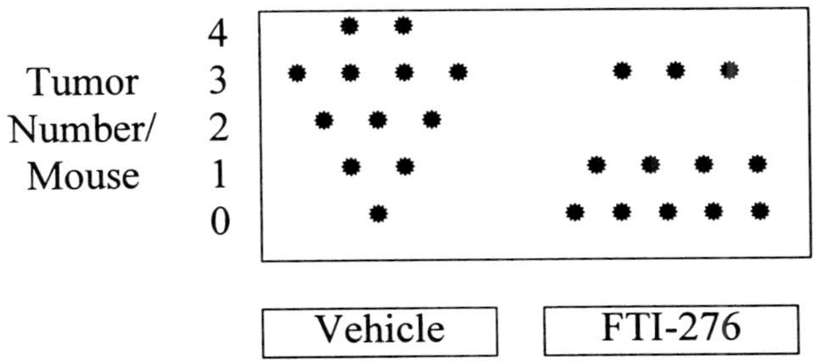

Fig. 6. Rank order of tumor incidence with 30-d FTI-276 treatment. Mice were initiated with NNK, 100 mg/kg ip. Each dot represents the tumor burden of an individual tumor ($p < 0.01$).

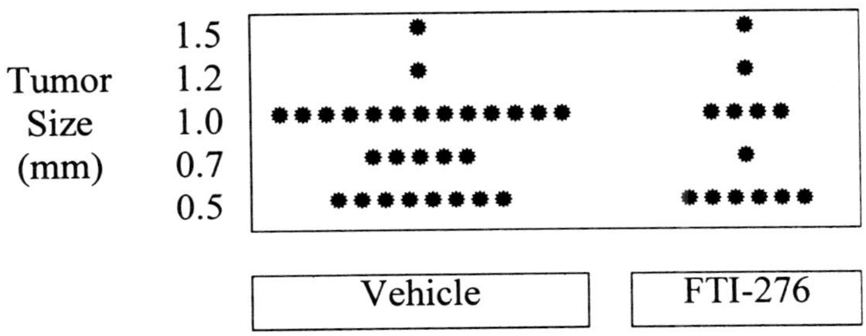

Fig. 7. Rank order of tumor incidence with 30-d FTI-276 treatment. Mice were initiated with NNK, 100 mg/kg ip. Each dot represents an individual tumor ($p < 0.01$).

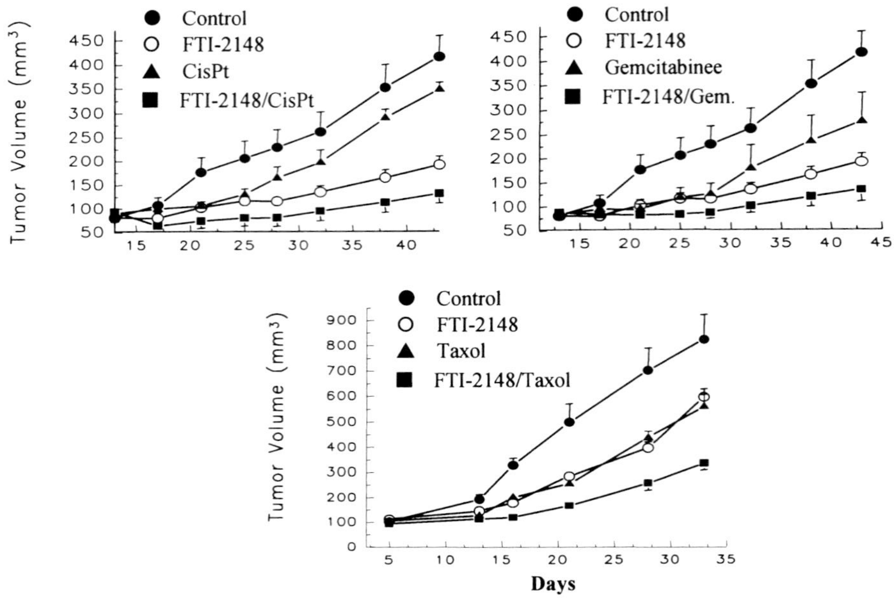

Fig. 8. Combination therapy with FTI-2148 and cytotoxic agents.

malformation in *Drosophila melanogaster*. This study demonstrated that these inhibitors when microinjected at a larvae stage of eye development, are very effective at reversing the Ras1-induced eye malformation without affecting normal eye development *(53)*. Experiments using cultured cells also suggest that FTIs lack toxicity to normal cells. FTI-276 was shown to not inhibit PDGF-dependent PDGFR tyrosine phosphorylation and subsequent MAPK activation in NIH3T3 cells *(52)*.

6. COMBINATION THERAPY OF FTIs OR GGTIs WITH CYTOTOXIC ANTICANCER DRUGS IS MORE EFFICACIOUS THAN MONOTHERAPY

Evidence from our research and that of others *(46–48)* indicated that FTIs' antitumor effect is often cytostatic and reversible. This suggested that combination therapy may be beneficial. Furthermore, there are several reasons why combination therapy with FTIs or GGTIs and other clinically used anticancer drugs with different mechanisms of action may prove to be more beneficial than monotherapy. First, because a tumor is made up of a heterogeneous population of cells with different genetic alterations, treatment with more than one agent may avoid resistance of the tumor to a single drug. Second, Ras has been shown to induce resistance to radiation and some cytotoxic agents. Therefore, inhibition of K-Ras prenylation may sensitize human tumors to cytotoxic agents or radiation (*see* Chapter 12). On the other hand, treatment with cytotoxic agents may also alter tumor cells such that they become even more sensitive to FTIs or GGTIs. To explore these possibilities we have implanted sc FTI-2148-containing mini-pumps (25 mg/kg/d × 14 d) (in mice bearing A-549 cells as in Fig. 4) 2 d prior to treatment with either gem-citabine (80 mpk [ip] every 4th d, 3 times), or cisplatin (5 mpk [ip] every 4th d, 3 times). Our studies have shown that combination therapy is more efficacious than monotherapy (Fig. 8) *(72)*. Similar results were obtained with taxol (12.5 mpk [ip] every 4th d, 3 times) and

FTI-2148 at a lower dose (12.5 mpk/d × 14 d minipump) (Fig. 8). In all cases the effect was additive rather than synergistic. We have also demonstrated that GGTI-2154 in combination with either gemcitabine, cisplatin, or taxol was more effective than monotherapy *(72)*. More extensive studies where dosing and scheduling are optimized are needed in order to determine the most efficacious protocols for these combination therapy studies.

7. INHIBITION OF PRENYLATION RADIOSENSITIZES HUMAN TUMORS

In collaboration with Drs. Bernhard, McKenna, and Muschel from the University of Pennsylvania, FTI-277 was shown to sensitize H-Ras transformed fibroblasts to radiation-induced apoptosis *(66)*. More recent studies have shown that human tumors with H-Ras mutations are also radiosensitized by FTI-277. However, human tumors with K-Ras mutations required both FTI-277 and GGTI-298 *(69)*. (For further details, please *see* Chapter 12.)

8. FTI-277 INDUCES APOPTOSIS BY BLOCKING THE PI-3 KINASE/AKT-2 SURVIVAL PATHWAY

In an attempt to understand the mechanism of antitumor activity of FTIs, we have evaluated the effects of FTI-277 on the PI-3kinase/AKT survival pathway. We have found that FTI-277 blocks growth factor and adhesion stimulated PI-3Kinase and AKT-2 activation and subsequent phosphorylation of the proapoptotic protein BAD *(81)*. Furthermore, in a set of nine human pancreatic and ovarian carcinomas, FTI-277 induced apoptosis only in those that overexpressed AKT-2. More importantly, FTI-277 induced-apoptosis was rescued by a constitutively activated AKT-2 *(81)*. These studies suggest that a farnesylated protein upstream of PI3-Kinase and AKT-2 mediates survival of human tumors by activating this kinase and that FTI-277 by inhibiting the farnesylation of such protein induces apoptosis. Unlike previous studies that demonstrated that FTIs induce apoptosis only when Ras-transformed cells are either deprived of serum or substratum attachment, our findings indicate that FTIs can induce apoptosis in attached human tumors and in the presence of serum *(81)*.

9. EFFECTS OF FTI-277 AND GGTI-298 ON CELL-CYCLE REGULATION

Our continued efforts to understand the mechanism by which FTIs and GGTIs inhibit human tumor growth lead us to investigate the effects of these inhibitors on the cell-division cycle. First, we have evaluated the effects of these inhibitors on cell cycle distribution by flow cytometry. We have shown that, depending on the human tumor cell line, FTI-277 can induce either a G1/G0 arrest, a G2/M enrichment or have no effect on cell-cycle distribution *(56,57)*. In contrast, GGTI-298 blocks all cells in the G0/G1 phase of the cell cycle and prevents entry into S phase *(55,57)*. In cells where FTI-277 enriches in G2/M, we found that FTI-277 and FTI-2153 block mitosis at pro-metaphase by inhibiting bipolar spindle formation and chromosome alignment *(87)*. On the other hand, GGTI-298 affected cell-cycle biochemical events in a manner consistent with its ability to block human tumors in the G1 phase of the cell cycle. GGTI-298 induces the expression of the cyclin-dependent kinase inhibitor p2l WAF at the transcriptional levels in a p53-independent manner and this involves an Sp1/TGF-β responsive element *(57,58)*. This induction of p21waf appears to be mediated by inhibition of RhoA geranylgeranylation *(57,58)*. This is consistent with the fact that Ras was shown to activate RhoA, which in turn suppresses p21 transcription *(82)*. We have also shown that dominant negative RhoA activates, whereas activated RhoA suppresses, p21waf expression in human pancreatic cell line (Panc-1) *(58)*. In addition to inducing p2lWAF expression, GGTI-298 also induced p21 and p27 partner switching from CDK6 to CDK2 *(83)*. Furthermore, GGTI-298 inhibited the activities of CDK2 and CDK4 but not CDK6 and accumulated pRb in its hypophosphorylated form *(83)*. Finally, we have shown that GGTI-298 inhibits PDGF-stimulated PDGF receptor tyrosine phosphorylation *(52)*. This suggests that a geranylgeranylated protein such as RhoA may be involved in early signaling events triggered by growth factor receptor tyrosine kinases, and that GGTI-298 effects on cell-cycle events are mediated by interfering with RhoA-dependent events at the receptor level.

10. GGTI-298 BUT NOT FTI-277 INHIBITS PROLIFERATION AND INDUCES APOPTOSIS IN VASCULAR SMOOTH MUSCLE CELLS (VSMC)

The aforementioned studies demonstrated that FTIs and GGTIs have major effects on growth, survival, apoptosis, and signaling pathways of cancer cells. We next evaluated the effects of these inhibitors on the biology of VSMC, the aberrant proliferation of which is at the heart of several cardiovascular diseases. First, we confirmed that FTI-277 and GGTI-298 are effective inhibitors of protein prenylation in VSMC by analyzing their effects on the processing of Ras and Rap1A in cultured VSMC from rat pulmonary artery *(59)*. Fig. 9A shows that control cells treated with vehicle contained only fully processed Ras and Rap1A proteins. FTI-277 inhibited the processing of Ras but did not inhibit the processing of Rap1A. GGTI-298 inhibited Rap1A processing with little effect on Ras processing (Fig. 9B).

Because small GTPases, such as Ras, Rho, and Rac have been shown to be involved in G_1 to S transition of the cell cycle *(76,77)*, we reasoned that disruption of their function could lead to suppression of DNA synthesis and cell-growth inhibition. To test this, we treated cells with a range of concentrations of FTI-277 or GGTI-298 and measured the effect on serum-stimulated [^{3}H] thymidine uptake after 2 d. GGTI-298 significantly decreased thymidine uptake at 10 and 20 μ*M* by 50 and 90%, respectively. FTI-277 was less effective and decreased thymidine uptake by only 25 and 39% at 10 and 20 μ*M*, respectively *(59)*. These results suggested that GGTI-298 had an antiproliferative effect in serum-stimulated VSMC. We therefore determined the effect of FTI-277 and GGTI-298 on cell number. GGTI-298 significantly inhibited (80%), whereas FTI-277 only reduced by 30%, VSMC proliferation after 3 d of treatment *(59)*.

Reduction in cell number could be caused by a negative effect on growth or to a loss of cells through programmed cell death. To evaluate the effects of the inhibitors on apoptosis, cells were harvested at 48 h, stained with Hoechst 33342, and examined microscopically under epi-illumination. Control cells exhibited a distinct nuclear membrane enclosing diffusely distributed, moderately fluorescent chromatin (Fig. 10). Cells treated with GGTI-298 showed areas of brightly fluorescent, condensed chromatin, and partial loss of the nuclear membrane. Furthermore, the nuclear membrane was lost and the chromatin

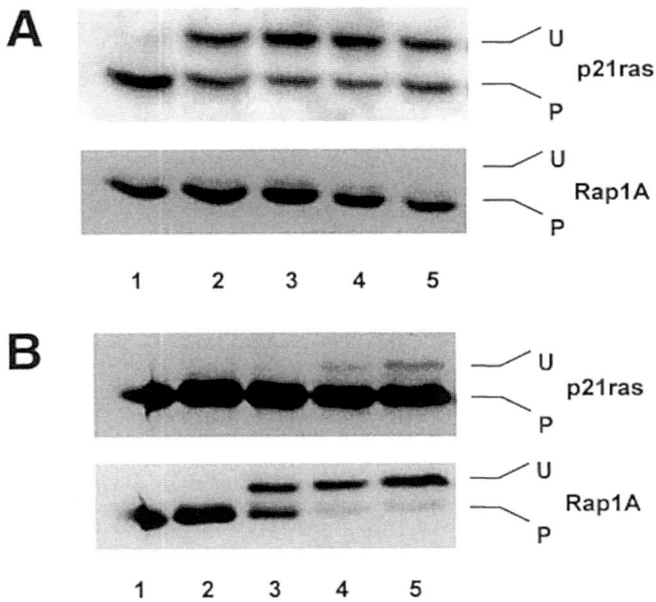

Fig. 9. Inhibiting protein prenylation blocks the processing of Ras and Rap1A in VSMC. Cells were treated with either FTI-277 or GGTI-p298 and the lysates were then immunoblotted with either anti-Ras or anti-Rap1A antibodies. Bottom band, P (processed) form of the protein; top band, U (unprocessed) form. (**A**) lanes 1–5, FTI-277 at 0, 2.5, 5, 10, and 20 μ*M*. (**B**) lanes 1–5, GGTI-298 at 0, 2.5, 5, 10, and 20 μ*M*, respectively

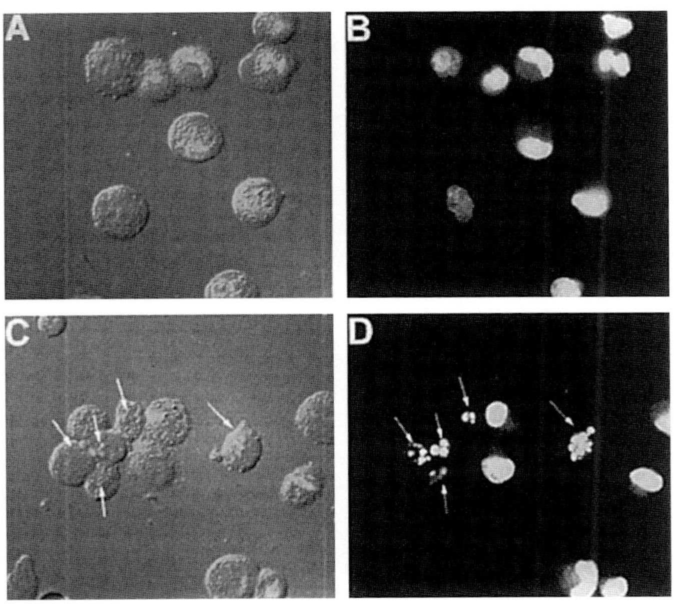

Fig. 10. Morphological assessment of apoptosis in VSMC. Photomicrograph of cells incubated in medium alone (**A**,**B**) or with GGTI-298 (**C**,**D**) for 48 h and viewed under a ×60 objective lens. (**A**) and (**C**) diffraction interference contrast images. (**B**) and (**D**) same cells stained with the DNA-binding fluorochrome Hoechst 33342. Cells considered apoptotic show areas of highly fluorescent, condensed chromatin (arrows) and partial loss of the nuclear membrane.

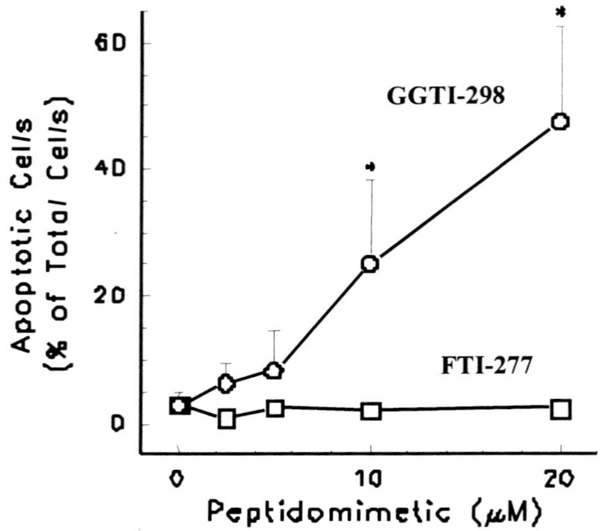

Fig. 11. GGTI-298 induces apoptosis in VSMC. Cells were treated for 48 h with a range of concentrations of FTI-277 or GGTI-298, and apoptosis was defined microscopically as described in Fig. 10.

dispersed into several small aggregates. Serum deprivation alone did not increase the incidence of apoptosis nor did treatment with FTI-277. Treatment with GGTI-298, but not FTI-277, increased apoptosis significantly and in a concentration-dependent manner (Fig. 11) *(59)*.

11. INHIBITION OF PROTEIN GERANYLGERANYLATION CAUSES SUPERINDUCTION OF NOS-2 PROTEIN AND NITRITE LEVELS BY IL-1β IN VSMC

The aforementioned studies showed that GGTI-298 inhibits VSMC growth and induces apoptosis. Because nitric oxide (NO) is known to inhibit VSMC proliferation, we investigated the effects of GGTI-298 on NO formation as a possible mechanism of action of GGTI-298 growth-inhibiting effects. To this end, we treated VSMC with GGTI-298 prior to treatment with IL-1β, a known inducer of NOS-2. We first analyzed the effects of GGTI-298 on IL-1β-induced nitrites as a measure of medium-released nitric oxide. Figure 12 shows that in the absence of inhibitors, IL-1β (0–10 ng/mL) induced a concentration-dependent, but modest, increase in the medium levels of nitrite. Pretreatment with GGTI-298 (10 μ*M*) caused a dramatic increase in IL-1β-induced nitrite formation (from 4 to 52 μ*M*). The medium of cells treated with 1 ng/mL IL-1β alone accumulated nitrite levels of 4 μ*M*, whereas the medium of cells treated with GGTI-298 prior to IL-1β accumulated levels of nitrites that were more than 10-fold higher (42 μ*M*) (Fig. 12). In the absence of IL-1β, GGTI-298 increased basal levels of nitrites to levels comparable to those obtained by stimulation of cells with 1 ng/mL IL-1β alone *(67)*. In contrast to the effects of GGTI-298, treatment of cells with FTI-277 inhibited IL-1β-stimulated nitrite production (Fig. 12).

The dramatic increase in the level of nitrite brought about by GGTI-298 could be owing to direct activation of NOS-2 enzymatic activity or superinduction of NOS-2 protein. To determine the effects of GGTI-298 on NOS-2 protein levels, VSMC were treated with GGTI-298 for 24 h and then stimulated with IL-1β for a further 24 h. Figure 13 shows

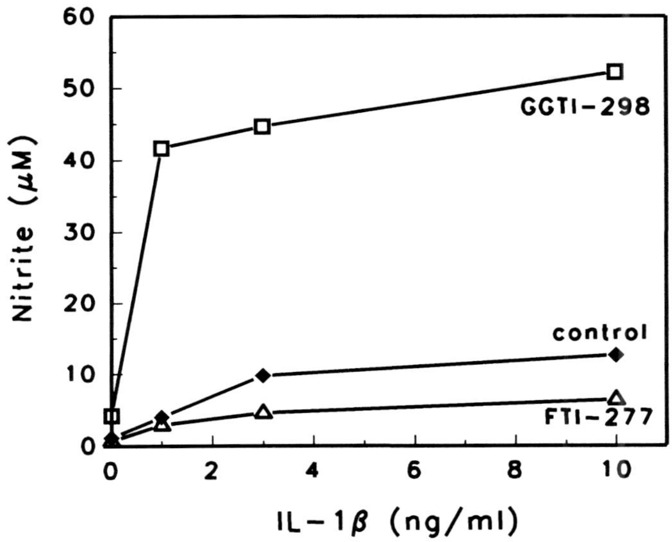

Fig. 12. Effect of FTI-277 and GGTI-298 on IL-1β-stimulated nitrite formation. Cells were pretreated for 24 h with FTI-277 (10 μ*M*) or GGTI-298 (10 μ*M*) or were left untreated. Fresh medium with or without FTI-277 or GGTI-298 and containing IL-1β at 0, 1, 3, or 10 ng/mL was then added. After a further 24 h, the medium was harvested and assayed for nitrites. The curves show the response of IL-1β of untreated cells (◆) or cells treated with FTI-277 (△) or GGTI-298 (□).

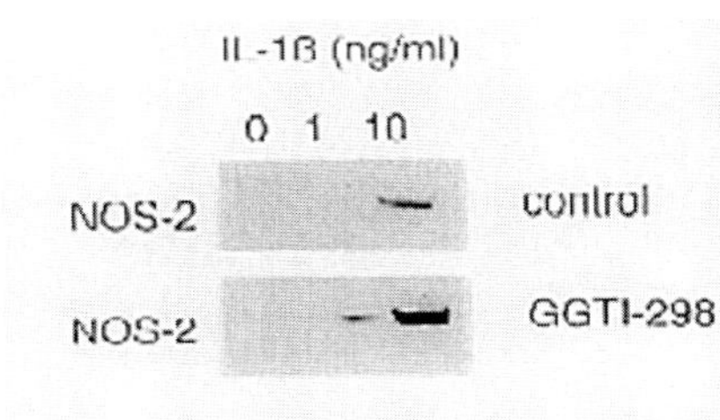

Fig. 13. Effect of GGTI-298 on IL-1β stimulated expression of NOS-2 protein levels in VSMC. Cells were treated with and without GGTI-298 in the manner described in the legend to Fig. 12. Following 24 h of IL-1β stimulation, cells were harvested, and NOS-2 protein levels were determined by Western analysis.

that in the absence of GGTI-298 (control), significant induction of NOS-2 protein levels in VSMC occurred only at 10 ng/mL IL-1β, whereas in the presence of inhibitor, NOS-2 was induced at concentrations as low as 1 ng/mL. In VSMC treated with 10 ng/mL IL-1β, GGTI-298 enhanced the ability of IL-1β to induce NOS-2 protein by fivefold.

12. FTI-277 BLOCKS IL-1β INDUCTION OF NOS-2 IN VSMC

The aforementioned results demonstrated that inhibition of protein geranylgeranylation causes a superinduction of NOS-2 by IL-1β. We next determined the consequences of inhibiting protein farnesylation. VSMC were pretreated for 24 h with FTI-277 prior to a 24 h treatment with various concentrations of IL-1β (0–10 ng/mL). Figure 14 shows that pretreatment of VSMC with FTI-277 (10 μ*M*) blocked IL-1β induction of NOS-2 protein.

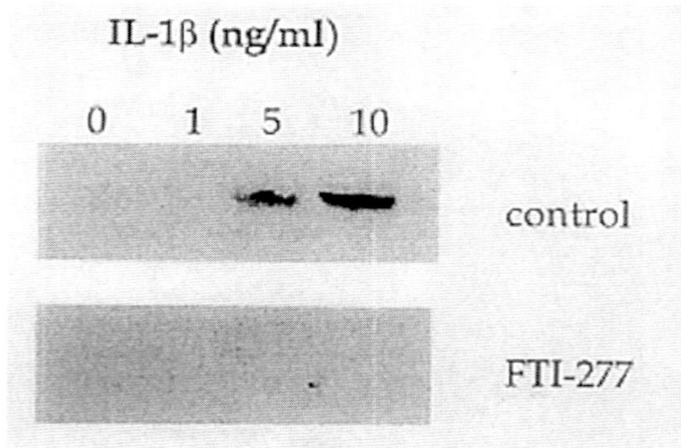

Fig. 14. Effect of FTI-277 on IL-1β-stimulated expression of NOS-2 protein. VSMC were treated with and without FTI-277 (10 μ*M*) as described in the legend to Fig. 12 and stimulated with IL-1β for 24 h at the concentrations shown. Cell lysates were then immunoblotted to determine the levels of NOS-2 proteins.

13. GGTI-298 ENHANCES, WHEREAS FTI-277 BLOCKS, THE ABILITY OF IL-1β TO INDUCE NOS-2 mRNA IN VSMC

We next determined whether the effects of GGTI-298 and FTI-277 on NOS-2 protein and nitrite levels were owing to alterations at the mRNA level. VSMC were treated with vehicle, IL-1β, FTI-277, GGTI-298, and IL-1β with either FTI-277 or GGTI-298, and the levels of NOS-2 mRNA determined by Northern blotting, as described previously *(67)*. Stimulation of cells with IL-1β shows a modest induction of NOS-2 mRNA (Fig. 15, lanes 1–3) while treatment of cells with FTI-277 or GGTI-298 alone had no detectable effect (lanes 4 and 7). Treatment of cells with FTI-277 blocked the ability of IL-1β to induce NOS-2 mRNA (compare lanes 1–3 to lanes 4–6). In contrast, GGTI-298 enhanced the ability of IL-1β to induce NOS-2 by 6- and 14-fold at 1 and 10 ng/mL IL-1β, respectively (compare lanes 1–3 to lanes 7–9) *(67)*.

14. GGTI-298 AND FTI-277 BLOCK IL-1β AND PDGF-STIMULATED SUPEROXIDE FORMATION IN VSMC

The aformentioned results show that inhibition of protein farnesylation blocks IL-1β induction of NOS-2, whereas inhibition of protein geranylgeranylation superinduces IL-1β-dependent NOS-2 and increases production of NO. Next we evaluated the effects of inhibition of protein prenylation on the formation of superoxide, an oxidant radical that is cogenerated with NO in response to IL-1β *(84)*. VSMC were pretreated with FTI-277 (5 μ*M*) or GGTI-298 (10 μ*M*) on each of two successive days. The cells were then treated with IL-1β (100 U/mL) for 2 h and harvested. The cell-conditioned medium collected over 2 h was analyzed for superoxide dismutase inhibitable-superoxide production by a cytochrome c reduction assay *(84)*. Control cells treated with vehicle-released basal superoxide levels that were stimulated 3.5-fold by IL-1β. However, both FTI-277 and GGTI-298 significantly reduced IL-1β stimulated superoxide levels to baseline *(84)*.

We next determined whether this effect of prenylation inhibitors on superoxide formation was specific to IL-1β. Therefore, we treated VSMC with a known smooth-muscle mitogen, PDGF, and determined the levels of superoxide in the culture medium. PDGF-BB caused a concentration-dependent increase in VSMC-derived superoxide anion

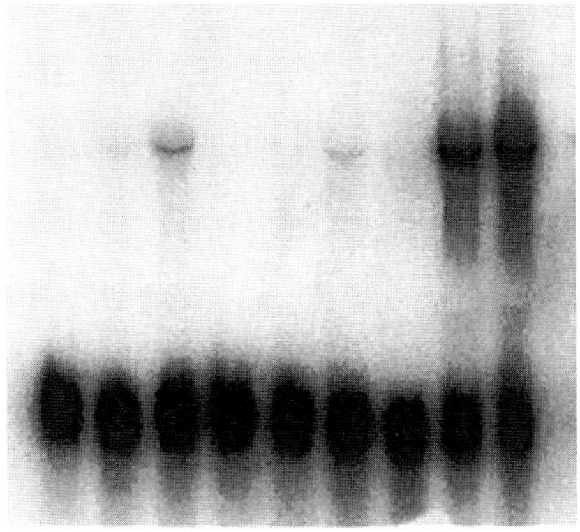

Fig. 15. Effects of GGTI-298 and FTI-277 on IL-1β induction of NOS-2 mRNA. VSMC were treated with and without GGTI-298 or FTI-277 and stimulated with IL-1β. Total RNA was isolated and the levels of NOS-2 and cyclophilin mRNAs were determined by Northern-blot analysis as described previously *(67)*. Lane 1, vehicle; lanes 2 and 3, IL-1β (1 and 10 ng/mL, respectively); lanes 7–9, GGTI-298 (10 μ*M*) and 1L-1β (0, 1, and 10 ng/mL, respectively).

during a 2-h incubation. Pretreatment of cells with FTI-277 and GGTI-298 blocked production of superoxide stimulated by PDGF. Thus, FTI-277 and GGTI-298 were able to block superoxide production in VSMC by both a cytokine (IL-1β) and a mitogen (PDGF) *(84)*.

The effect of FTI-277 and GGTI-298 implicated prenylated proteins in the signal transduction pathways mediating superoxide production. We considered H-Ras, a farnesylated protein, and RhoA and Rac1, two geranylgeranylated proteins, as likely candidates because they all have been implicated in growth factor (i.e., PDGF) signaling and proliferation. To this end we have started first by exploring the role of H-Ras in superoxide formation. We have stably transfected VSMC with a GTP-locked, constitutively active form of Ras (Val 12 H-Ras mutant). FTI-277 inhibited processing of both endogenous wild-type Ras and mutant H-Ras. GGTI-298 had little effect on processing of the endogenous or mutant forms of Ras. In the absence of mitogen or cytokine stimulation, the mutant H-Ras transfectants produced levels of superoxide similar to those of wild-type cells stimulated with PDGF-BB or IL-1β and significantly higher than the levels of superoxide produced by cells transfected with the pSV2 neo empty plasmid vector. Production of superoxide was suppressed by pre-treatment with FTI-277, consistent with a role for the exclusively farnesylated H-Ras mutant *(84)*. GGTI-298, however, also suppressed superoxide production, suggesting the involvement of a geranylgeranylated proteins (such as RhoA and Rac1) down stream of H-Ras *(84)*.

15. GGTI-297 IS A BETTER INHIBITOR THAN FTI-276 OF NEOINTIMAL HYPERPLASIA FOLLOWING BALLOON ANGIOPLASTY INJURY IN THE RAT CAROTID ARTERY MODEL

The aforementioned studies demonstrated that inhibition of protein geranylgeranylation results in VSMC growth inhibition and apoptosis. Inhibition of protein farnesylation also inhibited VSMC proliferation, although to a lesser degree, but did not induce apoptosis. We next evaluated whether this VSMC growth inhibition by prenyltransferase inhibitors could also occur in vivo. To this end, we determined the ability of GGTI-297 and FTI-276 to inhibit neointimal hyperplasia in vivo in a rat carotid artery injury model. Vascular injury was created by introducing a balloon catheter via external carotid artery. Two weeks after injury, the vehicle-treated animals had pronounced neointimal hyperplasia. In contrast, rats treated with GGTI-297 (12.5 mg/kg/d) delivered from a 7-d mini-pump through a catheter cannulated to the contralateral jugular vein at the time of carotid angioplasty, had reduced neointimal formation *(85)*. The intimal thickness from injured animals treated with DMSO vehicle was 32.8 ± 4.1 µm. GGTI-297 treatment had a pronounced effect on the development of neointimal hyperplasia reducing the neointimal thickening by 72% of that observed in the vehicle only treated groups *(85)*.

We next investigated the ability of the FTase inhibitor FTI-276 to inhibit neointimal formation in the same rat carotid artery injury model. In contrast to GGTI-297, the effects of FTI-276 on VSMC proliferation after angioplasty is not as pronounced. FTI-276 treatment resulted in only a 44% reduction in neointimal thickening *(85)*.

16. SUMMARY

We have designed, synthesized, and biologically evaluated tetrapeptide mimics of the CAAX box as selective FTase and GGTase I inhibitors. Although we have shown that both FTIs and GGTIs are effective at inhibiting aberrant proliferation of epithelial as well as VSMC in culture, in whole animal studies we found FTIs to be more effective as anti-cancer agents, whereas GGTIs are more effective in cardiovascular disease.

Our FTIs are highly selective for FTase over GGTase I both in vitro and in whole cells and antagonize oncogenic H-Ras signaling. We have also shown, in human tumors, that unlike the prenylation of H-Ras and N-Ras, which is inhibited by FTIs alone, the prenylation of K-Ras requires both FTIs and GGTIs. Yet, inhibition of the growth of these human tumors in soft agar and in nude mice requires only FTIs. This indicates that when FTIs inhibit K-Ras farnesylation, K-Ras becomes geranylgeranylated (but the tumor growth is still inhibited). Therefore, either geranylgeranylated K-Ras is unable to drive malignant transformation, or farnesylated proteins in addition to K-Ras are required for malignant transformation. We have also shown that combinations of FTIs with cytotoxic agents such as taxol, gemcitabine, and cisplatin are more efficacious than monotherapy at inhibiting human tumor growth in nude mice. Attempts to determine the mechanism by which FTIs inhibit tumor growth led to our discovery that these agents cause apoptosis by inhibiting the PI3-K/AKT-2 survival pathway.

As mentioned above, although GGTIs also inhibit human tumor growth in vitro and in vivo, their effects are more pronounced on aberrant proliferation associated with cardiovascular disease. For example, GGTIs are very effective at inhibiting cell growth and inducing apoptosis of VSMC. In whole animal studies, GGTIs reduced neointimal hyperplasia following balloon angioplasty-induced rat artery injury. Inhibition of protein farnesylation with FTI-276 was not as effective in reducing neointimal hyperplasia. One possible mechanism by which GGTIs inhibit aberrant VSMC proliferation is by enhancing the ability of cytokines and mitogens to stimulate the production of nitric oxide, a known inhibitor of VSMC growth. Other possible mechanisms could be related to the ability of GGTIs to block cells in G1 by inhibiting growth factor receptor tyrosine phosphorylation, inducing CDK inhibitor expression and inhibiting CDK activity and pRb phosphorylation. Thus, GGTIs are more potent than FTIs at inhibiting VSMC proliferation. This suggests that protein geranylgeranylation plays a critical role in aberrant VSMC proliferation and gives support to the idea of using GGTase I as a target for cardiovascular disease therapy.

17. IMPORTANT ISSUES TO BE ADDRESSED

Although we have achieved major accomplishments towards the development of FTIs and GGTIs as therapeutic agents for cancer and cardiovascular diseases, there are several important issues that remain to be addressed. We still do not know the precise mechanism by which these agents inhibit tumor growth. Although we have recently discovered the PI-3 kinase/AKT-2 survival pathway as a target for FTI-induced apoptosis in human tumors, we still do not know which farnesylated protein upstream of PI3-kinase in this pathway is targeted by FTIs *(81)*. The kinetics of inhibition of AKT-2 activity by FTI-277 are fast and suggest a farnesylated protein with a short half-life. Although RhoB has a short half-life, we have demonstrated that RhoB does not affect AKT-2 activity *(81)*. Furthermore, we have shown that both farnesylated and geranylgeranylated RhoB suppress human tumor growth in nude mice arguing against RhoB being a target for FTIs *(86)*. Downstream of AKT-2, we do know that phosphorylation of the proapoptotic protein BAD is inhibited by FTIs but whether this involves Bcl-X_L and cytochrome c release is not known. One key observation of our FTI-induced apoptosis work is that this occurs only in human tumors that overexpress AKT-2. If this is confirmed over a larger number of human tumors, the diagnostic and prognostic implications will be of great value in clinical settings.

Other critical issues that remain to be addressed in clinical settings relate to which biochemical correlates are best for assessing FTIs' efficacy in humans. Inhibition of farnesylation of such key proteins as Ras is reasonable but alternative prenylation by GGTase I adds a layer of complication. A more direct target for proof of concept in humans is to measure inhibition of FTase activity itself as a biochemical end-point for the FTI clinical trials. A critical question to answer here is whether complete inhibition of tumor FTase activity is required for inhibition of tumor growth. Our preliminary data indicate that this is not the case and that complete inhibition of human tumor growth in nude mice requires only 75% inhibition of FTase (data not shown). Whether this will also be the case in human clinical trials remains to be determined. Although the Ras mutation status does not appear to predict sensitivity of human tumors to FTIs in preclinical experiments, this issue must be addressed in clinical trials. A final issue relates to evaluating whether combination therapy of FTIs with other agents would be beneficial. Several in vivo studies have shown that combining FTIs with cytotoxic agents is additive but not synergistic. However, much more work is needed to optimize these combination therapies prior to investigations in the clinical arena. Obviously, it is also very important to evaluate the benefits of combining FTIs with other anti-signaling drugs.

Much work is required with GGTIs towards their proof of concept as therapeutic agents. With regards to the use of GGTIs as anti-cancer agents, a concern is that these molecules may be toxic owing to the wider spectrum of GGTase I substrates that have important physiological roles. This concern is somewhat dampened by the fact that in whole animal studies, doses of GGTIs that inhibit cancer growth as well as those that reverse restenosis are not toxic. More careful investigation is required to determine the actual therapeutic window of these agents. Another important issue to be addressed with

GGTIs is whether or not combination with FTIs is of value. Although data with our first generation GGTIs showed no benefit, these experiments should be repeated with our recently designed, highly potent, and highly selective agents *(72)*. This is particularly important in view of the fact that we now know that inhibition of K-Ras prenylation requires both FTIs and GGTIs. Finally, although GGTIs show great anti-proliferative activity and induce apoptosis of VSMC, as well as inhibit restenosis in a rat model, further work needs to be done to document their use in cardiovascular disease.

ACKNOWLEDGMENT

We would like to thank our collaborators for their contributions: Drs. Ian Pollack, Larry Shears, Timothy Billiar, Jonathan Finder, Paul Davies, Bruce Pitt, and Richard Carthew (University of Pittsburgh); Drs. Channing Der and Adrienne Cox (University of North Carolina); Dr. Saul Rosenberg (Abbott); Drs. Gillis McKenna, Ruth Muschel, and Eric Bernhard (University of Pennsylvania); Dr. Gille Favre (Universite Paul Sabatier); Drs. Jalila Adnane and Jin Cheng (University of South Florida); and Dr. Ming You (Medical College of Ohio). This work was partially supported by a grant from the National Cancer Institute (CA67771).

REFERENCES

1. Barbacid M. Ras genes. *Ann Rev Biochem* 1987; 56:779–828.
2. Barbacid M. Human oncogenes, in: *Important Advances in Oncology* (Devita and Rosenberg, eds.). Lippincott, Philadelphia, 1986, pp. 3–22.
3. Bar-Sagi D , Feramisco JR. Microinjection of the ras oncogene protein into PC12 cells induces morphological differentiation. *Cell* 1985; 42:841–848.
4. McCormick F. Ras GTPase activating protein: signal transmitter and signal terminator. *Cell* 1989; 56: 5–8.
5. McCormick F. Signal transduction. How receptors turn *ras* on. *Nature* 1993; 363:15,16.
6. Lowenstein EJ, Daly RJ, Batzer AG, Li W, Margolis B, Lammers R, et al. The SH2 and SH3 domain-containing protein GRB2 links receptor tyrosine kinases to *ras* signaling. *Cell* 1992; 70:431–442.
7. Egan SE, Giddings BW, Brooks MW, Buday L, Sizeland AM, Weinberg RA. Association of Sos *ras* exchange protein with Grb2 is implicated in tyrosine kinase signal transduction and transformation. *Nature* 1993; 363:45–51.
8. Stokoe D, Macdonald SG, Cadwallader K, Symons M, Hancock JF. Activation of Raf as a result of recruitment to the plasma membrane. *Science* 1994; 264:1463–1467.
9. Heldin C-H, Westermark B. Signal transduction by the receptors for platelet-derived growth factor. *J Cell Sci* 1990; 96:193–196.
10. Kazlauskas A, Cooper JA. Autophosphorylation of the PDGF receptor in the kinase insert region regulates interactions with cell proteins. *Cell* 1989; 58:1121–1133.
11. Fantl WJ, Escobedo JA, Martin GA, Turck CW, del Rosario M, McCormick F, Williams LT. Distinct phosphotyrosines on a growth factor receptor bind to specific molecules that mediate different signaling pathways. *Cell* 1992; 69:413–423.
12. Kumjian DA, Wahl MI, Rhee SG, Daniel TO. Platelet-derived growth factor (PDGF) binding promotes physical association of PDGF receptor with phospholipase C. *Proc Natl Acad Sci USA* 1989; 86: 8232–8236.
13. Kazlauskas A, Ellis C, Pawson T, Cooper JA. Binding of GAP to activated PDGF receptors. *Science* 1990; 247:1578–1581.
14. Coughlin SR, Escobedo JA, Williams LT. Role of phosphatidylinositol kinase in PDGF receptor signal transduction. *Science* 1989; 243:1191–1194.
15. Kazlauskas A, Cooper JA. Phosphorylation of the PDGF receptor beta subunit creates a tight binding site for phosphatidylinositol 3 kinase. *EMBO J* 1990; 9:3279–3286.
16. Gibbs JB, Sigal IS, Poe M, Scolnick EM. Intrinsic GTPase activity distinguishes normal and oncogenic ras p21 molecules. *Proc Natl Acad Sci USA* 1984; 81:5704–5708.

17. McGrath JP, Capon DJ, Goeddel DV, Levinson AD. Comparative biochemical properties of normal and activated human ras p21 protein. *Nature* 1984; 310:644–649.
18. Sweet RW, Yokoyama S, Kamata T, Feramisco JR, Rosenberg M, Gross M. The product of ras is a GTPase and the T24 oncogenic mutant is deficient in this activity. *Nature* 1984; 311:273–275.
19. Manne V, Bekesi E, Kung HF. Ha-ras proteins exhibit GTPase activity: point mutations that activate Ha-ras gene products result in decreased GTPase activity. *Proc Natl Acad Sci USA* 1985; 82:376–380.
20. Willumsen BM, Christensen A, Hubbert NC, Papageroge AG, Lowy DR. The p21 ras c-terminus is required for transformation and membrane association. *Nature* 1984; 310:583–586.
21. Willumsen BM, Norris K, Papageorge AG, Hubbert NC, Lowy DR. Harvey murine sarcoma virus p21 ras protein: biological and biochemical significance of the cysteine nearest the carboxy terminus. *EMBO J* 1984; 3:2581–2585.
22. Hancock JF, Magee AI, Childs JE, Marshall CJ. All ras proteins are polyisoprenylated but only some are palmitoylated. *Cell* 1989; 57:1167–1177.
23. Gutierrez L, Magee AI, Marshall CJ, Hancock JF. Post-translational processing of Ras is two-step and involves carboxyl-methylation and carboxy-terminal proteolysis. *EMBO J* 1989; 8:1093–1098.
24. Casey PJ, Solski PA, Der CJ, Buss JE. Ras is modified by a farnesyl isoprenoid. *Proc Natl Acad Sci USA* 1989; 86:8323–8327.
25. Jackson JH, Cochrane CG, Bourne JR, Solski PA, Buss JE, Der CJ. Farnesol modification of Kirsten-ras exon 4B protein is essential for transformation. *Proc Natl Acad Sci USA* 1990; 87:3042–3046.
26. Hancock JF, Paterson H, Marshall JC. A polybasic domain or palmitoylation is required in addition to the CAAX motif to localize Ras to the plasma membrane. *Cell* 1990; 63:133–139.
27. Casey PJ. Biochemistry of protein prenylation. *J Lipid Res* 1992; 33:1731–1740.
28. Sinensky M, Beck LA, Leonard S, Evans R. Differential inhibitory effects of lovastatin on protein isoprenylation and sterol synthesis. *J Biol Chem* 1990; 265:19,937–19,941.
29. Reiss Y, Goldstein JL, Seabra MC, Casey PA, Brown MS. Inhibition of purified Ras farnesyl: protein transferase by Cys-AAX tetrapeptides. *Cell* 1990; 62:81–88.
30. Goldstein JL, Brown MS. Regulation of the mevalonate pathway. *Nature* 1990; 343:425–430.
31. Kato K, Cox AD, Hisaka MM, Graham SM, Buss JE, Der CJ. Isoprenoid addition to Ras is the critical modification for its membrane association and transforming activity. *Proc Natl Acad Sci USA* 1992; 89: 6403–6407.
32. Moores SL, Schaber MD, Mosser SD, Rands E, O'Hara MB, Garsky VM, et al. Sequence dependence of protein isoprenylation. *J Biol Chem* 1991; 266:14,603–14,610.
33. Reiss Y, Seabra MC, Armstrong SA, Slaughter CA, Goldstein JL, Brown MS. Nonidentical subunits of p21H-ras farnesyltransferase. Peptide binding and farnesyl pyrophosphate carrier functions. *J Biol Chem* 1991; 266:10,672–10,677.
34. Yokoyama K, Goodwin GW, Ghomashchi F, Glomser JA, Gelb MH. A protein geranylgeranyltransferase from bovine brain: implications for protein prenylation specificity. *Proc Natl Acad Sci USA* 1991; 88:5302–5306.
35. Yokoyama K, Gelb MH. Purification of a mammalian protein geranylgeranyltransferase. Formation and catalytic properties of an enzyme-geranylgeranyl pyrophosphate complex. *J Biol Chem* 1993; 268: 4055–4060.
36. Seabra MC, Reiss Y, Casey PJ, Brown MS, Goldstein JL. Protein farnesyltransferase and geranylgeranyltransferase share a common alpha subunit. *Cell* 1991; 65:429–434.
37. Zhang FL, Diehl RE, Khol NE, Gibbs JB, Giros B, Casey PJ, Omer CA. cDNA cloning and expression of rat and human protein geranylgeranyl transferase type-I. *J Biol Chem* 1994; 269:3175–3180.
38. Armstrong SA, Hannah VC, Goldstein JL, Brown MS. CAAX geranylgeranyl transferase transfers farnesyl as efficiently as geranylgeranyl to RhoB. *J Biol Chem* 1995; 270:7864–7868.
39. Adamson P, Marshall CJ, Hall A, Tilbrook PA. Post-translational modifications of p21rho proteins. *J Biol Chem* 1992; 267:20,033–20,038.
40. Lebowitz PF, Casey PJ, Predergast GC, Thissen JA. Farnesyltransferse inhibitors alter the prenylation and growth-stimulating function of RhoB. *J Biol Chem* 1997; 272:15,591–15,594.
41. Lerner E, Qian Y, Hamilton AD, Sebti SM. Disruption of oncogenic K-Ras 4B processing and signaling by a potent Geranylgeranyltransferase I inhibitor. *J Biol Chem* 1995; 270:26,770–26,773.
42. Lerner EC, Zhang TT, Knowles D, Qian Y, Hamilton AD, Sebti SM. Inhibition of the prenylation of K-Ras but not H- or N-Ras is highly resistant to CAAX peptidomimetics and requires both a farnesyltransferase and a geranylgeranyltransferase I inhibitor in human tumor cell lines. *Oncogene* 1997; 15(11):1283–1288.

43. Whyte DB, Kirschmeier P, Hockenberry TN, Nunez-Oliva I, James L, Catino JJ, et al. K- and N-Ras are geranylgeranylated in cells treated with farnesyl protein transferase inhibitors. *J Biol Chem* 1997; 272:14,459–14,464.
44. Rowell CA, Kowalczyk JJ, Lewis MD, Garcia AM. Direct demonstration of geranylgeranylation and farnesylation of Ki-Ras in vivo. *J Biol Chem* 1997; 272:14,093–14,097.
45. James GL, Goldstein JL, Brown MS. Polylysine and CVIM sequences of K-RasB dictate specificity of prenylation and confer resistance to benzodiazepine peptidomimetic *in vitro*. *J Biol Chem* 1995; 270: 6221–6226.
46. Sebti SM, Hamilton AD. Inhibition of Ras prenylation: A novel approach to cancer chemotherapy. *Pharmacol Therapeut* 1997; 74:103–114.
47. Gibbs JB, Oliff A. The potential of farnesyltransferase inhibitors as cancer chemotherapeutics. *Annu Rev Pharmacol Toxicol* 1997; 37:143–166.
48. Cox AD, Der CJ. Farnesyltransferase inhibitors and cancer treatment: targeting simply Ras? *Biochim Biophys Acta* 1997; 1333:F51–F71.
49. Lerner E, Qian Y, Blaskovich SM, Fossum R, Vogt A, Cox A, et al. Ras CAAX peptidomimetic FTI-277 selectively blocks oncogenic Ras signaling by inducing cytoplasmic accumulation of inactive Ras/Raf complexes. *J Biol Chem* 1995; 270:26,802–26,806.
50. Sun Z, Qian Y, Hamilton AD, Sebti SM. Ras CAAX peptidomimetic FTI-276 selectively blocks in nude mice the growth of a human lung carcinoma with a K-Ras mutation and a p53 deletion. *Cancer Res* 1995; 55:4243–4247.
51. McGuire T, Qian Y, Blaskovich MA, Fossum RD, Sun J, Marlowe T, et al. CAAX pep-tidomimetic FTI-244 decreases platelet-derived growth factor receptor tyrosine phosphorylation levels and inhibits stimulation of phosphatidylinositol 3-kinase but not mitogen-activated protein kinase. *Biochem Biophys Res Comm* 1995; 214:295–303.
52. McGuire T, Qian Y, Hamilton AD, Sebti SM. Platelet derived growth factor receptor tyrosine phosphorylation requires protein geranylgeranylation and not farnesylation. *J Biol Chem* 1996; 271:27,402–27,407.
53. Kauffman R, Qian Y, Vogt A, Sebti SM, Hamilton AD, Carthew R. Activated drosophila Ras 1 is selectively suppressed by isoprenyltransferase inhibitors. *Proc Natl Acad Sci USA* 1995; 92:10,919–10,923.
54. Sun J, Qian Y, Hamilton AD, Sebti SM. Both farnesyltransferase and geranylgeranyltransferase I inhibitors are required for inhibition of oncogenic K-Ras prenylation but each alone is sufficient to suppress human tumor growth in nude mouse xenografts. *Oncogene* 1998; 16:1467–1473.
55. Vogt A, Qian Y, Hamilton AD, Sebti SM. Protein geranylgeranylation not farnesylation, is required for the G_1 to S phase transition in mouse fibroblasts. *Oncogene* 1996; 13:1991–1999.
56. Miguel K, Pradines A, McGuire TM, Hamilton AD, Sebti SM, Favre G. GGTI-298 induces GO/G1 block and apoptosis whereas FTI-277 causes G2/M enrichment in A549 cells. *Cancer Res* 1997; 57: 1846–1850.
57. Vogt A, Sun J, Qian Y, Hamilton AD, Sebti SM. The geranylgeranyltransferase I inhibitor GGTI-298 arrests human tumor cells in G0/G1 and induces p21WAF/CIP/SDII in a-p53 independent manner. *J Biol Chem* 1997; 272:27,224–27,229.
58. Adnane J, Bizouarn F, Qian Y, Hamilton AD, Sebti SM. p21WAF1/CIP1 is upregulated by the geranylgeranyltransferase I inhibitor GGTI-298 through a TGFβ/Sp1-responsive element: involvement of the small GTPase RhoA. *Mol Cell Biol* 1998; 18:6962–6970.
59. Stark WW, Blaskovich MA, Johnson BA, Vasudevan A, Hamilton AD, Sebti SM, Davies P. Inhibition of geranylgeranylation but not farnesylation promotes apoptosis in vascular smooth muscle cells. *Am J Physiol Lung Cell Mol Physiol* 1998; 275:L55–L63.
60. Nigam M, Seong CM, Qian Y, Hamilton A, Sebti SM. Potent inhibition of human tumor p21 *ras* farnesyl transferase by A_1A_2-lacking CA_1A_2X peptidomimetics. *J Biol Chem* 1993; 268:20,695–20,698.
61. Qian Y, Blaskovich M, Saleem M, Wathen S, Hamilton A, Sebti SM. Design and structural requirements of potent peptidomimetic inhibitors of p21ras farnesyl transferase. *J Biol Chem* 1994; 269:12,410–12,413.
62. Qian Y, Blaskovich MA, Seong C-M, Vogt A, Hamilton AD, Sebti SM. Peptidomimetic inhibitors of p21ras farnesyltransferase: hydrophobic functionalization leads to disruption of p21ras membrane association in whole cells. *Biorg Med Chem Lett* 1994; 4:2579–2584.
63. Vogt A, Qian Y, Blaskovich M, Fossum R, Hamilton A, Sebti SM. A non-peptide mimetic of Ras CAAX: selective inhibition of farnesyltransferase and Ras processing. *J Biol Chem* 1995; 270:660–664.
64. Vogt A, Sun J, Qian Y, Hamilton A, Sebti SM. Burkitt lymphoma Daudi cells contain two distinct farnesyltransferases with different divalent cation requirements. *Biochemistry* 1995; 34:12,398–12,403.
65. Qian Y, Vogt A, Sebti SM, Hamilton AD. Design and synthesis of non-peptide Ras CAAX mimetics as potent farnesyltransferase inhibitors. *J Med Chem* 1996; 39:217–233.

66. Bernhard EJ, Kao G, Cox AD Sebti SM, Hamilton AD, Muschel RJ, McKenna WG. The farnesyltransferase inhibitor FTI-277 radiosensitizes Ras^H-transformed rat embryo fibroblasts. *Cancer Res* 1996; 56: 1727–1730.
67. Finder JD, Litz JL, Blaskovich MA, McGuire TF, Qian Y, Hamilton AD, Davies P, Sebti SM. Inhibition of protein geranylgeranylation causes a superinduction of nitric oxide synthase-2 by IL-1β in pulmonary artery smooth muscle cells. *J Biol Chem* 1997; 272:13,484–13,488.
68. Clark GJ, Kinch MS, Rogers-Graham K, Sebti SM, Hamilton AD, Der CJ. The Ras-related protein Rheb is farnesylated and antagonizes Ras signaling and transformation. *J Biol Chem* 1997; 272:10,609–10,615.
69. Bernhard E, McKenna WG, Hamilton AD, Sebti SM, Qian Y, Wu J, Muschel RJ. Inhibiting Ras prenylation increases the radiosensitivity of human tumor cell lines with activating mutations of Ras oncogenes. *Cancer Res* 1998; 58:1754–1761.
70. Qian Y, Vogt A, Vasudevan A, Sebti SM, Hamilton AD. Selective inhibition of Type I geranylgeranyltransferase *in vitro* and in whole cells by CAAL peptidomimetics. *Biorg Med Chem Lett* 1998; 6:293–299.
71. Bredel M, Pollack IF, Freund JM, Hamilton AD, Sebti SM. Inhibition of Ras and related G-proteins as a novel therapeutic strategy for blocking malignant glioma growth. *Neurosurgery* 1998; 43(1):124–132.
72. Sun J, Blaskovich MA, Knowles D, Qian Y, Ohkanda J, Bailey RD, et al. Antitumor efficacy of a novel class of non-thiol-containing peptidomimetic inhibitors of farnesyltransferase and geranylgeranyltransferase I: combination therapy with the cytotoxic agents cisplatin, taxol and gemcitabine. *Cancer Res* 1999; 59:4919–4926.
73. James GL, Goldstein JL, Brown MS, Rawson TE, Somers TC, McDowell RS, et al. Benzodiazepine peptidomimetics: potent inhibitors of Ras farnesylation in animal cells. *Science* 1993; 260:1937–1941.
74. Kohl NE, Mosser SD, deSolms SJ, Giuliani EA, Pompliano DL, Graham SL, et al. Selective inhibition of ras-dependent transformation by a farnesyltransferase inhibitor. *Science* 1993; 260:1934–1937.
75. Kohl NE, Wilson FR, Mosser SD, Giuliani E, deSolms SJ, Conner MW, et al. Protein farnesyltransferase inhibitors block the growth of ras-dependent tumors in nude mice. *Proc Natl Acad Sci USA* 1994; 91: 9141–9145.
76. Khosravi-Far R, Solski PA, Clark GJ, Kinch MS, Der CJ. Activation of Rac1, RhoA and mitogen-activated protein kinases is required for Ras transformation. *Mol Cell Biol* 1995; 15:6443–6453.
77. Qiu RG, Chen J, McCormick F, Symons M. A role for Rho in Ras transformation. *Proc Natl Acad Sci USA* 1995; 92:11,781–11,785.
78. Lebowitz PF, Du W, Prendergast GC. Prenylation of RhoB is required for its cell transforming function but not its ability to activate serum repsonse element-dependent transcription. *J Biol Chem* 1997; 272: 16,093–16,095.
79. Pollack IF, Bredel M, Erff M, Hamilton AD, Sebti SM. Inhibition of Ras and related G-proteins as a therapeutic strategy for blocking malignant glioma growth II: preclinical studies in a nude mouse model. *Neurosurgery* 1999; 45:1208–1214.
80. Lantry LE, Zhang Z, Hu D, Yao R, Crist KA, et al. Effect of farnesyltransferase inhibitor FTI-276 on established lung adenomas from A/J mice induced by 4-(methyl-nitrosamino)-1-(3-pyridyl)-1-butanone. *Carcinogenesis* 2000; 21:113–116.
81. Jiang K, Coppola D, Crespo NC, Nicosia SV, Hamilton AD, Sebti SM, Cheng JQ. Phosphatidylinositol 3-kinase/AKT2 pathway as a critical target for farnesyltransferase inhibitor-induced apoptosis. *Mol Cell Biol* (in press).
82. Olson MF, Paterson HF, Marshall CJ. Signals from Ras and Rho GTPases interact to regulate expression of p21Waf1/Cip1. *Nature* 1998; 394:295–299.
83. Sun Z, Qian Y, Chen Z, Marfurt J, Hamilton AD, Sebti SM. The geranylgeranyltransferase I inhibitor GGTI-298 induces hypophosphorylation of Rb and partner switching of cyclin-dependent kinase inhibitors: a potential mechanism for GGTI-298 antitumor activity. *J Biol Chem* 1999; 274:6930–6934.
84. Boota A, Johnson BA, Lee K, Blaskovich MA, Hamilton AD, Pitt BR, Sebti SM, Davies P. Prenyltransferase inhibitors block superoxide production by pulmonary vascular smooth muscle. *Am J Physiol: Lung Cell Mol Physiol* 2000; 278:L329–L334.
85. Shears L, Sun Z, Qian Y, Billiard T, Hamilton AD, Sebti SM. Prenyltransferase inhibitors suppress neointima formation following rat artery angioplasty: geranylgeranyltransferase I inhibitors are more effective than farnesyltransferase inhibitors. *J Cell Mol Cardiol* (Submitted).
86. Chen Z, Sun J, Pradines A, Favre G, Adrane J, Sebti SM. Both farnesylated and geranylgeranylated RhoB inhibit malignant transformation and suppress human tumor growth in nude mice. *J Biol Chem* 2000; 275:974–978.
87. Crespo N, Yen T, Hamilton A, Sebti S. FTI-2153 inhibits bipolar spindle formation and chromosome alignment and induces prometaphase accumulation during mitosis. *J Biol Chem* (Submitted).

14 Protein Prenylation in Trypanosomatids

A New Piggy-Back Medicinal Chemistry Target for the Development of Agents Against Tropical Diseases

Michael H. Gelb, PHD, *Frederick S. Buckner,* PHD, *Kohei Yokoyama,* PHD, *Junko Ohkanda,* PHD, *Andrew D. Hamilton,* PHD, *Lisa Nguyen,* PHD, *Bartira Rossi-Bergmann,* PHD, *Kenneth D. Stuart,* PHD, *Saïd M. Sebti,* PHD, *and Wesley C. Van Voorhis,* PHD

CONTENTS

1. NEED FOR ANTI-TRYPANOSOMAL AND -LEISHMANIAL DRUGS

The trypanosomatids include a number of parasites that cause widespread disease in man and domestic livestock in most tropical parts of the world. Organisms that cause devastating diseases include *Trypanosoma brucei* (*T. brucei*), *Trypanosoma cruzi* (*T. cruzi*), and various leishmania species, including *Leishmania mexicana amazonensis* (*L. amazonensis*) and *Leishmania braziliensis* (*L. braziliensis*).

There is tremendous need for new drugs against trypanosomes and leishmania. The World Health Organization (WHO) estimates that 16–18 million people in Latin America are chronically infected with *T. cruzi*, the causative agent of Chagas' disease *(1)*. *T. cruzi* is transmitted by bloodsucking reduviid insects that live in areas of poor housing in Latin America or by blood transfusion from infected humans. The epimastigote form replicates in the insect gut and transforms into infective metacyclic trypomastigotes. Trypomastigotes enter a variety of human cells, replicate intracellularly as round amastigotes, and transform to the motile trypomastigote form, which lyse the host cell. Trypomastigotes

From: *Farnesyltransferase Inhibitors in Cancer Therapy*
Edited by: S. M. Sebti and A. D. Hamilton © Humana Press Inc., Totowa, NJ

circulate, invade other host cells, or infect reduviid insects. While in the host, *T. cruzi* spends most of its time as amastigotes. Parasites are detected in infected hosts for decades after infection, but the mechanisms that *T. cruzi* uses to evade the host immune responses are not well understood. As many as 30% of those chronically infected with *T. cruzi* develop debilitating or fatal manifestations of Chagas' disease, including cardiomyopathy, megaesophagus, or megacolon *(1)*. The pathogenesis of chronic Chagas' disease may involve damage owing to the presence of parasites in affected organs and/or infection-induced autoimmunity. The infection is usually not detected until the chronic phase. The two drugs that serve as the principle treatments for Chagas' disease, benznidazole and nifurtimox, are highly toxic and fail to cure most patients with chronic disease *(1)*.

The WHO estimates that over 20 million people are infected with *Leishmania spp. (1)*. Leishmaniasis is transmitted by the sand fly, which carries the promastigote form. Following the fly bite, promastigotes invade macrophages in the skin of the host, where they transform into amastigotes and replicate intracellularly. Amastigotes then lyse host cells and spread to other macrophages. *Leishmania spp.* cause visceral disease (*L. donovani* and *L. donovani chagasi*) or cutaneous disease (*L. tropica* and *L. major* in the Old World and *L. mexicana spp.* and *L. braziliensis spp.* in the New World). Species that cause cutaneous disease can recrudesce creating serious sequelae, e.g., mucocutaneous disease caused by *L. braziliensis spp.* or diffuse cutaneous leishmaniasis caused by *L. mexicana spp.* Recent outbreaks of visceral leishmaniasis in the Sudan and of visceralotropic leishmaniasis in Gulf War veterans have demonstrated that leishmaniasis is a re-emerging disease that can have serious morbidity and mortality. Drug therapy for leishmaniasis primarily depends on pentavalent antimonials, though this drug requires parenteral administration and is associated with cardiotoxicity *(2)*.

African trypanosomiasis is found in 36 countries in sub-Saharan Africa, and recently the WHO has estimated that 300,000 cases occur annually *(1)*. *Trypanosoma brucei spp.* are transmitted by the tsetse fly, which carries the procyclic form. In the mammalian host, the procyclics transform into trypomastigotes (bloodstream form), which replicate extracellularly in the bloodstream and cerebrospinal fluid of mammals. The infection is almost uniformly fatal once established in humans. Drugs against *T. brucei* are only active with parenteral administration and are quite toxic. The active drugs include pentamidine or suramin when the infection has not spread to the central nervous system (CNS) and melarsoprol (an arsenical) when it has *(2)*. The enzyme inhibitor Eflornithine is a less toxic alternative for CNS infection, but is administered intravenously, is expensive, and is not effective against *T. brucei rhodesiense (1)*.

2. PROTEIN PRENYLATION

2.1. Protein Prenylation in Mammalian Cells

In the late-1980s, the structures of prenyl groups attached to proteins in mammalian cells were determined by Gelb, Glomset, and their coworkers *(3)*. Protein prenylation involves the attachment of 15-carbon farnesyl or 20-carbon geranylgeranyl groups to the C-terminal cysteine residues of a specific set of proteins. Many of these prenylated proteins are small GTPases including Ras, Rab, Rac, and Rho that play a role in cellular signal transduction and intracellular vesicle trafficking *(4)*. The γ-subunits of heterotrimeic G proteins, nuclear lamins, multiple proteins in the phototransuction cascade, and viral antigens are other examples of prenylated proteins *(5)*. The functions of protein prenyl groups are not fully understood, but they play a role in binding proteins to membranes and to direct interactions with other proteins *(4)*.

The farnesylation of Ras is absolutely essential for the ability of this protein to transform mammalian cells *(6)*. Thus, there is currently intense medicinal interest in protein prenylation. The enzyme that attaches the farnesyl group to proteins, protein farnesyltransferase (FTase), is a target for anticancer drugs. In fact, several hundred potent FTase inhibitors (FTIs) have been reported over the past few years *(7)*. Clinical trials with FTIs for the treatment of cancer have been initiated because such compounds are much less toxic than expected to normal cells in culture and to experimental animals, and they cause shrinkage of tumors implanted into animals (Chapters 4–9 and 13 of this volume).

There are three protein prenyltransferases in mammals *(8,9)*. FTase transfers the farnesyl group from farnesyl pyrophosphate (FPP) to the cysteine SH of the C-terminal protein sequence CAAX (A is usually but not necessarily an aliphatic residue, and X is usually S, M, Q, A). Protein geranylgeranyltransferase I (GGTase I) transfers the 20-carbon geranylgeranyl group from geranylgeranyl pyrophosphate (GGPP) to CAAX (when X is L or F). Finally, GGTase II (also called Rab geranylgeranyltransferase) attaches two geranylgeranyl groups to both cysteines at the end of Rab proteins. The X-ray structures of rat FTase and the complex of the enzyme with substrates and inhibitors have been obtained *(10–12)*. After prenylation of the CAAX sequence, the last three residues (AAX) are removed by an endoprotease. The new C-terminal S-prenyl-cysteine is methylated on its a-carboxyl group. Both of these steps occur in the endoplasmic reticulum *(13,14)*.

2.2. Protein Prenylation in Trypanosomatids

We *(15)*—and, independently, Mark Field and coworkers *(16)*—showed that prenylation occurs in *T. brucei* (Fig. 1), and we found that it also occurs in *T. cruzi* and *Leishmania mexicana amazonensis* (*L. amazonensis*) *(17)*. Prenylation also occurs in *Giardia lamblia (18)* and *Schistosoma mansoni (19)*. In these parasites, protein prenylation was detected by showing that culturing cells with radiolabeled mevalonic acid leads to radiolabeling of a specific set of proteins when analyzed by gel electrophoresis. Mevalonic acid is the first committed precursor of isoprenoids in many organisms, although some microorganisms use an alternate pathway to biosynthesize isoprenoids, the deoxyxylulose phosphate pathway, and higher plants use both pathways *(20)*. Interestingly, it has recently been shown that the parasite that causes malaria (*Plasmodium falciparum*) uses the deoxyxylulose phosphate pathway for isoprenoid biosynthesis, and that the antibiotic fosmidomycin, which is a potent inhibitor of a reductoisomerase in this pathway, is a potent antimalarial agent *(21)*. Fosmidomycin has no effect on the growth of *T. brucei* or *T. cruzi* (Gelb and Buckner, unpublished observations) suggesting that mevalonate is the major, if not the only, isoprenoid precursor in these trypanosomatids. FTase and GGTase I activities have been recently detected in cytosolic extracts of *Plasmodium falciparum (22)*. Partial sequences of genes encoding putative protein prenyltransferase from the filarial parasites *Brugia malayi* and *Onchocerca volvulus* have been identified in genome databases (T. Egwang, Med Biotech Labs, Kampala, unpublished).

The cytosolic fraction of *T. brucei* was shown to contain FTase when assayed with the yeast RAS1 mutant protein with C-terminal sequence CVIM *(15)*. At that time, no sequences of CAAX-containing proteins from trypanosomatids were available. Over the past year, we have purified *T. brucei* FTase 60,000-fold to homogeneity *(23)*. *T. brucei* FTase was purified by screening a mixture of CAAX peptides (SSCALX, X is all 20 amino

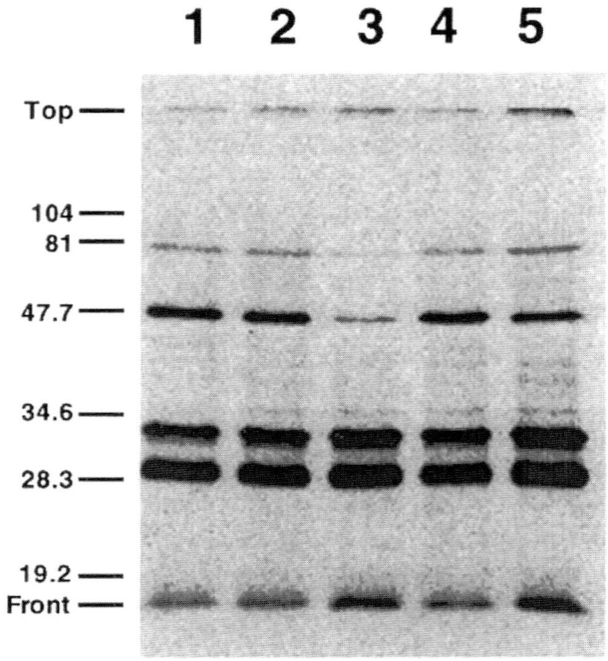

Fig. 1. Bloodstream form *T. brucei* (10^7 cells) was labeled for 24 h with 6.7 μ*M* (100 mCi) [^{3}H]mevalanolactone (hydrolyzed in cells to mevalonic acid) and 40 μ*M* of the hydroxymethylglutaryl-coenzyme A inhibitor simvastatin (to block endogenous mevalonate biosynthesis), and proteins were resolved on a 12.5% SDS-PAGE gel. Lane 1, absence of FTI; lane 2, 0.2 μ*M* FTI-277; lane 3, 5 μ*M* FTI-277; lane 4, 0.2 μ*M* mammalian GGTase I inhibitor GGTI-298; lane 5, 5 μ*M* GGTI-298. Adapted with permission from ref. *(23)*.

acids) for one that displays high affinity for the enzyme (ability to block competitively the prenylation of RAS1-CVIM farnesylation). The enzyme is a heterodimer composed of subunits of apparent molecular weights 61 and 65 kDa when analyzed by denaturing gel electrophoresis. Strong evidence that the 61 kDa band is the β-subunit of *T. brucei* FTase comes from the demonstration that a radiolabeled photoaffinity analog of farnesyl pyrophosphate containing an α-diazo-ester group *(24)* was able to radiolabel this protein band *(23)*. The same analog also labels the β-subunit of mammalian FTase *(25,26)*, and the corresponding geranylgeranyl pyrophosphate analog labels the β-subunit of mammalian GGTase I *(27)*. The subunits of mammalian FTases are considerably smaller (46 and 48 kDa for the β- and α- subunits, respectively). More recently, we have cloned both subunits of *T. brucei* FTase using degenerate PCR primers designed from partial amino-acid sequences obtained from the 61 and 65 kDa gel bands. The amino-acid sequences strongly suggest that these proteins are the subunits of *T. brucei* FTase. Attempts are underway to express large amounts of the enzyme for structural studies.

We showed that *T. brucei* FTase obeys different substrate specificity rules than the mammalian homolog *(17)*. Whereas H-Ras proteins with C-terminal sequences CVLS and CVLM are good substrates for mammalian FTase, only H-Ras-CVLM could be detectably farnesylated by the *T. brucei* enzyme *(15,23)*. Based on these initial results, it is anticipated that FTase inhibitors (FTIs) that are selective for the parasite vs human enzyme can be prepared (*see below*).

Using Ras-CVIM and [^{3}H]FPP as substrates, we have detected a single peak of FTase activity by anion exchange chromatography of cytosol from *T. cruzi* epimastigotes (insect form) and *L. amazonensis* promastigotes (insect form) (K. Yokoyama, F. Buckner, and M. H. Gelb, unpublished results). Efforts are underway to clone the FTase subunits from these trypanosomatids. Using a number of substrates (proteins and peptides) for mammalian GGTase I of the type CAAL, we have not been able to detect GGTase I activity in anion exchange chromatography fractions derived from the cytosol of bloodstream form *T. brucei*, *T. cruzi* epimastigotes, and *L. amazonensis* promastigotes. These results suggest that GGTase I may not be present in these trypanosomatids, and further studies are underway to examine this issue. Using a mammalian Rab protein with C-terminal motifs CXC, we have been able to detect GGTase II activity in column fractions derived from the cytosol of the three trypanosomatids (K. Yokoyama, F. Buckner, and M. H. Gelb, unpublished results). This was expected because several Rab GTP-binding proteins with CXC and CC motifs have been recently cloned from Trypanosomatids (*see below*).

3. FTASE AS A "PIGGY-BACK" DRUG TARGET

Drugs against trypanosomatid infections are not currently being developed in the pharmaceutical industry because of the limited potential for commercialization of such products. It is difficult to develop enzyme inhibitors as drugs because many compounds that are good in vitro enzyme inhibitors will not survive the next stages of development. Compounds must be bioavailable to parasites in vivo and have reasonable in vivo half-lives. Furthermore, compounds must have minimal toxicity to the host. Because CAAX mimetics are being extensively developed as anticancer agents, there is a considerable body of knowledge concerning their in vitro potency, pharmacokinetic properties, and toxicity profiles. If CAAX mimetics have the added ability to selectively kill parasites, the development of such agents as anti-parasite agents can occur in parallel with anti-cancer drug development. This "piggy-back" approach may be one of the few ways to develop drugs for use in developing countries.

Previously reported inhibitors of mammalian FTase and GGTase I (Fig. 2) were tested on *T. brucei* FTase in vitro, and results are summarized in Table 1. The CAAX mimetics L-745,631 *(28)* and FTI-276 *(29)* are potent inhibitors of both *T. brucei* and rat FTases but poorly inhibit rat GGTase I. GGTI-297 shows modest selectivity for mammalian GGTase I versus FTase but is considerably more potent against the *T. brucei* enzyme (Table 1). Thus, in contrast to mammalian FTase, which disfavors X = Leu in the CAAX motif, *T. brucei* FTase tolerates well this amino-acid residue. This is consistent with the notion that *T. brucei* may lack GGTase I (*see above*), and that *T. brucei* proteins with X = L may be good substrates for parasite FTase. The novel peptidomimetic SCH-44342 *(30)*, although quite potent against rat FTase, is 600-fold less active on parasite FTase. These results underscore the differences in the active site of the two FTases that can be exploited for the design of parasite-selective inhibitors. Note that the methyl esters of FTI-276 and GGTI-297, FTI-277 and GGTI-298, respectively, are much less potent FTIs.

The effect of FTI methyl ester prodrugs FTI-277 and GGTI-298 on protein prenylation in bloodstream *T. brucei* was examined in cells labeled with ^{3}H-mevalonic acid in the presence of simvastatin (to block endogenous mevalonic-acid production). As shown in Fig. 1, 5 μM FTI-277 significantly blocked tritium incorporation into a specific set of *T. brucei* proteins, with molecular weights of 48 and 77 kDa (as well as others). FTI-277

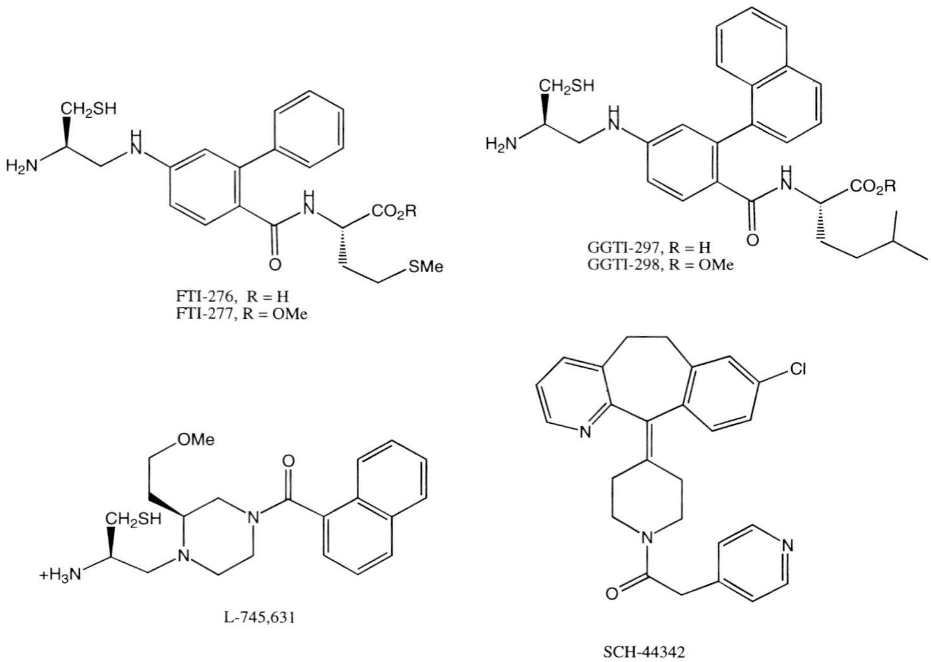

Fig. 2. Structures of the CAAX mimetic FTIs.

Table I
Inhibition of *T. brucei* FTase by CAAX Mimetics[a]

	IC_{50} *(nM)*[b]		
Inhibitor	*T. brucei FTase*	*Rat FTase*	*Rat GGTase I*
L-745, 631	52	23 (5)[c]	8,500 (10,000)[c]
FTI-276	1.7	4 (0.6)[c]	100 (50)[c]
FTI-277 (prodrug)	40	405	4,500
GGTI-297	3.2	35 (190)[c]	50 (50)[c]
GGTI-298 (prodrug)	46	500	305
SCH-44342	158,000	(250)[c]	(>114,000)[c]

[a] In vitro inhibition studies with FTase and GGTase I inhibitors were carried out with *T. brucei* and rat FTase using 5 μ*M* RAS-CVI*M* and 0.75 μ*M* [^{3}H]FPP for FTase or using 5 μ*M* H-Ras-CVLL and 1 μ*M* [^{3}H]GGPP as substrates for GGTase I.

[b] Estimated error for all IC_{50}s is <20%.

[c] Values in parentheses are previously reported IC_{50} values for L-745,631 *(28)* and for FTI-276 and GGTI-297 *(29)* (using H-Ras-CVLS and H-Ras-CVLL as substrates for human FTase and GGTase I, respectively), and for SCH-44342 using H-Ras-CVLS and H-Ras-CVLL as substrates and rat FTase and GGTase I, respectively *(30)*.

produced a stronger effect than did GGTI-298, which is consistent with its higher potency on *T. brucei* FTase activity in vitro and on cell growth. Radiolabeling of the prominent bands in the 28–35 kDa region was not significantly affected by these inhibitors. At least some of these proteins may be Rab proteins, which are present in trypanosomatids, do not contain CAAX sequences and are most likely doubly geranylgeranylated by GGTase II. CAAX mimetics are not expected to inhibit parasite GGTase II because the mammalian form of this enzyme does not recognize short peptides *(9)*.

Table 2
Inhibition of *T. brucei* and *T. cruzi* Growth by CAAX Mimetics[a]

	EC_{50} (μM)			
	T. brucei		*T. cruzi*	
Inhibitor	*Bloodstream form*	*Procyclic form*	*Amastigote form*	*Epimastigote form*
L-745, 631	25	36	11	80
FTI-277	0.7	17	8	~100
GGTI-298	1.7	18	3	45
SCH-44342	30	ND[b]	ND	ND

[a] Parasites were cultured in the presence of CAAX mimetics or vehicle for 3 d (*T. brucei* bloodstream and procyclic forms) or 7 d (*T. cruzi* amastigotes in mouse 3T3 cells and epimastigotes).
[b] ND, not determined.

FTI-277 and GGTI-298, and L-745,631 and SCH-44342 were tested for their effect on the growth of the bloodstream and insect (procyclic) forms of *T. brucei*. Cells were cultured for 3 d in the presence of various amounts of inhibitors, at which time potencies of the inhibitors were determined. The concentrations of inhibitors required to reduce the cell number relative to control culture by twofold (EC_{50}) are listed in Table 2. Both FTI-277 and GGTI-298 are highly potent anti-parasite agents (EC_{50} = 0.7 and 1.7 μM, respectively). Growth curves for bloodstream form *T. brucei* in the presence of different concentrations of these compounds are shown in Fig. 3. Growth of bloodstream parasites was completely blocked with 1 μM FTI-277 and with 5 μM GGTI-298. GGTI-298 has been shown to be useful for selective inhibition of mammalian GGTase I in vivo and for studying the consequence of this inhibition *(31)*. However, growth inhibition of *T. brucei* caused by GGTI-298 might be owing to inhibition of protein farnesylation but not geranylgeranylation, because both this compound and FTI-277 are potent inhibitors of *T. brucei* FTase in vitro (Table 1). The bloodstream form is more sensitive to these compounds than the procyclic form (Table 2), possibly because of poor penetration of CAAX mimetic through the dense protein layer that covers the plasma membrane of procyclic parasites. L-745,631 and SCH-44342, which are considerably less potent at inhibiting *T. brucei* FTase in vitro, are also less potent at stunting parasite growth. For all tested inhibitors, cell shape deformation was observed within 24 h after adding inhibitor, and significant cell lysis occurs during the course of the treatment. Thus, these CAAX mimetics seem to be cytocidal rather than cytostatic.

The effect of CAAX mimetics on the growth of *T. cruzi* amastigotes inside of 3T3 host cells was assayed using the Tulahuen strain of this parasite that is stably transfected with the *Escherichia coli* β-galactosidase gene (*LacZ*) *(32)*. This procedure permits the number of amastigotes to be readily quantified using an enzyme-linked immunoassay (ELISA) plate reader and a colorimetric substrate for *E. coli* β-galactosidase. Figure 4 shows that FTI-277, GGTI-298, and L-745,631 block growth of *T. cruzi* amastigotes in a dose-dependent manner (measured after 7 d of culture). The EC_{50}s are listed in Table 2. Of the three CAAX mimetics tested, only GGTI-298 was toxic at low micromolar amounts to noninfected 3T3 host cells. On the other hand, concentrations of L-745,631 and FTI-277 sufficient to block completely amastigote growth did not affect the growth of noninfected host cells. GGTI-298 but not the FTase inhibitor FTI-277 has been reported to block cell-cycle phase progression from G_0/G_1 to S phase in 3T3 cells *(31)*. As shown in Table 2, the CAAX mimetics also inhibit the growth of insect form *T. cruzi* (epimastigote)

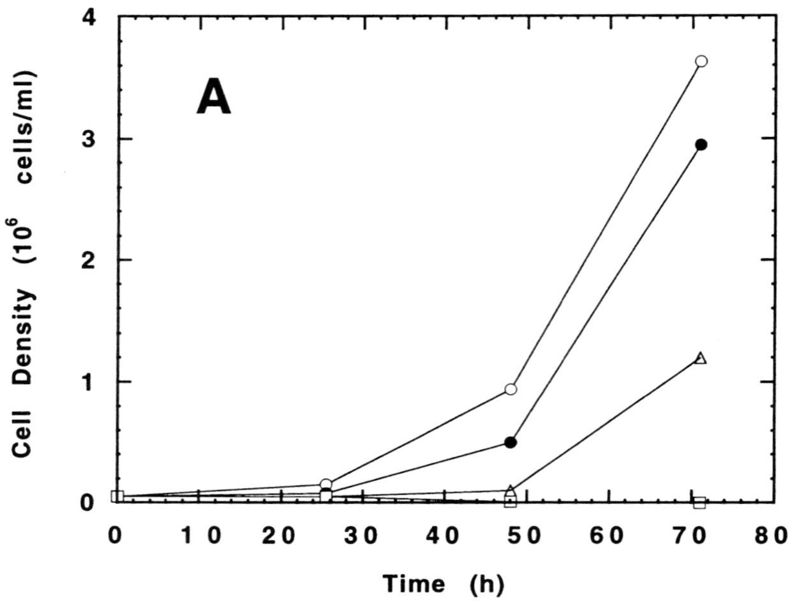

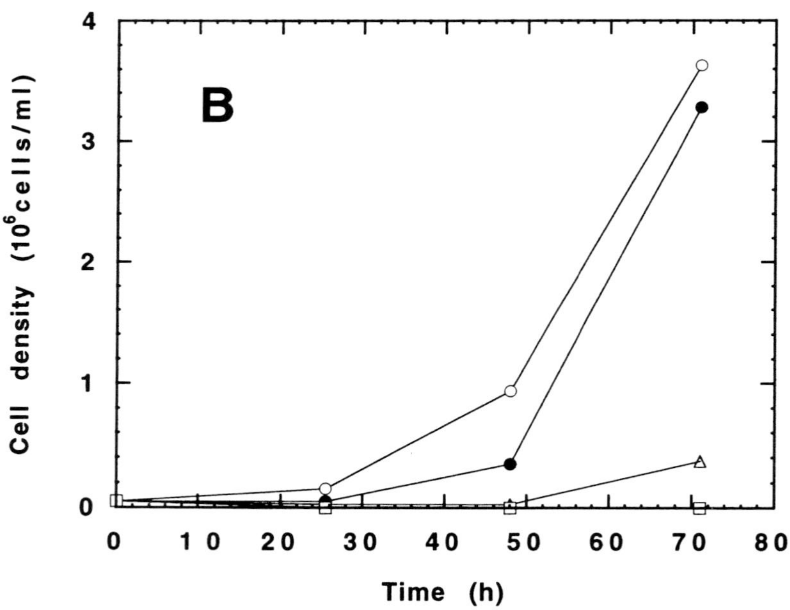

Fig. 3. Growth inhibition of bloodstream form *T. brucei* by CAAX analogs. (**A**) Bloodstream form *T. brucei* (5×10^4 cells/mL) was cultured with 0 (○), 0.5 (●), 0.75 (△), 1 μ*M* (□) FTI-277, and cells were counted daily. (**B**) Same as (**A**) except with GGTI-298, 0 (○), 1 (●), 2.5 (△), or 5 μ*M* (□). Adapted with permission from ref. *(23)*.

in a dose-dependent manner. The compounds are roughly an order of magnitude less potent against epimastigotes compared to amastigotes. Studies are underway to test the CAAX mimetics as inhibitors of partially purified FTases from *T. cruzi*, *L. amazonensis*, and *L. braziliensis* as well as to test these compounds for their ability to block the growth of various Leishmania species in vitro (preliminary results look encouraging).

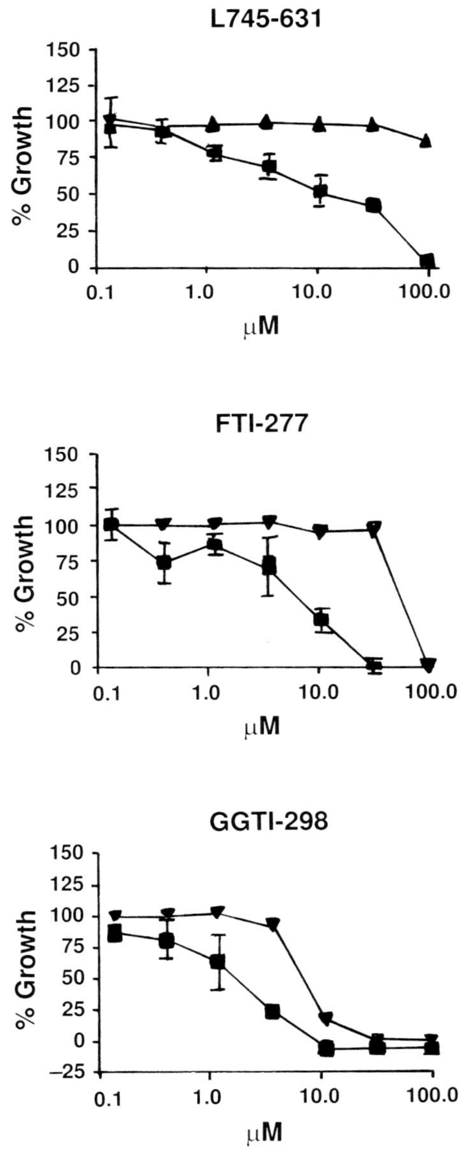

Fig. 4. Growth inhibition of *T. cruzi* amastigotes and of murine 3T3 fibroblasts by CAAX mimetics. Amastigotes (■) in 3T3 host cells were cultured in the presence of the indicated inhibitors for 7 d, and growth was measured by the colorimetric assay of β-galactosidase *(17,32)*. Separately, 3T3 fibroblasts (▲) were cultured in the presence of the inhibitor, and growth was measured with Alamar-Blue after 5 d. Adapted with permission from ref. *(23)*.

FPP analogs, which are low- to sub-micromolar inhibitors of mammalian and *T. brucei* FTases, are also able to block the incorporation of ^{3}H-mevalonic acid into bloodstream *T. brucei* proteins *(17)*. Low micromolar concentrations of these compounds are also toxic to *T. brucei* bloodstream and procyclic forms *(17)*. Although FPP analogs are not being developed as anti-cancer drugs, these results provide useful data showing that inhibitors of FTase that have structures very different from CAAX mimetics are also lethal to parasites.

4. TRYPANOSOMATID GTPASES

GTPases in mammalian cells control critical processes including growth regulation (Ras family), cytoskeletal organization and cell-cycle progression (Rho family), and vesicle trafficking (Rab proteins). Because most GTPases are prenylated, and because most prenylated proteins in eukaryotic cells are GTPases *(4)*, the study of protein prenylation and GTPases go hand in hand. Only a few CAAX-containing trypanosomatid GTPases have been cloned. Engman's lab *(33)* has cloned CAAX-containing DnaJ chaperone proteins from *T. cruzi*, and Hide's lab *(34)* cloned a *T. brucei* Ras/Rap-like protein. Eleven members of the Rab GTPase family (with CAC and CC termini) have been cloned from *T. brucei* and judging from their intracellular localizations are involved in endocytosis and exocytosis *(35)*. The cloning of several additional trypanosomatid Rab GTPases with C-terminal motifs CAC and CC *(36,37)* is helping to define secretory pathways in Leishmania, *T. cruzi*, *Giardia lamblia*, *Entamoeba histolytica*, and *Toxoplasma gondii*. A more detailed description of trypanosomatid GTPases can be found in a recent review article by Field and coworkers *(35)*.

5. FUTURE DIRECTIONS

With the cloning of *T. brucei* FTase subunits almost complete, it should be possible to overexpress the enzyme and to obtain its X-ray crystal structure. Such a structure together with that of mammalian FTase *(10)* will enable a structure-based approach to improve further the potency and selectivity of trypanosomatid FTIs. Given that many mammalian FTIs also inhibit the *T. brucei* enzyme, only minor structural modifications to the CAAX mimetics should be needed to optimize their affinity for the parasite enzyme. Efforts are also underway to clone the FTases from *T. cruzi* and *L. amazonensis* and to overexpress these enzymes for structural studies.

Gene disruption experiments can be routinely carried out with trypanosomatids. Such studies will allow the requirement of protein farnesylation for parasite survival to be rigorously tested. Continued studies are needed to determine if trypanosomatids contain singly geranylgeranylated proteins through the action of GGTase I. Our working hypothesis is that the selective toxicity of FTIs to trypanosomatids vs mammalian cells may be owing to the ability of these compounds to block the farnesylation of vital cell proteins such as farnesylated GTPases—which are geranylgeranylated in mammalian cells—that control processes including cytoskeleton structure and signal transduction pathways. Indeed, GGTase I inhibitors are toxic to mammalian cells (for example, *see* ref. *38*).

CAAX mimetics are sufficiently potent at killing trypanosomatids in vitro to warrant the testing of these compounds for their ability to reduce parasitemia in trypanosomatid infected animal models. Such studies will be carried out with those CAAX mimetics that have already been shown to display acceptable pharmacokinetic properties and that are minimally toxic in animals. We remain optimistic that CAAX mimetic-based FTIs will prove useful for the treatment of cancer and parasitic infections.

REFERENCES

1. World Health Organization. *Tropical Disease Research: Progress 1995–1996*. World Health Organization, Geneva. 1997.
2. Croft SL. The current status of antiparasitic chemotherapy. *Parasitology* 1997; 114:S3–S15.

3. Glomset JA, Gelb MH, Farnsworth CC. Prenyl protein in eukaryotic cells: a new type of membrane anchor. *Trends Biochem Sci* 1990; 15:139–142.
4. Glomset JA, Farnsworth CC. Role of protein modification reactions in programming interactions between ras-related GTPases and cell membranes. *Annu Rev Cell Biol* 1994; 10:181–205.
5. Clarke S. Protein isoprenylation and methylation at carboxy-terminal cysteine residues. *Annu Rev Biochem* 1992; 61:355–386.
6. Hancock JF, Magee AI, Childs JE, Marshall CJ. All ras proteins are polyisoprenylated but only some are palmitoylated. *Cell* 1989; 57:1167–1177.
7. Leonard DM. Ras farnesyltransferase: a new therapeutic target. *J Med Chem* 1997; 40:2971–2990.
8. Yokoyama K, Goodwin GW, Ghomashchi F, Glomset J, Gelb MH. Protein prenyltransferases. *Biochem Soc Trans* 1992; 20:479–484.
9. Casey PJ, Seabra MC. Protein prenyltransferases. *J Biol Chem* 1996; 271:5289–5292.
10. Park H-W, Boduluri SR, Moomaw JF, Casey PJ, Beese LS. Crystal structure of protein farnesyltransferase at 2.25 angstrom resolution. *Science* 1997; 275:1800–1804.
11. Strickland CL, Windsor WT, Syto R, Wang L, Bond R, Wu Z, et al. Crystal structure of farnesyl protein transferase complexed with a CaaX peptide and farnesyl diphosphate analogue. *Biochemistry* 1998; 37: 16,601–16,611.
12. Strickland CL, Weber PC, Windsor WT, Wu Z, Le HV, Albanese MM, et al. Tricyclic farnesyl protein transferase inhibitors: crystallographic and calorimetric studies of structure-activity relationships. *J Med Chem* 1999; 42:2125–2135.
13. Schmidt WK, Tam A, Fujimura-Kamada K, Michaelis S. Endoplasmic reticulum membrane localization of Rce1p and Ste24p, yeast proteases involved in carboxyl-terminal CAAX protein processing and amino-terminal a-factor cleavage. *Proc Natl Acad Sci USA* 1998; 95:11,175–11,180.
14. Romano JD, Schmidt WK, Michaelis S. The *Saccharomyces cerevisiae* prenylcysteine carboxyl methyltransferase Ste14p is in the endoplasmic reticulum membrane. *Mol Biol Cell* 1998; 9:2231–2247.
15. Yokoyama K, Lin Y, Stuart KD, Gelb MH. Prenylation of proteins in *Trypanosoma brucei*. *Mol Biochem Parasitol* 1997; 87:61–69.
16. Field H, Blench I, Croft S, Field MC. Characterisation of protein isoprenylation in procyclic form *Trypanosoma brucei*. *Mol Biochem Parasitol* 1996; 82:67–80.
17. Yokoyama K, Trobridge P, Buckner FS, Scholten J, Stuart KD, Van Voorhis WC, Gelb MH. The effects of protein farnesyltransferase inhibitors on trypanosomatids: inhibition of protein farnesylation and cell growth. *Mol Biochem Parasitol* 1998; 94:87–97.
18. Lujan HD, Mowatt MR, Chen G-Z, Nash T.E. Isoprenylation of proteins in the protozoan *Giardia lamblia*. *Mol Biochem Parasitol* 1995; 72:121–127.
19. Chen G-Z, Bennett JL. Characterization of mevalonate-labeled lipids isolated from parasite proteins in *Schistosoma mansoni*. *Mol Biochem Parasitol* 1993; 59:287–292.
20. Eisenreich W, Schwarz M, Cartayrade A, Arigoni D, Zenk MH, Bacher A. The deoxyxylulose phosphate pathway of terpenoid biosynthesis in plants and microorganisms. *Chem Biol* 1998; 5:R221–R233.
21. Jomaa H, Wiesner J, Sanderbrand S, Altincicek B, Weidemeyer C, Hintz M, et al. Inhibitors of the nonmevalonate pathway of isoprenoid biosynthesis as antimalarial drugs. *Science* 1999; 285:1573–1576.
22. Chakrabarti D, Azam T, DelVecchio C, Qui L, Park YI, Allen CM. Protein prenyl transferase activities of *Plasmodium falciparum*. *Mol Biochem Parasitol* 1998; 94:175–184.
23. Yokoyama K, Trobridge P, Buckner FS, Van Voorhis W C, Stuart K D, Gelb MH. Protein farnesyltransferase from *Trypanosoma brucei*: a heterodimer of 61 and 65 kDa subunits as a new target for antiparasite therapeutics. *J Biol Chem* 1998; 273:26,497–26,505.
24. Allen CM, Baba T. Photolabile analogs of the allylic pyrophosphate substrate of prenyltransferases. *Methods Enzymol* 1985; 110:117–124.
25. Bukhtiyarov YE, Omer CA, Allen CM. Photoreactive analogues of prenyl diphosphates as inhibitors and probes of human protein farnesyltransferase and geranylgeranyltransferase type I. *J Biol Chem* 1995; 270:19,035–19,040.
26. Omer CA, Kral AM, Diehl RE, Prendergast GC, Powers S, Allen CM, et al. Characterization of recombinant human farnesyl-protein transferase: cloning, expression, farnesyl diphosphate binding, and functional homology with yeast prenyl-protein transferases. *Biochemistry* 1993; 32:5167–5176.
27. Yokoyama K, McGeady P, Gelb MH. Mammalian protein geranylgeranyltransferaes-I: substrate specificity, kinetic mechanism, metal requirements, and affinity labeling. *Biochemistry* 1995; 34:1344–1354.
28. Williams TM, Ciccarone TM, MacTough SC, Bock RL, Conner MW, Davide JP, et al. 2-Substituted piperazines as constrained amino acids. Application to the synthesis of potent, non carboxylic inhibitors of farnesyltransferase. *J Med Chem* 1996; 39:1345–1348.

29. McGuire TF, Qian Y, Vogt A, Hamilton AD, Sebti SM. Platelet-derived growth factor receptor tyrosine phosphorylation requires protein geranylgeranylation but not farnesylation. *J Biol Chem* 1996; 271: 27,402–27,407.
30. Njoroge FG, Doll RJ, Vibulbhan B, Alvarez CS, Bishop WR, Petrin J, et al. Discovery of novel non-peptide tricyclic inhibitors of Ras farnesyl protein transferase. *Bioorg Med Chem* 1997; 5:101–113.
31. Vogt A, Qian Y, McGuire TF, Hamilton AD, Sebti SM. Protein geranylgeranylation, not farnesylation, is required for the G1 to S phase transition in mouse fibroblasts. *Oncogene* 1996; 13:1991–1999.
32. Buckner FS, Verlinde CLM, La Flamme AC, Van Voorhis WC. Efficient technique for screening drugs for activity against *Trypanosoma cruzi* using parasites expressing β-galactosidase. *Antimicr Agents Chemother* 1996; 40:2592–2597.
33. Tibbetts RS, Jensen JL, Olson CL, Wang FD, Engman DM. The DnaJ family of protein chaperones in *Trypanosoma cruzi*. *Mol Biochem Parasitol* 1998; 91:319–326.
34. Sowa, MP, Coulter LJ, Tait A, Hide G. A novel gene encoding a ras-like GTP-binding protein from *Trypanosoma brucei*: an evolutionary ancestor of the ras and rap genes of higher eukaryotes? *Gene* 1999; 230:155–161.
35. Field MC, Ali BRS, Field H. GTPases in protozoan parasites: tools for cell biology and chemotherapy. *Parasitol Today* 1999; 15:365–371.
36. El-Sayed NM, Alarcon CM, Beck JC, Sheffield VC, Donelson JE. cDNA expressed sequence tags of *Trypanosoma brucei* rhodesiense provide new insights into the biology of the parasite. *Mol Biochem Parasitol* 1995; 73:75–90.
37. Mendonca SM, Campos CB, Gueiros FFJ, Lopes UG. Identification of GTPase genes in the protozoa parasites *Trypanosoma cruzi* and *Leishmania amazonensis*. *Biol Res* 1993; 26:3–9.
38. Vogt A, Sun J, Qian Y, Hamilton AD, Sebti SM. The geranylgeranyltransferase-I inhibitor GGTI-298 arrests human tumor cells in G0/G1 and induces p21(WAF1/CIP1/SDI1) in a p53-independent manner. *J Biol Chem* 1997; 272:27,224–27,229.

15 Early Clinical Experience with Farnesyl Protein Transferase Inhibitors

From the Bench to the Bedside

Amita Patnaik, MD *and Eric K. Rowinsky,* MD

CONTENTS

1. INTRODUCTION

The ability to target anticancer therapies to specific molecular components of the signal transduction cascade involved in malignant transformation represents a major therapeutic advance in the development of cytostatic antineoplastic agents. Inhibitors of Ras farnesylation are examples of such molecularly targeted therapies and are the culmination of rational drug design and dedicated high-throughput screening of natural products and libraries. Protein farnesyltransferase (FTase) inhibitors (FTIs) are now entering early phase clinical investigations alone and in combination, with encouraging preliminary safety and pharmacologic data.

Ras proteins are guanine nucleotide-binding proteins that play pivotal roles in the control of normal and transformed cell growth. Following stimulation by various growth factors and cytokines, Ras activates several downstream effectors, including the Raf-1/MAP kinase pathway and the Rac/Rho pathway. *Ras* mutations represent the most common dominant oncogene mutations and are present in approx 30% of human cancers, including a substantial proportion of pancreatic and colorectal adenocarcinomas. Mutated *ras* genes give rise to mutated proteins that remain locked in an active state, thereby relaying uncontrolled proliferative signals. Ras undergoes several post-translational modifications that facilitate its attachment to the inner surface of the plasma membrane. The first and most critical modification is the addition of a farnesyl isoprenoid moiety to cysteines of proteins ending at the carboxyl terminal with a CAAX motif where C is cysteine, A is

From: *Farnesyltransferase Inhibitors in Cancer Therapy*
Edited by: S. M. Sebti and A. D. Hamilton © Humana Press Inc., Totowa, NJ

aliphatic, and X is any amino acid but preferably methionine or serine. This farnesylation reaction is catalyzed by the enzyme protein farnesyltransferase (FTase). Inhibition of FTase prevents Ras from maturing into its biologically active form, and thus FTase is of considerable interest as a potential therapeutic target. A closely related family member, geranylgeranyltransferase I (GGTase I) catalyzes the transfer of geranylgeranyl to cysteines of proteins that end with CAAX where X is leucine. Different classes of FTIs have been identified that block the farnesylation of Ras, reverse Ras-mediated cell transformation in human cell lines, and inhibit the growth of human tumor cells in nude mice. FTIs have been well-tolerated in animal studies and do not produce the generalized cytotoxic effects in normal tissues that are a major limitation of most conventional anticancer agents. Clinical evaluations of FTIs to determine the feasibility of administration schedules similar to those that result in optimal therapeutic indices in preclinical studies are ongoing. It is the purpose of this chapter to review the different classes of FTIs, the early clinical experience with these compounds to date, and discuss the challenges in designing disease-directed Phase II and III evaluations of their effectiveness.

2. TYPES OF FTIs

The acquisition of detailed kinetic information about the FTase reaction and the physicochemical nature of FTase substrates has led to the rational design of FTIs *(1–7)*. Three general approaches have been used:

1. The design and synthesis of farnesyl diphosphate (FDP) analogs that compete with the substrate FDP for FTase;
2. Peptidomimetics or CAAX mimetics that compete with the CAAX portion of Ras for FTase; and
3. Bisubstrate analogs that combine the features of both FDP analogs and peptidomimetics (Fig 1).

Still other approaches have resulted in the development of several types of structurally and functionally unrelated compounds that are nonpeptidomimetics of FTase. The recent elucidation of the crystal structure of FTase most likely will further our understanding of the binding of specific classes of inhibitors and provide insight into the optimal design of FTIs *(8)*.

2.1. FDP Analogs

Over the past decade, a number of structurally diverse molecules have been isolated that have demonstrated specific activity against FTase. These compounds have been identified as a result of two drug discovery strategies: 1) high-throughput screening of natural products or libraries; and 2) rational principles of drug design. The former approach has identified several natural products that are competitive inhibitors of FDP, including chaetomellic acids and manumycin analogs *(9–12,14,15)*. These compounds selectively inhibit FTase, with substantially less inhibitory activity against GGTase I, and have potencies in the submicromolar to micromolar range *(1,16)*. The affinity of manumycin for FTase is 10 times less than FDP, however, it also inhibits the growth of several human pancreatic cancer cell lines, with IC_{50}s ranging from 3.5–7.5 μmol/L *(17)*. One of the manumycin analogs, UFC1-C was shown to inhibit the growth of K-Ras-transformed fibrosarcoma at a dose of 6.3 mg/kg, administered intraperitoneally (i.p.) for 5 d from d 0 to d 4; however, inhibition of FTase in mammalian cells has not been directly demonstrated, and thus

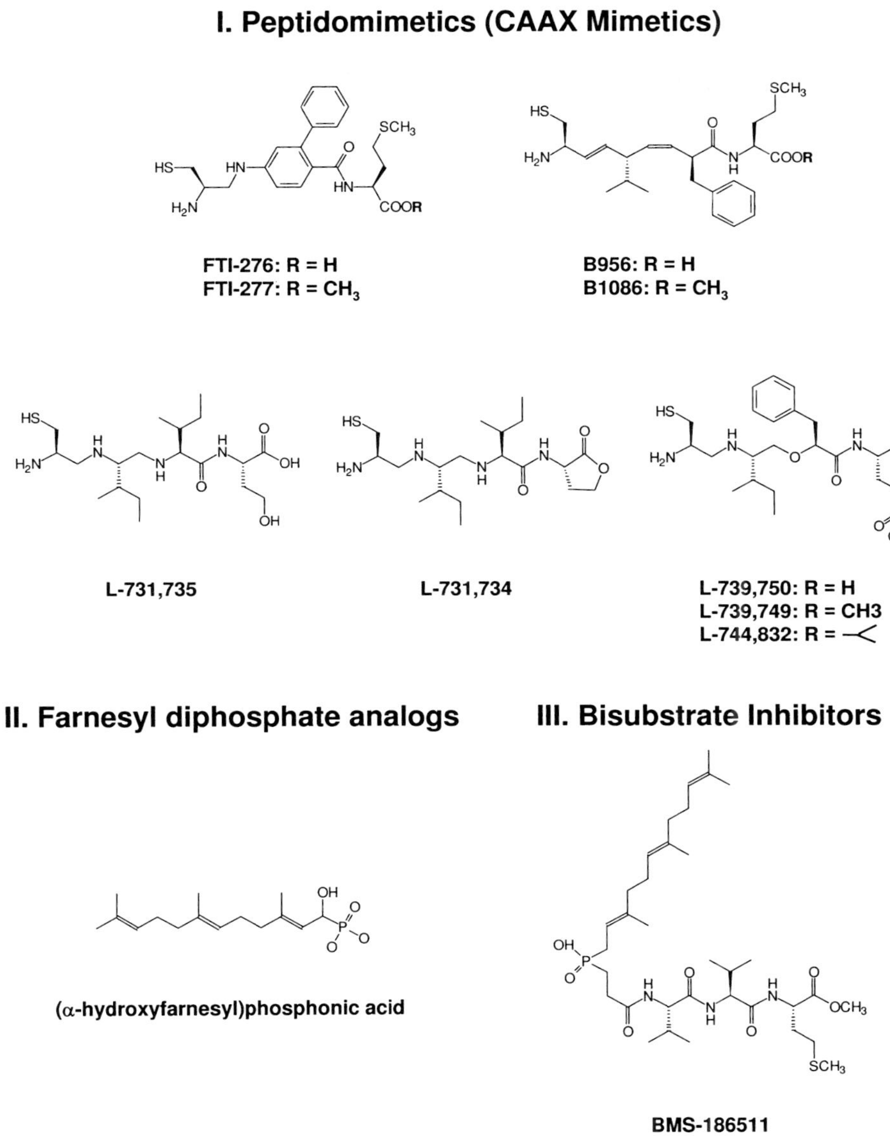

Fig. 1. Representative compounds from three major classes of FTIs.

it is not clear whether the biological effects of this compound are owing to FTIs *(14)*. Inhibitors of FTase have also been designed based on the farnesyl moiety of the FDP substrate. Among the first FTIs to demonstrate activity in cell-culture systems was the nonhydrolyzable FDP α-hydroxyfarnesyl-phosphonic acid, which inhibits FTase with an inhibitory constant (K_i) of 5 nmol/L *(1,4,18)*. The agent inhibited Ras processing in H-*ras*-transformed NIH 3T3 fibroblasts at concentrations as low as 1 μmol/L *(19)*. Other, more highly selective FDP analogs that inhibit FTase at submicromolar concentrations in vitro have also been synthesized and have been shown to inhibit H-Ras processing in whole cells at concentrations of approx 1 μmol/L *(1,18)*. These FDP analogs have also been demonstrated to block H-Ras-mediated transformation of NIH 3T3 fibroblasts at concentrations of 100 μmol/L, and none were toxic to untransformed cells at concentrations up to 250 μmol/L *(1)*. However, these synthetic FDP analogs have not yet demonstrated relevant antitumor activity in animal models.

Although FDP binds to FTase at low nanomolar affinity, intracellular FDP concentrations are near micromolar, which means that most FDP binding sites on FTase in the cell are occupied *(4)*. Thus, FDP analogs will likely need to possess higher affinity than FDP for FTase. Further, FDP is a ubiquitous cofactor required for enzymes involved in many cellular processes, which implies that FDP analogs may produce significant toxicity and therefore clinically useful compounds will have to be much more selective for FTase than other FDP-utilizing enzymes in the cell.

2.2. Peptidomimetics

The finding that CAAX tetrapeptides contain the primary determinants for enzyme recognition led to the synthesis of a number of peptides, such as FTIs, using the principles of rational drug design. The demonstration that tetrapeptides with aromatic amino-acid substitutions at the second aliphatic amino-acid position two residues away from the cysteine group were nonsubstrate FTIs aroused interest in developing low-molecular weight CAAX peptidomimetics as a principal strategy for FTase inhibition *(20,21)*.

Although CAAX peptides are potent FTIs in acellular systems, several physicochemical characteristics of peptides limit their usefulness against tumor cells growing in tissue culture and in animals, and these compounds generally lose two or three logs of potency in whole cells. First, the free C-terminal carboxylate residue of CAAX mimetics is negatively charged, which makes them relatively impermeable to the plasma membrane. To mask the negative charge, a prodrug strategy has been used to synthesize ester or lactone derivatives, with the assumption that the ester or lactone would be hydrolyzed to the more active acid in the cell. These prodrugs are, however, susceptible to cleavage by esterases and other hydrolytic enzymes in plasma, and thus the challenge has been to develop prodrugs that are resistant to hydrolysis in plama but still sensitive to the intracellular hydrolysis required to generate the active FTIs. Second, the labile peptidic bonds of these compounds are rapidly degraded by intracellular proteases, and additional chemical modifications to enhance compound stability are required. A pseudopeptide strategy, whereby peptide bonds in CAAX are reduced to their methyleneamino forms, has been used to create several potent and stable peptidomimetics. For example, reduction of the first and second amide linkages and substitution of homoserine for methionine has been used to synthesize L-731,735, which is relatively stable in the cell *(22)*. L-731,735 is a potent inhibitor of FTase (IC_{50}, 18 nmol/L); the IC_{50} of its prodrug, L-731,734, is much greater (IC_{50}, 282 nmol/L). A further application of this approach involves the synthesis of the methyleneoxy-isostere L-738-750, which is a potent FTI (IC_{50}, 1.8 nmol/L) prepared by replacing the amide linkages between the two central amino acids in the CAAX motif with an oxyether bridge *(23)*. Both L-738,750 and its prodrug methyl ester derivative, L-739,749 inhibit H-Ras processing at concentrations of 0.1–1.0 μmol/L and suppress the growth of mutated H-Ras transfected tumors in nude mice *(23,24)*. A similar prodrug, L-744,732, has been demonstrated to inhibit the growth of more than 70% of tumor cell lines in vitro at concentrations of 2–20 μmol/L. L-778,123 is another peptidomimetic FTI with a benzylimidazole core and low nanomolar activity against FTase, which is in early clinical evaluation. It inhibits the prenylation of Ras proteins and anchorage-independent growth of *ras*-transformed cells in vitro at low micromolar concentrations.

A more recent approach to developing peptidomimetic FTIs is to eliminate the prodrug strategy. One permutation of this approach involves deletion of the X residue in the CAAX box, followed by further modifications of the resultant C-terminal elements *(25)*. This strategy has produced cell-permeable compounds that are pure competitive inhibitors of the protein substrate but are not themselves substrates of FTase. These agents also possess in vitro potencies for FTase in the range of 25–500 nmol/L. In addition, despite deletion of the X residue, which determines prenylation specificity, these pseudopeptides retain >100-fold selectivity for FTase vs GGTase I. The development of these agents has been limited by nonmechanism-based cytotoxicity.

Another related approach involves replacing the peptidic features of the two central amino acids of the CAAX tetrapeptide with stable hydrophobic spacers. This approach, which uses 4-aminobenzoic acid and its derivatives to replace the amino acids, has been used to synthesize FTI-276, which is one of the most potent compounds in its class, and its prodrug FTI-277 *(26)*. The IC_{50} value for FTI-276 against FTase in vitro is 0.5 nmol/L, and FTI-277 inhibits H-Ras processing in vivo with an IC_{50} of 100 nmol/L. Still another pseudopeptidomimetic approach, in which other spacers are used to replace the central two amino acids in the CAAX tertrapeptide, led to the synthesis of B956 and its prodrug, B1086 *(27)*. B956 inhibits both H-Ras and K-Ras processing (IC_{50}, 0.5, and 25 μmol/L, respectively). These agents have been shown to inhibit the growth of transformed cell lines without Ras mutations at concentrations ranging from 16–80 μmol/L and to inhibit tumor growth in nude mice *(27)*.

2.3. Nonpeptidomimetic FTIs

Random high-volume screening of histamine-receptor antagonists from compound libraries led to the identification of a class of novel nonpeptidic, nonsulfhydryl tricyclic inhibitors of FTase that does not depend on a prodrug strategy *(1,6)*. The prototypical tricyclic FTI SCH44342 (Fig. 2) actively competes with the CAAX substrate. This agent inhibits human FTase (IC_{50} ~ 250 nmol/L) and Ras precessing in Cos-7 monkey kidney cells that transiently expressed H-Ras (IC_{50}, 3 μmol/L) *(28–31)*. SCH66336 (Fig. 3), a subsequent lead compound in this series, resulted from blocking of susceptible metabolic sites on SCH44342, with the finding of greatly improved oral pharmacokinetic properties. This compound is an 11-piperidinyl trihalogenated analog that possesses excellent oral bioavailability in both the mouse and monkey and also displays improved metabolic stability ($t_{1/2}$ of 1.4 h in the mouse and 3 h in the monkey). In addition to the improved pharmacokinetics of SCH66336, the intrinsic potency (IC_{50}, 1.9 nmol/L) is also significantly improved compared to earlier compounds in this series and this agent is currently in Phase I clinical trials *(28)*. The pentapeptide PD083176 (Fig. 4) was also identified by high-volume screening of a compound library, and further structure-activity studies led to a series of potent derivatives *(32)*. PD083176 lacks the cysteine residue common to most potent FTIs and was shown to be competitive with FDP. Although this agent inhibits human FTase (IC_{50}, 10 nmol/L), it does not penetrate cells. However, when 5 pmol was microinjected into *Xenopus* oocytes, PD083176 inhibited insulin-induced cell maturation, a Ras-mediated process, but not progesterone-induced maturation, a process not dependent on Ras.

R115777 (Fig. 5) is a nonpeptidomimetic FTI that is an oral quinolone analog of imidazole-containing heterocyclic compounds that was initially developed as an antifungal agent. In vitro, using isolated human FTase, R115777 inhibited farnesylation of a

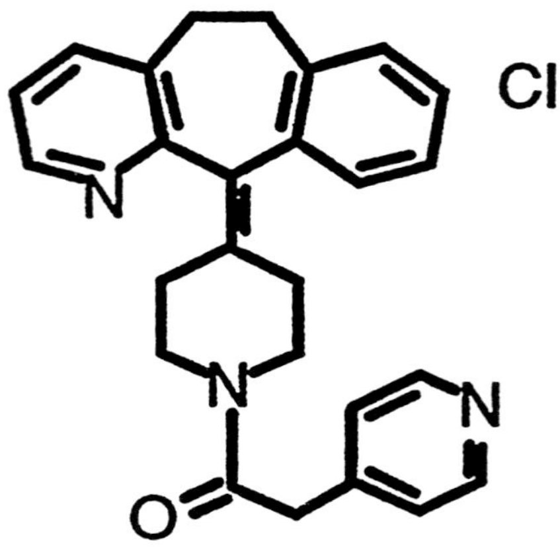

SCH 44342

Fig. 2

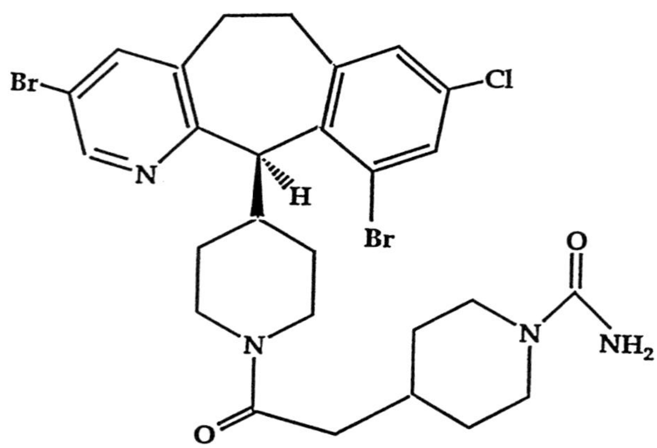

SCH 66336

Fig. 3

Figs. 2 and 3. Examples of two nonpeptidic tricyclic FTIs identified from high-volume screening of histamine-receptor antagonists. SCH 66336 is a lead compound presently in clinical investigations.

lamin B peptide substrate with an IC_{50} of 0.86 nmol/L and also inhibited the farnesylation of the resistant K-*ras* B peptide substrate with an IC_{50} of 7.9 nmol/L. This compound was the first FTI to be studied in human clinical trials. In vitro evaluations of the activity of R115777 against a panel of human tumor cell lines demonstrated that 80% were sensitive to R115777, and significant growth inhibition at concentrations ≤120 nmol/L occurred in 100% of tumors. The growth of CAPAN-2, HCT-116, and LoVo tumor cells was inhibited at IC_{50} values ranging from 16–22 nmol/L *(33)*. In vivo, R115777 inhibited the growth of tumors bearing K-ras mutations in a dose-dependent manner, with accompanying antiangiogenic and apoptotic effects *(34)*.

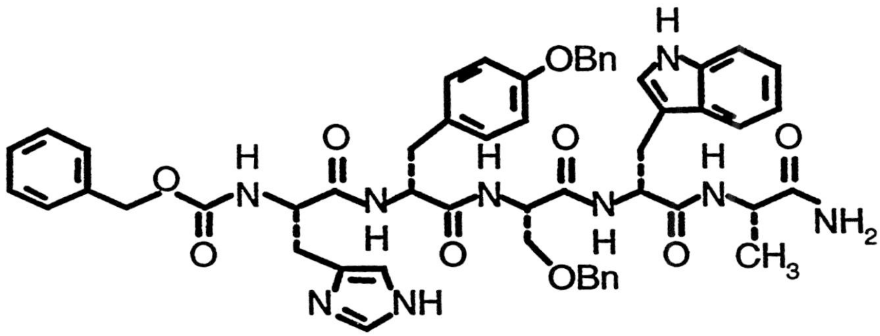

PD083176

Fig 4. A nonpeptidic pentapeptide FTI identified by high-volume screening of a compound library.

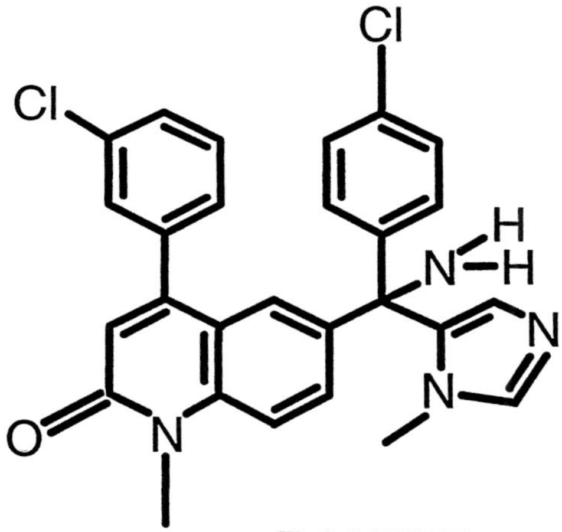

R115777

Fig. 5. An oral nonpeptidic methyl-quinolone FTI currently in clinical development.

2.4. Bisubstrate Analogs

Structural and kinetic analyses of FTase have revealed a sequential mechanism whereby an enzyme-FDP-CAAX ternary complex is formed before catalysis and raised the possibility that bisubstrate analogs that mimic the transition state of the enzyme might be both potent and specific inhibitors of the enzyme. Instead, bisubstrate analogs that incorporate the structural motifs of both FDP and the CAAX tetrapeptide motif are highly potent in vitro *(35)*. The bisubstrate analog BMS-186511 is 2000-fold more specific for FTase than GGTase and has minimal effect on normal cells *(36)*. The compound also inhibits Ras signaling and growth in H-*ras*-transformed and K-*ras*-transformed NIH 3T3 cells at concentrations as low as 0.1 μmol/L, with farnesylation of Ras almost completely inhibited at 100 μmol/L. Furthermore, the agent has been shown to inhibit the anchorage-independent growth of ST88-14, a malignant Schwannoma cell line that is deficient in the expression of neurofibromin *(36)*. Because neurofibromin has intrinsic Ras GTPase-activating activity and cells deficient in it have elevated levels of Ras-GTP, it is conceivable that inhibitors of FTase will be useful in treating patients with type I neurofibromatosis *(37,38)*. BMS-214662 is also a bisubstrate analog associated with growth-inhibitory effects and phenotype reversion in cells transformed by H-*ras*, K-*ras*, and wild-type *ras*. It induces growth inhibition and regression of some established tumors in vivo, including multi-drug resistant tumors. Efficacy has been demonstrated with both intermittent and chronic dosing schedules in vivo and the agent in currently in Phase I clinical trials.

3. CLINICAL EVALUATION OF FTIs

3.1. Phase I and Feasibility Studies

Several FTIs have entered early phase clinical investigations. A major challenge in developing such compounds is the selection of an optimal dose for subsequent disease-directed studies, because antitumor activity may not correlate with toxicity (unlike conventional cytotoxic therapy, in which toxicity and antitumor activity are related, albeit weakly). Toxic effects in normal tissues may not be evident at doses that inhibit Ras farnesylation, or may not be quantifiable, or even related to FTase inhibition. Pharmacologically guided studies may be used to assess whether biologically relevant plasma concentrations associated with maximal inhibition of Ras farnesylation and antitumor activity in preclinical studies are being achieved in patients. Interspecies differences in tissue distribution of drug, protein binding, pharmacokinetics, and metabolic processes, however, may preclude extrapolating pharmacologic parameter targets from animals to humans, thereby limiting the usefulness of pharmacologic studies. The development and validation of assays of protein prenylation in accessible tissues that may be a surrogate for farnesylation of Ras in tumors will facilitate the ability to define the optimal doses of FTIs in Phase I evaluations. Protein prenylation can be assessed by using a diverse series of assays. For example, the specific protein prenylation (e.g., nuclear lamins), or global protein prenylation can be measured by labeling cellular proteins with [^{3}H]mevalonic acid, the precursor of the isoprenoids, or metabolically labeling in vitro with [^{3}H] mevalonic acid, [^{3}H]FDP, or [^{3}H]GGDP *(39)*. Alternatively, inhibition of prenylation of marker proteins can be quantified using gel mobility shift assays. These assays may be helpful in selecting doses of FTIs that achieve maximal inhibition of prenylation of marker proteins validated to correlate with a desirable target effect.

Determination of the ideal mode and schedule of adminstration of FTIs is an important clinical issue. There is experimental evidence indicating that continuous drug exposure, perhaps optimally achieved with continuous treatment, is required to achieve maximal efficacy. The use of protracted dosing schedules, however, raise concerns about both acquired drug resistance and toxicity. Acquired drug resistance has been noted with the FTIs both in vitro and in vivo *(27,40,41)*. In addition, the most likely long-term toxic effects of protracted continuous treatment may not be fully appreciated on the basis of the standard procedures used in preclinical toxicology studies of new anticancer agents in animals. In both preclinical and early clinical investigations, it will be important to rigorously monitor physiologic processes, such as vision and muscular function, that require essential farnesylated proteins. Because many other farnesylated proteins await identification, it will also be prudent to monitor patients for unexpected toxicity, particularly long-term effects.

Preliminary clinical safety data for several different FTIs administered on various chronic treatment schedules are available. Tables 1 and 2 provide a summary of toxicities

Table 1
Summary of Toxicities for FTIs Currently in Clinical Trials

Drug	*Neutrophil*	*Platelets*	*Fatigue*	*Nausea*	*Vomiting*	*Diarrhea*	*Neuro*	*Confusion*
L-778,123	+	+	+	+	+	–	–	+
R115777	+	+	+	+	+	–	+	+
SCH66336	+	+	+	+	+	+	+	–
BMS214662	–	–	+	+	+	+	+	

Table 2
Summary of Pharmacokinetics for FTIs Currently in Clinical Trials

Drug	*Route*	T_{max} *(h)*	$T_{1/2}$ *(h)*	*Exceeding biologically relevant concentrations*
L-778,123	iv/oral	NA[a]	2	>>>
R115777	Oral	0.5–4	5	>>>
SCH66336	Oral	1.5–12	3.6–17.6	>>>
BMS214662	iv/oral5	NA[a]	2–4	Unknown

[a]NA, not available

and pharmacokinetics of FTIs currently in clinical trials. R115777, an orally bioavailable methyl-quinolone, which is a nonpeptidomimetic inhibitor sharing structural similarities to the CAAX motif of Ras, was the first FTI to enter clinical evaluations. R115777 has been administered on several different schedules in patients with advanced malignancies. In a Phase I study of patients with advanced solid malignancies where R115777 was administered twice-daily for 5 consecutive days every 2 wk, it was administered orally at escalating doses as a solution (25–850 mg twice orally) or as pellet capsules (500–1300 mg twice orally). In this study, there was evidence of rapid gastrointestinal absorption, biphasic elimination with an initial $t_{1/2}$ of approx 5 h, and achievement of biologically relevant steady-state plasma concentrations within 2 to 3 d of initiating drug treatment *(42,43)*. Peak plasma concentrations were achieved within 0.5–4 h after oral drug administration and there was little drug accumulation. Pharmacokinetics were dose proportional in the 25–325 mg dose range for the oral solution. Urinary excretion of unchanged R115777 was <0.1% of the oral dose. At doses below 1300 mg twice daily, R115777 was well-tolerated, although an unacceptably high rate of dose-limiting toxicity, consisting of grade 3 neuropathy (1/6 patients), grade 2 fatigue with a decrease in two performance status levels (4/6 patients), and gastrointestinal complaints, were observed at the 1300 mg twice daily dose level. The most frequent clinical grade 2 or 3 adverse events in any cycle were nausea, vomiting, headache, fatigue, anemia, and hypotension. One patient with metastatic colon cancer treated at the 500 mg twice daily dose level had a 46% decrease in CEA levels, associated with clinical improvement in symptoms and stable disease for 5 mo *(43)*. In contrast, neutropenia and thrombocytopenia were the principal dose limiting side effects when R115777 was administered twice daily for 21 d at doses of 60–420 mg/m^2 with the projected maximum tolerated dose (MTD) being 240 mg/m^2 twice daily, and achievement of biologically relevant plasma steady-state concentrations. Peak plasma concentrations were reached 0.8–3 h after oral dosing. Day 1 values of R115777 C_{max} and AUC_{0-12h} increased linearly over the dose range studied. The plasma elimination of R115777 was biphasic, with an initial $t_{1/2}$ value of approx 5 h. Steady-state was reached in the first 2–3 d of the 21-d dosing period *(44,45)*. Grade 3–4 neutropenia and thrombocytopenia, evident by d 15, with duration approx 7 d were the principal dose-limiting toxicities. Grade 3 confusion (1 patient) and grade 3 bilirubin elevation (1 patient) were also dose-limiting. Stable disease of >6 mo duration was observed in two patients (parotid carcinoma, hormone-refractory carcinoma) *(44,45)*.

The membrane permeable peptidomimetic inhibitor, L-778,123 has been administered as a continuous 7-d intravenous infusion in Phase I clinical trials, with a view towards proceeding to a protracted administration schedule *(46,47)*. This FTI has a benzylimidazole core, low nanomolar activity against FTase, and inhibits prenylation of Ras proteins and anchorage-independent growth of *ras*-transformed cells in vitro at low micromolar concentrations. Continuous intravenous administration of this agent at doses ranging from 35–1120 mg/m^2 for 7 d, resulted in dose-limiting toxicities consisting of QTc prolongation (1/4 patients) and grade 4 thrombocytopenia (1/4 patients) at the 1120 mg/m^2 dose level. Grade 2–3 thrombocytopenia was seen in four courses at dose levels of 35, 560, and 1120 mg/m^2, whereas grade 3 neutropenia was seen in four courses at the 560 and 1120 mg/m^2 dose levels. Grade 2–3 somnolence and confusion was observed in two patients. Mild to moderate nausea, vomiting, and fatigue were also observed. Pharmacokinetic studies have demonstrated that steady-state concentrations are achieved within 3 h, with $t_{1/2\alpha}$ and $t_{1/2\beta}$ of 0.6 and 3.3 h, respectively. Biologically relevant concentrations of 3–5 μmol/L, which are capable of inhibiting Ras processing and growth of tumors with *ras* mutations in preclinical studies were seen at the 280 mg/m^2/d dose level. Inhibition of farnesylation of a marker protein hDJ2 in peripheral blood mononuclear cells, was related to dose and plasma L-778,123 concentrations. Maximal inhibition of farnesylation was achieved on d 4 and 8 and returned to pretreatment levels by d 17 *(46,47)*. Extensive testing of retinal signal transduction was carried out in this trial with electroretinograms performed at baseline and prior to each course of treatment. Asymptomatic, reversible decrements in parameters indicative of rod and cone function were seen in three patients across all dose levels.

Similarly, the tolerability and pharmacokinetic profile of SCH66336, an orally bioavailable tricyclic inhibitor of FTase, is also being evaluated in Phase I studies *(28,48–50)*. In a Phase I study where SCH66336 was administered twice daily for 7 d, every 3 wk, the dose-limiting toxicities were diarrhea, nausea, vomiting, and fatigue at the 400 mg BID dose-level. The drug was absorbed slowly with a T_{max} of 1.5–12 h post-dose. $T_{1/2}$ values ranged from 4.8–17.6 h. Mean AUC and plasma steady-state trough concentrations increased in a greater than dose-proportional manner. In treated patients, inhibition of prelamin A farnesylation was demonstrated in buccal mucosal cells, using double-label immunohistochemistry. There was one partial response lasting 9 mo in a patient with previously treated nonsmall cell lung carcinoma. The recommended Phase II dose on this schedule was 350 mg twice daily *(48)*. In another Phase I trial where SCH66336 was administered twice daily for 2 wk every month (doses of 25–300 mg twice daily), the dose limiting toxicities were also diarrhea, nausea, vomiting, and fatigue. Pharmacokinetics were nonlinear, with approximately two- to fivefold drug accumulation at d 14. The recommended Phase II dose was 200 mg twice daily *(49)*. In Phase I studies of SCH66336 administered on a twice daily oral continuous dosing schedule, vomiting, diarrhea, myelosuppression (grade 3 leukopenia and neutropenia), fatigue, and neurocortical toxicity (grade 3 confusion and disorientation) were the principal toxicities, at 300 and 400 mg twice daily and the recommended Phase II dose was 240 mg twice daily *(48–50)*. Pharmacokinetics, were nonlinear, and $t_{1/2}$ values ranged from 3.6–12.3 h. Stable disease of >9 mo duration was seen in two patients (pseudomyxoma peritonei and thyroid carcinoma) *(50)*.

Phase I studies of the FTI inhibitor, BMS-214662 are ongoing, examining dosing schedules of single intravenous administration every 3 wk or daily ×5 intravenous treatments every 3 wk. To date, toxicities have included, fatigue, nausea, vomiting, anorexia, hepatic transaminase elevation and ataxia. The $t_{1/2}$ is short (range: 2–4 h) and biologic correlative studies demonstrate dose-related inhibition of FTase catalytic activity in peripheral blood mononuclear cells, with abrupt regeneration within 24 h *(50a,50b)*.

3.2. Combination Studies with Other Agents and Therapeutic Modalities

Given the importance of multiple pathways in the malignant transformation process, it is likely that combination therapies would have greater effectiveness than single-agent regimens. An example of enhanced antitumor activity arising from the combination of a signal transduction inhibitor with a cytotoxic agent comes from recent studies with the recombinant human monoclonal antibody (MAb), rhuMAb HER2 (Herceptin®) directed against a transmembrane tyrosine/kinase receptor (p185^{HER2}), which has partial homology with other members of the EGFR family. Herceptin has demonstrated efficacy in patients with HER2-overexpressing metastatic breast cancer. Recently, preclinical work combining Herceptin with paclitaxel or doxorubicin showed greater inhibition of growth in in vitro and in vivo models than that observed with any agent alone and this synergistic effect has been corroborated in clinical studies *(51,52)*. Similarly, the FTIs may complement the activity of other anticancer agents that may or may not affect Ras-mediated pathways. Additionally, although FTIs have the capacity rapidly to reduce and nearly ablate tumors in some preclinical studies (rather than simply preventing tumor growth), which is the principal effect preclinically, residual tumors proliferated following withdrawal of the agents. Therefore, combinations of FTIs and classical cytotoxic chemotherapeutic agents may result in greater cytoreduction and may reduce the need for protracted therapy. The overlapping antitumor spectra and nonoverlapping toxicity profiles of FTIs and cytotoxic agents provide a rationale for assessing the efficacy and feasibility of combination regimens. Although the precise chemotherapeutic agents to evaluate in combination with inhibitors of FTase will ultimately be related to the logistics and appropriateness of the agent for the particular clinical setting, the selection may also be based on a unique mechanistic rationale. For example, the FTI L-744,832 and antimicrotubule agents that prevent tubulin depolymerization, such as the taxanes and epothilones, have been demonstrated to inhibit the growth of several breast cancer cell lines in vitro in a synergistic manner, whereas interactions between the FTI and antimicrotubule agents that induce tubulin depolymerization are much less pronounced, but still additive *(28,53)*. Further, the results of mechanistic studies have indicated that L-744,832 enhances the mitotic block induced by agents that prevent tubulin polymerization. The combination of paclitaxel or cisplatin with minimally effective concentrations of R115777 was demonstrated to produce additive antiproliferative activity against human MCF-7 breast, CAPAN-2 (pancreatic), and C32 melanoma cells growing in tissue culture as well as established tumor xenografts *(54)*. The interaction between R115777 and paclitaxel was additive irrespective of the order of drug administration and the duration of the response to R115777 was not enhanced by paclitaxel. In another study, the combination of the FTI SCH66336 and paclitaxel demonstrated either synergistic or additive activity against a broad panel of human tumor cell lines, except for one breast cancer cell line in which the combination demonstrated antagonism *(55)*. The results were independent of p53 mutational status, *ras* mutational status, or tissue of origin. Additive interactions have also been noted between FTIs and cisplatin, cyclophosphamide, doxorubicin, and 5-fluorouracil (5-FU) *(28,53)*.

The combination of FTIs with cytotoxic agents have recently entered Phase I clinical trials. The peptidomimetic FTI L-778,123 has been combined with paclitaxel in a Phase I trial with the starting dose of L-778,123 being 280 mg/m^2/d (7 d continuous infusion), and that of paclitaxel 175 mg/m^2 as a 3-h infusion. Initially, paclitaxel was administered on d 4 of the L-778,123 infusion; however, this resulted in the occurrence of QTc prolongation, tachycardia, and hypotension, as well as somnolence in several patients. No pharmacologic interactions between L-778,123, paclitaxel, and diphenhydramine (used in premedication) was noted to explain the occurrence of these toxic effects at doses which are tolerable for each agent alone. The use of a diphenhydramine challenge on d 1 of L-778,123 administration failed to reproduce the toxicities. The sequence of drug administration was subsequently changed to paclitaxel on d 1 prior to initiation of L-778,123, followed by the FTI given continuously i.v. over 7 d. This schedule of drug administration has not resulted in cardiovascular or neurologic toxicities; however, there have been heterogenous dose-limiting toxicities, including neutropenic sepsis, fatigue, and peripheral neuropathy. The MTD of this combination is projected to be L-778,124 at 280 mg/m^2 with paclitaxel of 175 mg/m^2 *(56)*.

The combination of R115777 with 5-FU and leucovorin has been examined in patients with advanced colorectal and pancreatic cancers in a Phase I trial. Patients received R115777 at doses ranging from 200-500 mg twice daily, with bimonthly fixed dose 5-FU/leucovorin (de Gramont regimen: Leucovorin 200 mg/m^2/2 h, 5-FU 400 mg/m^2 intravenous bolus, 5-FU 600 mg/m^2 over 22 h on d 1 and 2). Dose-limiting toxicity, consisting of grade 4 hematological toxicity, (two patients) was observed at the 500 mg twice daily dose level. There has been 1 episode of grade 4 hematological toxicity in the 400 mg twice daily cohort, with further accrual continuing at this dose level. No drug-related grade 3 nonhematological toxicities have been observed, and pharmacokinetic analysis of 5-FU in the presence and absence of R115777 is planned *(57)*.

The combination of R115777 with gemcitabine is also being examined in a Phase I trial. Patients are receiving R115777 at escalating doses from 100–300 mg twice daily, with gemcitabine at a fixed dose (1000 mg/m^2, d 1, 8, and 15 every 4 wk). Dose-limiting toxicity has been grade 4 neutropenia for greater than 5 d in 2 of 5 patients at the 300 mg twice daily level and 1 of 13 patients at the 200 mg twice daily level. Grade 3 thrombocytopenia has been observed at all dose levels, typically occurring concomitant with neutropenia. Nonhematologic toxicities have been mild and the recommended Phase II dose for this regimen is R115777 at 200 mg twice daily with gemcitabine 1000 mg/m^2 on d 1, 8, and 15. No pharmacologic interaction has been observed in this trial between R115777 and gemcitabine, and all patients, with the exception of one have received full single-agent doses of gemcitabine. Farnesylation of a marker protein, HSDJ, is also being determined by a gel shift assay and preliminary results suggest the progressive inhibition of protein prenylation with successive weeks of treatment *(58)*.

FTIs may also augment the responsiveness of tumors to other therapeutic modalities, such as agents targeting angiogenesis and ionizing irradiation. Oncogenic Ras is known to be involved in pathways of angiogenesis, and FTIs are capable of inhibiting angiogenesis *(46,55,59,60)*. In one study, L-739,749 was shown to block the expression of vascular endothelium-derived growth factor in H-*ras* transformed cells, and it is conceivable that FTIs will be used with therapeutics that principally target malignant angiogenesis *(61)*. H-*ras* and other oncogenes have also been demonstrated to confer resistance to the cytotoxic effects of ionizing radiation, and the inhibitors of FTase have demonstrated radiation-sensitizing properties in tumors growing in tissue culture and animals *(62,63)*. The augmentation of radiation may be attributed to the enhancement of irradiation-induced apoptosis by FTIs *(63)*. Furthermore, the radiosensitivity of normal cells is not enhanced, indicating a selective radiosensitizing effect, which provides a rationale for clinical evaluations of FTIs and ionizing radiation *(63,64)*.

4. FUTURE DISEASE-DIRECTED CLINICAL EVALUATIONS

The challenge in developing Phase II/III disease-directed studies of FTIs arises from the difficulty in selection of appropriate endpoints for their evaluation. Traditionally, the goal of Phase II trials have been the determination of efficacy, on the basis of objective tumor responses as described using standard criteria (e.g., World Health Organization), although it is clear that response rates are at best a surrogate for efficacy. The screening of compounds in Phase II studies using objective tumor response has evolved from the subsequent demonstration of efficacy as measured by prolonged survival in pivotal Phase III trials *(65)*. Although FTIs have induced regression of tumors in some animal models, the principal therapeutic effect clinically may be tumor growth inhibition or "cytostasis," rather than an appreciable cytoreductive response, which would be anticipated for a conventional chemotherapeutic agent. Therefore, a developmental plan that provides for a clinical trial situation with adequate sensitivity to detect and measure tumor growth inhibition will be essential for disease-directed evaluations. Although experimental evidence exists indicating that FTIs may inhibit the growth of tumors with and without *ras* mutations, Phase III and earlier, exploratory (Phase II) evaluations may have the greatest likelihood of detecting meaningful clinical activity if the studies are performed in tumor types that are highly likely to have *ras* mutations. Following rigorous "proof of principle" trials, the scope of disease-directed evaluations can be broadened, and patient eligibility requirements can be less restrictive. The principal endpoints of such pivotal trials ought to be median survival, percentage of patients alive at relevant intervals, time to progression, clinical benefit (e.g., performance status, weight loss, pain control), and improvement in quality of life.

Practically, however, some type of "lead" or indication that the FTIs possess relevant clinical activity, with the ability to modify the natural history of disease progression will ultimately need to be observed before resource-intensive Phase III studies are commenced. One such way of obtaining a "lead" prior to launching large randomized studies is to measure the relative time to tumor progression of patients receiving single-agent treatment with the FTI (period B) against that produced by treatment with a relevant standard therapy or supportive care, measured just before administration of the experimental agent (period A). On the basis of experience with agents that were later shown to have relevant clinical activity in randomized trials (i.e., 30% of patients had a longer time to progression on the new agent in earlier, single-arm studies than on the agent that they received prior to receiving the new agent), a 30% prolongation in the time to progression may be a reasonable threshold to use before proceeding to Phase III studies. Alternatively, "exploratory" single-arm or randomized Phase II studies that are designed with sufficient power to detect and quantify the relevant indices of tumor growth inhibition may provide meaningful leads about activity before randomized evaluations *(66)*. For example, in advanced pancreatic cancer, the percentage of patients surviving for at least 1 yr in exploratory nonrandomized studies may be considered a reasonable endpoint to use in gauging whether or not to proceed with randomized Phase III evaluations *(66, 67)*. Considering the results of Phase II and III studies of gemcitabine in patients with advanced pancreatic cancer, an FTI demonstrating a 1-yr survival rate with a lower limit of a 95% confidence interval of at least 20% might be viewed encouragingly as a candidate for Phase III development. Similarly, the proportion of patients having progressive disease as their best response appears to relate inversely to the ultimate utility of any particular agent in a specific clinical setting, and a maximum acceptable threshold of patients with progressive disease as their best response may be used to predict the potential usefulness of the agent. A retrospective analysis of The National Cancer Institute of Canada, Clinical Trials Group Phase II studies of new agents indicates that the rates of disease progression of agents felt to be most promising in breast, lung carcinomas, and gliomas appear to be less than 20, 30, and 40%, respectively *(65)*. The use of such thresholds, once validated, may be useful in screening FTIs before undertaking large randomized Phase III trials. Other surrogate endpoints which may be considered for efficacy in Phase II trials include assessment of target inhibition, the use of PET scanning to evaluate changes in tumor metabolism and, following changes in tumor markers. Although all of these potential end points remain intriguing possibilities for future Phase II trials, no proposed alternative endpoints have in fact been validated, and thus the challenge will be to integrate them successfully as we search for new paradigms for evaluating novel cytostatic agents *(65)*.

5. SUMMARY

FTIs represent a novel class of anticancer agents that target the signal transduction cascade by preventing the activation of mutated Ras proteins. The prospect of developing such target-specific agents on the basis of understanding the primary molecular defects that underlie the malignant transformation process presents the intriguing possibility of enhancing antitumor efficacy while sparing normal tissue toxicity. However, despite encouraging results from preclinical studies, it is still unclear as to whether FTIs as a class of agents can inhibit tumor growth in patients with advanced disease, and many challenges exist in their clinical development. An ongoing concern in the development of FTIs relates to the possibility that K-*ras* inhibition can be circumvented by the occurrence of cross-prenylation by GGTase I. Given, however, the number of physiologic proteins that are known to be substrates for GGTase I, the development of any inhibitors would have be highly selective in nature. Selective inhibitors of GGTase I, including GGTI-2154 and GGTI-298 *(68)*, have in fact been synthesized, and it may be possible to combine these agents with FTIs or cytotoxic agents, to effectively target cells harboring K-*ras* mutations *(68)*.

There are many unresolved questions pertaining to the development of FTIs; however, some of the most clinically pointed ones pertain to optimal schedules of administration, selection of biologically relevant doses, appropriate endpoints for evaluation in disease-directed studies, which combinations of cytotoxic therapies to pursue, and whether to restrict initial studies to *ras*-bearing tumors. Regardless of the obstacles that lie ahead, the accumulated biological data obtained thus far indicate that FTIs possess remarkable potential as components of the therapeutic armamentarium against malignant diseases and, possibly, nonmalignant disorders involving aberrant cellular proliferation.

REFERENCES

1. Leonard DM. Ras farnesyltransferase: a new therapeutic target. *J Med Chem* 1997; 40:2971–2990.
2. Cox AD, Der CJ. Farnesyltransferase inhibitors and cancer treatment: targeting simply Ras? *Biochem Biophys Acta* 1997; 1333:551–571.
3. Omer CA, Anthony NJ, Buser-Doepner CA, et al. Farnesyl-protein transferase inhibitors as agents to inhibit tumor growth. *Biofactors* 1997; 6:359–366.
4. Gibbs JB, Oliff A. The potential of farnesyltransferase inhibitors as cancer chemotherapeutics. *Annu Rev Pharmacol Toxicol* 1997; 37:143–166.
5. Yamane HK, Farnsworth CC, Xie HY, et al. Brain G protein gamma subunits contain all-transgeranyl-geranylcystein methyl ester at their carboxyl terminal. *Proc Natl Acad Sci USA* 1990; 87:5868–5674.
6. Sebti SM, Hamilton AD. New approaches to anticancer drug design based on the inhibition of farnesyl-transferase. *DDT* 1998; 3:26–33.
7. Heimbrook DC, Oliff A. Therapeutic intervention and signaling. *Curr Opin Cell Biol* 1998; 10:284–288.
8. Park HW, Boduluri SR, Moomaw JF, et al. Crystal structure of protein farnesyltransferase at 2.25 angstrom resolution. *Science* 1997; 275:1800–1804.
9. Gibbs JB, Pompliano DL, Mossner SD, et al. Selective inhibition of farnesyl-protein transferase blocks Ras processing in vivo. *J Biol Chem* 1993; 268:7617–7620.
10. Singh SB, Zink DL, Liesch JM, et al. Isolation and structure of chaetomellic acids A and B from *Chaetomella acutiseta*: farnestl pyrophosphate mimic inhibitors of Ras farnesyl-protein transferase. *Tetrahedron* 1993; 49:5917–5926.
11. Singh SB. Synthesis of chaetomellic acid A: a potent inhibitor of Ras farnesyl-protein transferase. *Tetrahedon Lett* 1993; 34:6521–6524.
12. Lingham RB, Silverman KC, Bills GF, et al. *Chaetomella acutiseta* produces chaetomellic acids A and B which are reversible inhibitors of farnesyl-protein transferase. *Appl Microbiol Biotechnol* 1993; 40: 370–374.
13. Singh SB, Liesch JM, Lingham RB, et al. Actinoplanic acid A: a macrocyclic polycarboxylic acid which is a potent inhibitor of Ras farnesyl-protein transferase. *J Am Chem Soc* 1994; 116:11,606–11,607.
14. Hara M, Akasaka K, Akinaga S, et al. Identification of Ras farnesyltransferase inhibitors by microbial screening. *Proc Natl Acad Sci USA* 1993; 90:2281–2285
15. Tamanoi F. Inhibition of Ras farnesyltransferase. *Trends Biochem Sci* 1993; 18:349–353.
16. Lowy DR, Willumsen BM. Rational cancer therapy. *Nature Med* 1995; 1:792–797.
17. Kainuma O, Asano T, Hasegawa M, et al. Inhibition of growth and invasive activity of human pancreatic cancer cells by a farnesyltransferase inhibitor, manumycin. *Pancreas* 1997; 15:379–383.
18. Pompliano DL, Rands E, Schaber MD, Mosser SD, Anthony NJ, Gibbs JB. Steady-state kinetic mechanism of Ras farnesyl:protein transferase. *Biochemistry* 1992; 31:3800–3807.
19. Santillo M, Mondola P, Gioielli A, Seru R, Iossa S, Annella T, et al. Inhibitors of Ras farnesylation revert the increased resistance to oxidative stress in K-Ras transfomed NIH 3T3 cells. *Biochem Biophys Res Commun* 1996; 229:739–745.
20. Symons M. The Rac and Rho pathways as a source of drug targets for Ras-mediated malignancies. *Curr Opin Biotechnol* 1995; 6:668–774.
21. Brown MS, Goldstein JL, Paris KJ, et al. Tetrapeptide inhibitors of protein farnesyltransferase: amino-terminal substitution in phenylalanine-containing tetrapeptides restores farnesylation. *Proc Natl Acad Sci USA* 1992; 89:8313–8316.
22. Kohl NE, Mosser SD, de Solms SJ. Selective inhibition of *ras*-dependent transformation by a farnesyl-transferase inhibitor. *Science* 1993; 260:1934–1937.
23. Kohl NE, Mosser SD, deSolms SJ et al. Protein farnesyltransferase inhibitors block the growth of ras-dependent tumors in nude mice. *Proc Natl Acad Sci USA* 1994; 91:9141–9145.
24. Sepp-Lorenzino L, Ma Z, Rands E, et al. A peptidomimetic inhibitor of farnesly protein transferase blocks the anchorage-dependent and -independent growth of human tumor cell lines. *Cancer Res* 1995; 55:5302–5309.
25. deSolms SJ, Deana AA, Giulian EA, et al. Pseudodipeptide inhibitors of protein farnesyltransferase. *J Med Chem* 1995; 38:3967–3971.

26. Lerner EC, Qian Y, Blaskovich MA, et al. Disruption of oncogenic K-Ras4B processing and signaling by a potent geranylgeranyl transferase-I inhibitor. *J Biol Chem* 1995; 270:26,770–26,773.
27. Nagasu T, Yoshimatsu K, Rowell C, et al. Inhibition of human tumor xenograft growth by treatment with the farnesyl transferase inhibitor B956. *Cancer Res* 1995; 55:5310–5314.
28. Liu M, Bryant MS, Chen J, et al. Antitumor activity of SCH66356, an orally bioavailable tricyclic inhibitor of farnesyl protein transferase, in human xenograft models and wap-ras transgenic mice. *Cancer Res* 1998; 58:4947–4956.
29. Mallams AK, Njorge FG, Doll RJ, et al. Antitumor 8-chlorobenzocycloheptapyridines: a new class of selective nonpeptidic nonsulfhydryl inhibitors of ras farnesylation. *Bioorg Med Chem* 1997; 5:93–99.
30. Njoroge FG, Doll RJ, Vibulbhan B, et al. Discovery of novel nonpeptidic tricyclic inhibitors of Ras farnesyl protein transferase. *Bioorg Med Chem Lett* 1997; 5:101–113.
31. Bishop WR, Bond R, Petrin J, et al. Novel tricyclic inhibitor of farnesyl protein transferase: biochemical characterization and inhibition of Ras modification in transfected Cos cells. *J Biol Chem* 1995; 270: 30,611–30,618.
32. Leonard DM, Shuler KR, Poulter CJ, et al. Structure-activity relationships of cysteine-lacking pentapeptide derivatives that inhibit ras farnesyltransferase. *J Med Chem* 1997; 40:192–200.
33. End D, Skrzat S, Devine A, et al. R115777, a novel imidazole farnesyl protein transferase (FTI): biohemical and cellular effects in H-ras and K-ras dominant systems. *Proc Amer Assoc Cancer Res* 1998; 39:1847
34. Janssen Research Foundation. R115777 investigator brochure, 4th Ed. Janssen Research Foundation, Titusville, NJ, 1998.
35. Manne V, Yan N, Carboni JM, et al. Bisubstrate inhibitors of farnesyltransferase: a novel class of specific inhibitors of ras transformed cells. *Oncogene* 1995; 10:1763–1779.
36. Yan N, Ricca C, Fletcher J, et al. Farnesyltranferase inhibitors block the neurofibromatosis type I (NF1) malignant phenotype. *Cancer Res* 1995; 55:3569–3575.
37. Basu TN, Gutmann DH, Fletcher JA, et al. Aberrant regulation of ras proteins in malignant tumor cells from type I neurofibromatosis patients. *Nature* 1992; 356:713–715.
38. DeClue JE, Papageorge AG, Fletcher JA, et al. Abnormal regulation of mammalian $p21^{ras}$ contributes to malignant tumor growth in von Recklinghausen (type 10) neurofibromatosis. *Cell* 1992; 69:265–273.
39. James GL, Goldstein JL, Brown MS, Rawson TE, Somers TC, McDowell RS, et al. Benzodiazepine peptidomimetics: potent inhibitors of Ras farnesylation in animal cells *Science* 1993; 260:1937–1942.
40. Prendergast GC, Davide JP, Leboqitz PD, et al. Resistance of a variant *ras*-transformed cell line to phenotypic reversion by farnesyl transferase inhibitors. *Cancer Res* 1996; 56:2626–2632.
41. Kohl NE, Omer CA, Conner MW, et al. Inhibition of farnesyltransferase induces regression of mammary and salivary carcinomas in *ras* transgenic mice. *Nature Med* 1995; 1:792–797.
42. Zujewski J, Horak ID, Woestenborghs R, et al. Phase I trial of farnesyl-transferase inhibitor, R115777, in advanced cancer. *Proc Amer Assoc Cancer Res* 1998; 39:1848
43. Zujewski J, Horak ID, Bol CJ, et al. A phase I and pharmacokinetic study of farnesyl protein transferase inhibitor R115777 in advanced cancer. *J Clin Oncol* 2000; 18:927–941.
44. Schellens JHM, de Klerk G, Swart M, et al. Phase I and pharmacologic study with the novel farnesyltransferase inhibitor (FTI) R115777. *Proc Amer Assoc Cancer Res* 1999; 39:724.
45. Hudes GR, Schol J, Baab, et al. Phase I clinical and pharmacokinetic trial of the farnesyltransferase inhibitor R115777 on a 21-day dosing schedule. *Proc Am Soc Clin Oncol* 1999; 18:601.
46. Soignet S, Yao S-L, Britten D, et al. Pharmacokinetics and pharmacodynamics of the farnesyl protein transferase inhibitor (L-778,123) in solid tumors. *Proc Am Assoc Can Res* 1998; 40:517.
47. Britten CD, Rowinsky E, Yao S-L, et al. A phase I and pharmacologic study of the farnesyl protein transferase inhibitor L-778,123 in patients with solid cancers. *Proc Am Soc Clin Oncol* 1999; 18:597.
48. Adjei AA, Erlichman C, Davis JN, et al. A phase I and pharmacologic study of the farnesyl protein transferase inhibitor SCH 66336 in patients with locally advanced or metastatic cancer. *Proc Am Soc Clin Oncol* 1999; 18:598.
49. Hurwitz HI, Colvin OM, Petros WP, et al. A phase I and pharmacokinetic study of SCH 66336, a novel FPTI using a 2-week on, 2-week off schedule. *Proc Am Soc Clin Oncol* 1999; 18:599.
50. Eskens F, Awada A, Verweij, et al. Phase I and pharmacologic study of continuous daily oral SCH 66336, a novel farnesyl transferase inhibitor, in patients with solid tumors. *Proc Am Soc Clin Oncol* 1999; 18:600.
50a. Rose WC, Arico MA, Burke CL, et al. Preclinical antitumor activity of BMS-214662, a novel farnesyl transferase inhibitor. *Proc Am Assoc Cancer Res* 2000; 41:2835A.
50b. Sonnichsen D, Damle B, Manning J, et al. Pharmacokinetics and pharmacodynamics of the farnesyltransferase inhibitor BMS-214662 in patients with advanced solid tumors. *Proc Am Soc Clin Oncol* 2000; 19:720A.

51. Baselga J, Norton L, Albanell J, et al. Recombinant humanized anti-HER2 antibody (Herceptin) enhances the antitumor activity of paclitaxel and doxorubicin against HER2/neu overexpressing human breast cancer xenografts. *Cancer Res* 1998; 58, 2825–2831.
52. Goldenberg MM. Trastuzumab, a recombinant DNA-derived humanized monoclonal antibody, a novel agent for the treatment of metastatic breast cancer. *Clin Ther* 1999; 21:309–318.
53. Moasser MM, Sepp-Lorenzino L, Kohl NE, et al. Farnesyl transferase inhibitors cause enhanced mitotic sensitivity to taxol and epothilones. *Proc Natl Acad Sci USA* 1998; 95:1369–1374.
54. Skrzat SG, Bowden CR, End DW. Interaction of the farnesyl protein transferase inhibitor R115777 with cytotoxic chemotherapeutics in vitro and in vivo. *Proc Am Assoc Cancer Res* 1999; 40:523.
55. Shi B, Gurnani M, Yaremko B, et al. Enhanced efficacy of the farnesyl protein transferase inhibitor SCH 66336 in combination with paclitaxel. *Proc Am Assoc Cancer Res* 1999; 40:524.
56. Sharma S, Britten C, Spriggs D, et al. A phase I and PK study of farnesyl transfterase inhibitor L-778,123 administered as a seven day continuous infusion in combination with paclitaxel. *Proc Am Soc Clin Oncol*, 2000; 19:719A.
57. Peeters M, Van Cutsem E, Marse H, et al. Phase-I Combination Trial of the Farnesyltransferase Inhibitor (FTI) R115777 with A 5FU/LV Regimen in Advanced Colorectal (CRC) or Pancreatic (PC) Cancer. *Proc Am Soc Clin Oncol* 1999; 18:859.
58. Patnaik A, Eckhardt SG, Izbicka E, et al. A phase I and pharmacologic (PK) study of the farnesyltransfernase inhibitor, R115777 (R11) in combination with Gemcitabine (Gem). *Proc Am Soc Clin Oncol*, 2000; 19:5A.
59. Smets G, Xhonneux B, Cornelissen F, et al. R115777, a selective farnesyl protein transferase inhibitor (FTI), induces anti-angiogenic, apoptotic and anti-proliferative activity in CAPAN-2 and LoVo tumor xenografts. *Proc Am Assoc Cancer Res* 1998; 39:2170A.
60. Grugel S, Finkenzeller G, Weindel K, et al. Both v-Ha-Ras and v-Faf stimulate expression of the vascular endothelial growth factor in NIH 3T3 cells. *J Biol Chem* 1995; 270:25,915–25,919.
61. Rak J, Mitsuhashi Y, Bayko L, et al. Mutant *ras* oncogenes upregulate VEGF/VPF expression: implications for induction and inhibition of tumor angiogenesis. *Cancer Res* 1995; 55:4575–4580.
62. Bernhard EJ, McKenna WG, Hamilton AD, Sebti SM, Qian Y, Wu JM, Muschel RJ. Inhibiting Ras prenylation increases the radiosensitivity of human tumor cell lines with activating mutations of ras oncogenes. *Cancer Res* 1998; 58:1754–1761.
63. Bernhard EJ, Kao G, Cox AD, et al. The farnesyl transferase inhibitor FTI-277 radiosensitizes H-Ras transformed rat embryo fibroblasts. *Cancer Res* 1996; 56:1727–1730.
64. McKenna WG, Weiss MC, Endlich B, et al. Synergistic effects of the v-myc oncogene with H-Ras on radioresistance. *Cancer Res* 1990; 50:97–102.
65. Eisenhauer EA. Phase I and II trials of novel anti-cancer agents: endpoints, efficacy and existentialism. *Ann Oncol* 1998; 9:1047–1052.
66. Von Hoff DD. There are no bad anticancer agents, only bad clinical trial designs: twenty-first Richard and Hinda Rosenthal lecture. *Clin Cancer Res* 1998; 4:1079–1086.
67. Burris HA III, Moore MJ, Anderson J, et al. Improvements in survival and clinical benefit with gemcitabine as first-line therapy for patients with advanced pancreatic cancer: a randomized clinical trial. *J Clin Oncol* 1997; 15:2403–2413.
68. Sun J, Blaskovich MA, Knowles D, Qian Y, Ohkanda J, Bailey RD, Hamilton AD, Sebti SM. Antitumor efficacy of a novel class of non-thiol-containing peptidomimetic inhibitors of farnesyltransferase and geranylgeranyltransferase I: combination therapy with the cytotoxic agents cisplatin, taxol and gemcitabine. *Cancer Res* 1999; 59:4919–4926.

16 Phase I Trial of Oral R115777 in Patients with Refractory Solid Tumors

Preliminary Results

Gary R. Hudes, MD *and Jessie Schol,* RN

CONTENTS

1. INTRODUCTION

Several protein farnesyltransferase (FTase) inhibitors (FTIs) are completing evaluation in Phase I clinical trials *(1)*. R115777, an orally bioavailable substituted quinone, is among the first FTI to undergo evaluation in humans. This chapter summarizes the preliminary results of a Phase I trial of R115777 administered orally, twice daily (bid) for 21 consecutive days, conducted at Fox Chase Cancer Center *(2)*.

2. *Preclinical Activity of R115777*

A potent and selective inhibitor of FTase, R115777 inhibited farnesylation of lamin-B and K-RasB peptides with IC_{50} values of 0.86 n*M* and 7.9 n*M*, respectively. By contrast, inhibition of geranylgeranylation occurred only at concentrations > 50 μ*M*. In cell proliferation assays, growth of T24 bladder carcinoma (H-ras mutant), MCF-7 breast carcinoma, pancreatic CAPAN-2 (K-rasB mutation), and colon LoVo tumor cells (K-RasB) were inhibited with IC_{50}s ranging from 1.7–22 n*M*. Similar to other inhibitors of FTase, inhibition of K-Ras mutant cells required approx 10-fold higher concentrations of the drug. Growth delay rather than tumor regression was observed in T24, LoVo, and CAPAN-2 xenografts (Janssen Research Foundation, unpublished reports). Toxicity studies in beagle dogs showed that maximum tolerated dose (MTD) was dependent on duration of treatment. Myelosuppression affecting granulocytes and platelets was reversible after 4 and 7 d of treatment at 40 mg/kg, whereas doses exceeding 10 mg/kg were not tolerated on longer dosing schedules. Renal and gastrointestinal toxic effects were also observed, more severe with higher doses and longer duration of treatment.

From: *Farnesyltransferase Inhibitors in Cancer Therapy*
Edited by: S. M. Sebti and A. D. Hamilton © Humana Press Inc., Totowa, NJ

Based on these data, the initial Phase I clinical trial employed a 5-d on, 9-d off schedule, with cycles repeated every 2 wk *(3)*. While this trial was in progress, additional Phase I trials employing continuous oral dosing were initiated in the US and Europe.

3. FOX CHASE CANCER CENTER PHASE I TRIAL (USA-3)

3.1. Patients and Methods

Because continuous administration would be preferable for an agent with cytostatic rather than cytotoxic properties, a 21-d treatment schedule was selected for the second US Phase I trial, with cycles repeated every 28 d. The starting dose of R115777 was 60 mg/m^2 or approx 100 mg twice daily, based on the tolerability of this dose and fivefold higher doses in the earlier NCI Phase I trial. A Bayesian dose escalation design, Escalation with Overdose Control (EWOC), provided rules for dose escalation or de-escalation, depending upon the overall toxicity experience in all patients *(4)*. This dose escalation plan was designed with parameters θ = probability of dose-limiting toxicity at the MTD, and α = probability that any dose escalation exceeds the MTD, set at 0.33 and 0.3, respectively. Additional patients were accrued to dose levels thought to be at or near the MTD.

A total of 22 patients were enrolled and treated at doses ranging from 100 to 800 mg bid. Dose was calculated according to body surface area for the initial 12 patients. Based on pharmacokinetic data derived from USA-3 and other R115777 Phase I trials, treatment of subsequent patients was based on total dose regardless of body surface area.

Patients selected for the trial had incurable solid tumors for which no better treatment options existed. Also required were ECOG performance status of 0 or 1, adequate nutritional status and oral intake, and normal hematologic, renal, and hepatic function. Prior chemotherapy and radiotherapy must have been completed at least 4 wk prior to registration. Patients who had received radiotherapy encompassing 25% or more of the bone marrow-containing skeleton, or who had received high-dose chemotherapy with bone marrow or stem-cell transplantation, were excluded. Also excluded were patients with known hypersensitivity to imidazole drugs (e.g., ketoconazole and miconazole). All patients were required to give written informed consent to trial participation, and the treatment protocol was approved by the Fox Chase Cancer Center Institutional Review Board.

All patients registered participated in pharmacokinetic studies. Blood samples were obtained before and at multiple time points following the first and last doses of R115777, on d 1 and 22, respectively. Urine was not analyzed because urinary excretion of R115777 was determined to be <0.1% in the earlier Phase I trial. Measurement of R115777 plasma concentrations utilized reversed-phase high-pressure liquid chromatography (HPLC) with ultraviolet (UV) detection at 240 nm.

3.2. Preliminary Results

Of the initial 22 patients enrolled, 14 were male and 8 were female, with median age of 59 yr (range 35–73 yr). The tumor types represented were colorectal (7 patients), pancreatic carcinoma (4 patients), nonsmall cell lung cancer (2 patients), and one patient each ($n = 9$) with a variety of other tumors including renal, prostate, salivary gland, and hepatocellular cancers. Six of the 22 patients received two or more cycles of R115777, with 7 cycles received by one patient and 8 cycles received by another patient. The reason for discontinuing treatment was progression of disease in all but one case, a patient who refused further treatment after experiencing severe fatigue after 2 wk of R115777.

Single patient dose escalation proceeded from 100 mg bid up to 800 mg bid. Dose-limiting neutropenia and thrombocytopenia were encountered in patients treated at the 650 and 800 mg bid, and a third patient treated at 600 mg bid experienced dose-limiting fatigue. A total of 7 patients were accrued at the 400 mg level. Two of these patients experienced grade 4 thrombocytopenia, one with grade 4 neutropenia. Grade 4 granulocyte and platelet toxicity typically were detected at d 15, with nadirs at d 18–21 and recovery by d 28. Other toxic effects observed at 400 mg bid were reversible grade 3 bilirubin elevation in a patient with high volume liver metastases, grade 2 creatinine elevation in one patient, and grade 2 skin rash in another patient.

Based on the toxicity observed at 400 mg bid, additional patients were treated at the 300 mg bid dose. None of these patients experienced grade 3 or 4 neutropenia or thrombocytopenia, including two patients treated for 7 and 8 cycles, respectively, without toxicity. Grade 2 maculopapular rash, predominantly involving chest, abdomen, and back, occurred in one patient, and grade 3 fatigue developed in one other patient. Based upon these results, the provisional MTD and recommended dose of R115777 for Phase II study on the 21-d schedule is 300 mg bid.

Other toxic effects were mild and uncommon. Grade 1 diarrhea occurred in one patient each treated at 250, 300, 400, and 800 mg bid. Nausea or vomiting of grade 1 (six patients) or grade 2 (two patients) were reported at doses of R115777 ranging from 200 to 550 mg bid. Although no objective tumor regressions have been observed, three patients have received treatment for ≥6 mo without progression of their tumor.

3.3. Pharmacokinetic Studies

Plasma R115777 concentration data for all 22 patients were analyzed using noncompartmental methods *(5)*. C_{max} ranged from 357 to 3604 ng/mL over the dose range of 100–800 mg bid. Time to C_{max} (T_{max}) was a mean of 1.7 h (range 0.8–3.0) after R115777 administration. Corresponding values for area under the plasma concentration-time curve from times 0–12 h (AUC_{0-12h}) were 1024–17,283 ng-h/mL. Significant interpatient variability in C_{max} and AUC_{0-12h} was observed, with a trend toward increasing values of both C_{max} and AUC with increasing dose of R115777 *(2)*. Comparison of C_{max} and AUC values following the first and last doses did not show significant accumulation of R115777 over the 21-d dosing period. Mean elimination half-life for R115777 was 4.4 h (range 3.3 –5.8).

4. SUMMARY AND CONCLUSIONS

The preliminary findings of this Phase I trial indicate that R115777 can be administered safely for 21 consecutive days at doses that produce plasma concentrations capable of inhibiting FTase. The wide interpatient pharmacokinetic variability of R115777 is similar to that observed with other orally administered agents. Unlike the 5-d intermittent dosing schedule *(3)*, granulocytopenia and thrombocytopenia were dose-limiting toxicities for the 21-d chronic dosing schedule. Fatigue, nausea, and diarrhea were common, but usually mild. Rash, creatinine elevation, and hyperbilirubinemia were sporadic, reversible toxicities. Considering all the Phase I trials together *(2,3,6)*, the clinical toxicity of R115777 in humans has mirrored the preclinical toxicity profile in dogs.

Pending accrual of additional patients, 300 mg bid appears to be a well-tolerated starting dose for further clinical trials that employ 21-d dosing of R115777. An alternative dosing schedule being investigated in ongoing Phase I trials utilizes continuous bid dosing without the 7-d break *(6)*. Phase II studies employing the 21-d, bid schedule of R115777 in a variety of tumor types will commence in the year 2000.

In companion laboratory studies, we are measuring the activity of FTase and the prenylation of several proteins in peripheral blood mononuclear cells obtained from patients before and during treatment with R115777. It is hoped that these studies will provide additional information that may guide selection of an optimal schedule or dose of R115777 in future studies. These methods may prove to be feasible in tumor tissue obtained from small biopsies and thus may help elucidate mechanisms of sensitivity or resistance to FTIs in humans.

REFERENCES

1. Rowinsky EK, Windle JJ, Von Hoff DD. Ras protein farnesytransferase: a strategic target for anticancer therapeutic development. *J Clin Oncol* 1999; 17:3631–3652.
2. Hudes GR, Schol J, Ranganathan S, et al. Phase I clinical and pharmacokinetic trial of the farnesyltransferase inhibitor R115777 on a 21-day dosing schedule. *Proc Am Soc Clin Oncol* 1999; 18:156a (Abstract 601).
3. Zujewski J, Horak ID, Bol CJ, et al. Phase I pharmacokinetic study of farnesyl protein transferase inhibitor R115777 in advanced cancer. *J Clin Oncol* 2000; 18:927–941.
4. Baab J, Rogatko A, Zacks S. Cancer phase I clinical trials: efficient dose escalation with overdose control. *Stat Med* 1998; 17:1103–1120.
5. Gibaldi M, Perrier D. Pharmacokinetics, 2nd Ed. Marcel Dekker, New York, 1982.
6. Schellens JHM, de Klerk G, Swart M, et al. Phase I and pharmacologic study with the novel farnesyltransferase inhibitor (FTI) R115777. *Proc Am Assoc Can Res* 1999; 40:724 (Abstract 4780).

17 Farnesyltransferase and Geranylgeranyltransferase Inhibitors

The Saga Continues

Adrienne D. Cox, PHD, *L. Gerard Toussaint III*, MD, *James J. Fiordalisi*, PHD, *Kelley Rogers-Graham*, BS, *and Channing J. Der*, PHD

CONTENTS

1. INTRODUCTION

Since 1982, when mutated and oncogenic forms of *ras* genes were first identified in human tumor cells, their protein products have attracted considerable interest as a target for anticancer drug development. Researchers were inspired to delineate the functions of Ras proteins in normal cells and to determine how mutated Ras proteins were altered in these functions. The impressive accumulation of information about the genetics, biochemistry, biology, and structure of Ras proteins over the last 17 years has provided important clues to how anti-Ras drugs may be developed.

From: *Farnesyltransferase Inhibitors in Cancer Therapy*
Edited by: S. M. Sebti and A. D. Hamilton © Humana Press Inc., Totowa, NJ

Although a variety of approaches have been considered and pursued, the greatest progress to date has come from the understanding that Ras function—normal or oncogenic—is critically dependent on its post-translational modification by a farnesyl isoprenoid lipid *(1–3)*. The demonstration that genetically engineered mutants of oncogenic Ras lacking this lipid modification both failed to associate with the plasma membrane and were rendered completely nontransforming *(4,5)* confirmed that this was potentially an ideal approach to blocking oncogenic Ras function. In 1990, the key discovery and isolation of farnesyltransferase (FTase) *(6)*, the enzyme that catalyzes the farnesylation of Ras proteins, established a crucial target for the development of a new class of anti-Ras drugs. The further demonstration that a tetrapeptide CAAX (C = cysteine, A = aliphatic, X = any amino acid) sequence was both necessary and sufficient to signal Ras farnesylation, and that such a peptide was a potent inhibitor of FTase activity *(6,7)*, provided a platform for the rational development of FTase inhibitors (FTIs). A flurry of efforts by nearly two dozen pharmaceutical companies and academic universities has led to the development and testing of a wide range of FTIs as anti-Ras and anticancer agents.

Further intensity was added to these efforts with the first remarkable observations in 1993 that FTIs were potent inhibitors of Ras farnesylation and transformation in cell-culture assays, with surprisingly little toxicity for normal cells *(8–10)*. Subsequent studies revealed an equally remarkable ability of FTIs to block the growth of tumors in animal models *(11–16)* and encouraged the belief that FTIs would be quickly reduced to practice in the clinic. However, the path of these FTIs to clinical utility has taken a few unexpected turns and several significant bumps in the road have been encountered. Although FTIs have indeed proven to be potent antitumor agents in cell-culture and animal models, one of the unexpected outcomes of these drug discovery efforts is a deepening doubt as to whether this is owing solely to the anti-Ras actions of FTIs. Although several FTIs are now under assessment in Phase I and II clinical trials, the development and availability of FTIs have revealed complexities of Ras and tumor cell biology that were unexpected, and have generated as many questions as answers.

In addition, the findings that FTase shares a subunit with the highly related enzyme, geranylgeranyltransferase (GGTase I) *(17)*, and that GGTase-I modifies many proteins of interest in signal transduction and transformation *(18)*, have stimulated interest in developing GGTase I inhibitors. Although the enzymology is more complicated than that of FTase, and such inhibitors have proven more difficult to come by than FTIs, geranylgeranyltransferase inhibitors (GGTIs) are now also under investigation.

The chapters in this volume document the considerable progress that has been made in the discovery and development of FTIs and GGTIs, the biochemical and structural characterization of their enzyme targets, and their evaluation as anti-Ras, anticancer, and anti-cardiovascular drugs in cell-culture and animal tumor models. In this closing chapter, we first provide a brief history of the evolution and development of FTIs as an anti-Ras strategy. We then highlight some of the key mechanistic questions that have arisen regarding how FTIs and GGTIs mediate their antitumor action; answers to these questions will be critical as FTIs approach clinical approval. We also pinpoint some of the unresolved issues that are currently being debated in the field. Finally, we discuss some of the complications in the progress of FTIs to the clinic, which reflect fundamental differences between the evaluation of conventional cytotoxic drugs and the assessment of a new generation of drugs targeting the actions of specific genes whose products play key roles in promoting oncogenesis.

2. TARGETING RAS FOR CANCER TREATMENT

Despite their importance in cancer treatment, the clear therapeutic limitations of our current repertoire of cytotoxic anticancer drugs are well appreciated. Thus, it is widely embraced that novel drug development approaches will be required to accelerate our progress towards curing cancer. Much hope is placed in studying cancer genetics (the identification of specific genes whose aberrant function may promote the development of human cancers). There are high expectations that rational drug discovery efforts will be able to correct the biologic consequences of malfunctioning oncogenes and of non-functioning tumor suppressor genes. The 30% prevalence of *ras* mutations in human cancers has marked it as one of the most promising oncogene targets for such efforts. In particular, the high prevalence of *ras* gene mutations in carcinomas of the pancreas, lung, and colon lends anti-Ras drug strategies importance in the treatment of solid tumors for which current therapy is limited or ineffective.

There are three closely related *ras* genes in the human genome, and mutated versions of H-*ras*, K-*ras*, and N-*ras* have been found in a wide spectrum of human cancers *(19)*. Although H-*ras* has been the most heavily studied *ras* isoform, it is the least frequently mutated *ras* gene. Instead, it is K-*ras* mutations that are most frequently associated with human carcinomas, and N-*ras* mutations that are associated with hematopoietic neoplasms. In light of the strong sequence identity shared by the gene products of the three *ras* genes (>90% identity), and their essentially identical biochemical and biological properties, until very recently it has been widely assumed that all Ras proteins are functionally identical. As will be described in this chapter, this assumption has provided one of the major bumps in the road to the clinic for FTIs.

That the first promising anti-Ras agents were finally developed in 1992, a decade after Ras was first linked to human cancer development, reflects the long process of taking discoveries from the bench to the clinic. Although our knowledge of Ras function is now considerable, some of the first clues to how anti-Ras drugs may be developed proved difficult to translate to novel therapies. A number of approaches have been considered, including gene therapy targeting both mutated *ras* genes and Ras protein function.

At least four approaches have been considered for developing chemotherapeutic agents that target Ras activity as a molecular relay switch in cell signaling. Ras proteins function as small GTPases whose biological activity is controlled by their association with guanine nucleotides *(20,21)*. Ras proteins cycle between an active GTP-bound form and an inactive GDP-bound form. Ras GDP/GTP cycling is controlled by two classes of regulatory proteins. Guanine nucleotide exchange factors (GEFs) promote GDP dissociation from Ras, thereby allowing Ras to bind GTP *(20)*. GTPase activating proteins (GAPs) stimulate the hydrolysis of the bound GTP to cycle Ras back to its inactive GDP-bound state *(21)*. The mutated *ras* genes that are present in human tumors encode Ras proteins (mutated most commonly at residues 12, 13, or 61) that are insensitive to GAP stimulation, rendering them chronically active. Consequently, two approaches to anti-Ras drug development have been either to restore GAP-responsiveness to mutated Ras proteins or to recognize specifically and interfere with the activated, GTP-bound form of Ras. To date, these efforts have yielded no promising drugs.

Ras is positioned at the inner face of the plasma membrane, where it serves to relay the messages from a wide array of extracellular stimuli, via diverse cell-surface receptors, to cytoplasmic signaling networks *(1–3,22)*. These signaling networks terminate at various cytoplasmic and nuclear targets and initiate events that control normal cell proliferation and death. These aspects of Ras function have identified two promising options for blocking Ras function, which in turn have led to the best yield of candidate anti-Ras drugs. First, because Ras function relies on its ability to stimulate downstream signaling cascades, agents that block any of these signaling components may serve as effective inhibitors of Ras function. In particular, Ras activation of a cascade of protein kinases, including the Raf, MEK, and ERK kinases, has made these kinases targets for inhibition. For example, inhibitors of MEK can effectively block Ras transformation in cell-culture model systems *(24,25)*. Hence, targeting MEK and other Ras signaling components is currently a promising and active area of anti-Ras drug discovery. However, recent revelations that Ras signaling is much more complex than simply triggering the Raf/MEK/ERK cascade, and that there are significant cell type differences in the importance of particular signaling pathways in Ras function, have complicated this active area of anti-Ras drug discovery.

The observation by Lowy, Willumsen and colleagues *(26,27)* in 1984 that oncogenic Ras transforming activity was absolutely dependent on its association with the plasma membrane, suggested that preventing this interaction is a fourth means of modulating Ras function. The subsequent determination in 1989 that farnesylation (as distinct from palmitoylation) is the first crucial post-translational modification of Ras that promotes its association with membranes *(28–30)* focused efforts on understanding protein prenylation. Then, when the FTase enzyme was isolated and characterized in 1990 *(6)*, FTase was immediately established as a target for anti-Ras drug development. A frenzy of activity followed quickly in both the pharmaceutical industry and academia.

As summarized in the preceding chapters, a wide spectrum of chemically distinct FTIs has now been developed. Some have been identified by conventional random high throughput screening of chemical and natural product sources for inhibitors of FTase-mediated prenylation of Ras. In addition, yeast genetic approaches have also identified natural products (e.g., manumycin and so forth) as FTIs *(31)*. However, the majority of FTIs have been developed by exploiting the CAAX tetrapeptide sequence at the carboxyl-terminus of all Ras proteins that signals FTase-catalyzed addition of the isoprenoid. The important demonstration by Brown, Goldstein, and colleagues *(6)* that the CAAX tetrapeptide could serve as a potent inhibitor of FTase in vitro prompted the rational design of CAAX peptidomimetics as cell permeable FTIs.

An important goal of FTI development has been to increase selectivity for FTase. This is owing to the fact that some proteins terminating in CAAX tetrapeptides could be recognized by a second related enzyme, geranylgeranyltransferase I (GGTase I). GGTase I is a structurally related enzyme that shares a common α-subunit with FTase and a homologous β-subunit (30% homology) *(17)*. Whereas FTase catalyzes the addition of a C15 farnesyl isoprenoid, GGTase I catalyzes the addition of a C20 geranylgeranyl isoprenoid to the cysteine of a subset of CAAX sequences. FTase preferentially recognizes CAAX sequences where the terminal X residue is serine, methionine, cysteine, alanine, or glutamine *(7)*. GGTase I recognizes CAAX sequences terminating in leucine *(32)*. A major goal of FTI development has been to identify inhibitors that are selective for FTase vs GGTase I, and it has proven relatively easy to design CAAX peptidomimetics that are highly specific inhibitors of FTase. Although it was once believed that strong FTase selectivity was a desired property of FTIs, some uncertainty regarding this issue has arisen. We will discuss later whether increasing selectivity for FTase has been the best approach and whether GGTase I inhibitors (GGTIs) might also have importance as inhibitors of the isoprenylation of Ras and/or other important target proteins.

3. UNDERSTANDING FTASE BIOCHEMISTRY AND STRUCTURE: REFINING INHIBITORS

Medicinal chemistry efforts have resulted in the design of cell permeable and FTase-selective CAAX peptidomimetics. Concurrently, biochemical, genetic, and structural analyses of wild-type and mutant forms of FTase have provided insight into how FTase recognizes both the CAAX tetrapeptide substrate and farnesyl pyrophosphate (FPP), the donor for the transferred farnesyl group. For example, Tamanoi and colleagues utilized a yeast genetic screen to identify specific residues in the β-subunit of FTase that, if altered, allow FTase to recognize GGTase I CAAX substrates *(33)*.

Biochemical studies evaluating the catalytic functions of FTase and the related GGTase I have provided insight into how these structurally and biochemically related enzymes recognize different CAAX substrates and catalyze the addition of distinct isoprenoid groups *(34–38)*. Although still not fully elucidated, this information has also provided some appreciation of how and when alternative prenylation occurs (*see below*). The determination of the crystal structure of FTase *(39,40)* and its co-crystal structure with both FPP and CAAX tetrepeptides *(41)* has provided further understanding of enzyme function and illuminated methods for increasing FTI potency. In particular, co-crystal structures of FTase with specific FTIs will be informative in this regard.

Although the majority of FTIs act as CAAX tetrapeptide competitive inhibitors of FTase, FPP-competitive inhibitors have also been developed. In light of the ubiquitous involvement of FPP in cellular metabolism, it has been generally assumed that such inhibitors will be compromised by a much greater toxicity problem. However, some FPP-competitive inhibitors are selective for FTase and exhibit potency against a wide range of human tumor cells in vitro and in vivo *(15,42,43)*. Bisubstrate inhibitors that antagonize FTase interaction with both FPP and CAAX have also been described *(44)*. Of these FTIs, to date only CAAX-competitive compounds have been reported to advance to clinical trials.

In addition to farnesylation of the cysteine residue of the CAAX motif, Ras proteins undergo additional CAAX-mediated post-translational modifications (Fig. 1). The farnesylation step is followed by endoprotease cleavage of the AAX residues and carboxymethylation of the farnesylated cysteine residue *(2,45)*. H-Ras, K-Ras4A, and N-Ras undergo an additional modification where cysteine residue(s) upstream of the CAAX sequence are covalently modified by addition of the fatty acid palmitate *(2,46)*. Now that the respective enzymes have been cloned and characterized *(47,48)*, inhibitors of the -AAX endoprotease and carboxymethylation steps are also currently under evaluation as anti-Ras and anticancer agents. We observed previously that a K-Ras4B CAAX motif mutant able to undergo farnesylation, but not the subsequent modifications, could still potently transform NIH 3T3 cells *(5)*. However, these studies involved overexpression of the mutant protein in a highly transformation-sensitive rodent fibroblast cell line. Thus, it remains a reasonable possibility that drugs that block AAX proteolysis and/or carboxylmethylation may inhibit the activity of mutated Ras proteins expressed at endogenous levels in human tumors.

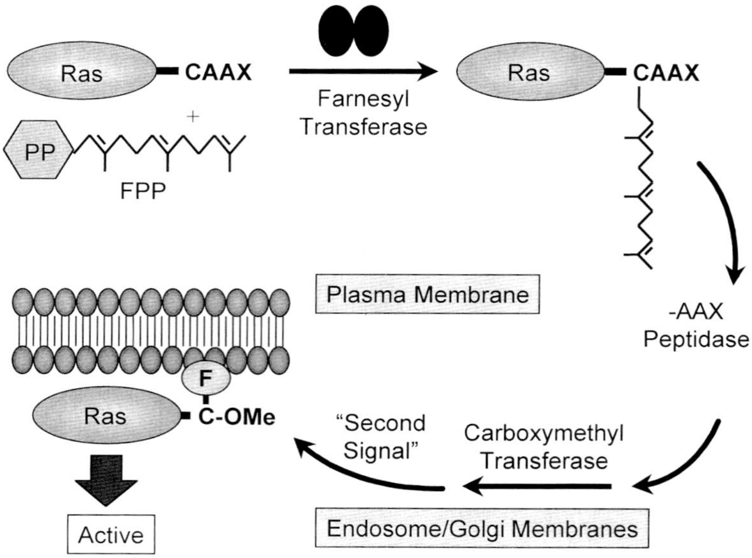

Fig. 1. Ras processing and targeting to the plasma membrane. The cytosolic farnesyltransferase (FTase) $\alpha\beta$ heterodimer catalyzes the covalent addition of farnesol from farnesylpyrophosphate (FPP) to the cysteine residue of the carboxyl terminal CAAX tetrapeptide sequence present in all Ras proteins. The peptidase and carboxymethyl transferase enzymes are located at the endosome/Golgi membranes, where they catalyze the removal of the AAX residues and the methylation of the resulting farnesyl-cysteine residue, respectively. Ras proteins that undergo only these modifications remain associated with endosomal membranes. In addition to the CAAX modifications, a "second signal" at the carboxyl terminus (palmitoylation for H-Ras and N-Ras; polylysine sequences for K-Ras4B) is required to complete the translocation of Ras from endosomal membranes to the plasma membrane. Sequence information is also required for correct plasma membrane localization.

4. FTIs AS ANTI-RAS DRUGS

Initial analyses of FTIs using cell-based transformation assays revealed the exciting promise of these drugs as anti-Ras and anticancer drugs. These studies revealed the potent ability of various FTIs to block the prenylation, membrane association, and transforming activity of oncogenic forms of the H-Ras protein when expressed in NIH 3T3 and other rodent fibroblast cell lines *(1–3)*. In contrast, the FTIs did not alter the growth of the untransformed counterparts of these H-Ras-transformed cell lines. Furthermore, these compounds did not block the growth of rodent fibroblasts that were transformed by oncoproteins whose transforming action is independent of Ras function, such as the Raf-1 serine/threonine kinase. When these studies were extended into animal models, FTIs showed the same remarkable ability to inhibit the growth of H-Ras-transformed rodent fibroblasts and to cause tumor regression of H-Ras transgene-induced mouse mammary tumors. No systemic toxicity was observed at the doses effective in reducing tumor growth. Thus, these FTIs showed the desired mechanism-based inhibitory activity and selectivity for Ras-transformed cells.

Although cell culture analyses showed that FTIs are cytostatic when assayed in a wide variety of cells, H-ras transgenic mouse studies demonstrated that FTIs can be cytotoxic under other conditions *(13,49)*. The question of whether FTIs are cytostatic or cytotoxic agents is further confused by recent observations that the response to FTIs in vitro is also dependent on culture conditions. Prendergast and colleagues *(50)* showed that when H-Ras transformed Raf-1 fibroblasts are deprived of substratum attachment, FTI treatment induces apoptosis. Similarly, Tamanoi and colleagues *(51)* found that a reduction in serum growth factors also changes the consequences of FTI treatment from inhibition of mitosis to initiation of programmed cell death. Furthermore, this apoptotic response was specific for K-Ras-transformed normal rat kidney cells and was not seen with their normal counterparts *(51)*. At present, the mechanisms underlying altered responses to FTIs in cells deprived of substratum attachment or growth factors, and the relation of these phenomena to tumor growth, are not understood. Nevertheless, such dramatic alterations in FTI action owing to heterogeneity in the tumor cell environment may partly explain the distinct consequences of FTIs in xenografts vs transgene-induced tumors.

In light of the fact that there are farnesylated proteins other than Ras, the general lack of normal cell toxicity was also an unexpected feature of FTIs. Further, because normal Ras function is believed to be both crucial for normal cell proliferation and dependent on farnesylation, why are FTIs tolerated by normal cells? Although sometimes referred to as anti-Ras drugs, these compounds target FTase and hence potentially act as antagonists of the function of all farnesylated proteins. The nuclear lamins (A and B) are involved in the assembly of the nuclear envelope, yet blocking their farnesylation does not have deleterious effects on cell growth *(52)*. Three proteins that mediate retinal signaling are farnesylated *(2)*, yet impairment of vision has not been documented. The basis for this lack of toxicity is presently not understood. Possible explanations include the retention of function owing to incomplete inhibition of farnesylation, the existence of FTI-insensitive proteins that facilitate redundant pathways, and alternative prenylation. The last possibility will be addressed in Section 5.

5. NOT ALL RAS PROTEINS ARE THE SAME

Because mutated alleles of K-*ras* (which encodes K-Ras4A and K-Ras4B owing to alternative splicing) and N-*ras*, are found in human tumors more commonly than H-*ras* *(19)*, FTI analyses were extended to rodent fibroblast cells transformed by oncogenic forms of K-Ras4B and N-Ras. In light of the relative ease with which FTIs could block H-Ras prenylation, both K-Ras and N-Ras proteins showed surprising resistance to FTI-mediated inhibition of prenylation *(53–55,88)*. Furthermore, although inhibition of their transforming activities could be achieved at much higher concentrations of FTI, this inhibitory activity did not coincide with inhibition of K-Ras or N-Ras prenylation *(55,88)*. Thus, a major and unexpected difference between Ras proteins was revealed. The higher affinity of the CAAX tetrapeptide sequences of K-Ras and N-Ras for FTase *(6)* may partially account for this resistance to FTIs. Also, a stretch of six consecutive lysines (the polybasic domain) unique to K-Ras4B and located just upstream from the CAAX motif contributes separately to functional FTI resistance of this Ras isoform *(54,56)*, probably by increasing FTase affinity via ionic interactions with an acidic surface near the FTase active site.

However, the more surprising additional explanation came from the unexpected finding that, in the presence of FTIs, K-Ras4A, K-Ras4B, and N-Ras proteins can be substrates for GGTase I *(57)* and become geranylgeranylated ("alternatively prenylated") in vivo *(58,59)*. Only the prenylation of H-Ras, which does not become alternatively prenylated, can be blocked effectively by FTI treatment (Fig. 2).

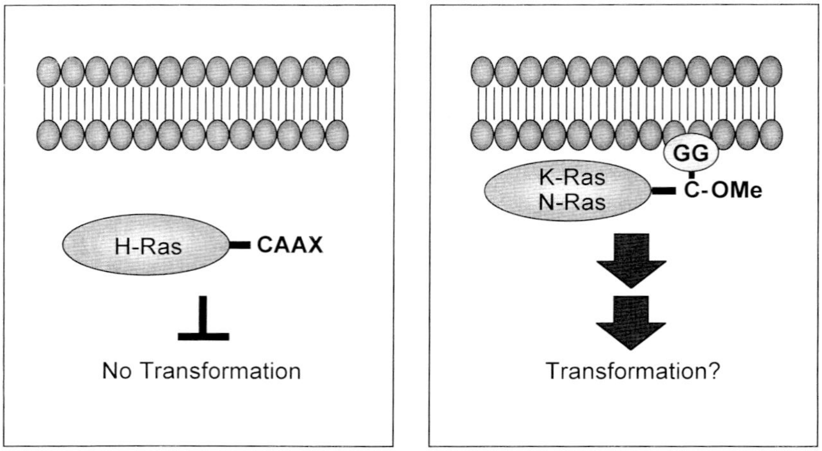

Fig. 2. FTase-mediated alterations in the prenylation state of Ras proteins. All Ras proteins are normally farnesylated (F). FTI treatment results in the accumulation of nonprenylated and biologically inactive H-Ras protein, whereas K-Ras and N-Ras proteins become substrates for GGTase I and are thus alternatively modified by geranylgeranylation (GG). Experimental studies suggest that GG-modified oncogenic Ras is functional when overexpressed, but whether alternatively processed normal or oncogenic Ras proteins fully mimic farnesylated Ras has not been clearly established.

The ability of K-Ras and N-Ras proteins to become alternatively prenylated in the presence of FTIs raises several questions that remain unanswered. First, we and others showed previously that geranylgeranyl- modified forms of activated H-Ras and K-Ras4B proteins retain their transforming activities in rodent fibroblasts and epithelial cells *(5,60)*. Is this also true in human tumors? Second, we observed that a geranylgeranylated version of normal H-Ras was a potent inhibitor of NIH 3T3 cell proliferation *(61)*. However, the accumulation of geranylgeranylated K-Ras and N-Ras proteins in normal tissue is not growth-inhibitory*(56)*. Thus, H-Ras and K-Ras may be functionally distinct such that geranylgeranylated K-Ras actually protects against toxicity in normal cells. Both the inhibition of farnesylation and the formation of alternatively prenylated proteins must be considered when assessing the biological actions of FTIs. Finally, it should be emphasized that many of our current concepts regarding Ras prenylation and biology have been derived from experimental studies where Ras proteins are overexpressed. The ability of these models to portray accurately the consequences of FTI treatment in human tumors where Ras is expressed at physiologic levels remains an area of valid concern.

Alternative prenylation of some Ras proteins has also prompted debate regarding whether GGTIs will serve as a necessary complement to FTI treatment. Additionally, it has created uncertainty regarding the original emphasis on FTase selectivity in drugs meant to target all Ras proteins. Despite the alternative prenylation concern, there remains a belief that inhibiting both FTase and GGTase I may result in too much toxicity to justify completely inhibiting Ras prenylation. Many more mammalian proteins are substrates for GGTase I than for FTase, and these include many proteins with vital cellular functions *(3)*. Examples are members of the Rho family of proteins (e.g., Rac1, RhoA, and Cdc42) whose functions in normal cellular physiology include modulation of actin cytoskeletal organization, gene expression, and cell cycle progression. However, since these same proteins have also been implicated as critical for Ras transformation *(62)*, one can envision that the blocking of their activities may further aid in blocking these malignant changes.

In contrast to FTase-specific inhibitors, development of GGTase I-specific inhibitors has been more difficult. Some highly selective GGTase I inhibitors have been engineered and assessed in cell-culture and animal models *(63,64)*. GGTI treatment has been shown to cause cell-cycle arrest of a variety of human tumor cells *(65–67)* and also to prevent growth and induce apoptosis of vascular smooth muscle cells, suggesting potential utility in averting restenosis *(68)*. Current speculation is that such arrest occurs via altering RhoA prenylation, thereby altering p21WAF regulation *(69–70)*. Surprisingly, although GGTIs alone inhibit cell growth, there is no clear evidence that they provide a useful complement to FTIs in blocking tumor growth in xenograft studies *(71)*. The anticipated toxic side effects of FTIs were not realized; perhaps the same will be seen for GGTIs.

Aside from the potential to undergo alternative prenylation, do Ras proteins differ in other ways? Although such biological differences are anticipated, to date only subtle differences in function have been described. The lack of a phenotype seen in H-*ras* or N-*ras* knockout mice *(72)* argues that these Ras proteins are functionally redundant with each other or with other, presumably related, proteins. In contrast, the embryonic lethal nature of K-*ras* knockout mice suggests that K-Ras may have a distinct function *(73,74)*. However, this lethality may also reflect the possibility that K-*ras* could be the sole *ras* gene expressed in some tissues at specific and critical developmental stages. Ras guanine nucleotide exchange factors (GEFs) and GTPase-activating proteins (GAPs), which modulate Ras activity, do not affect all Ras isoforms equivalently *(75,76)*, suggesting the possibility of their subtle modulation of Ras-related signaling pathways. Finally, the finding that H-, N-, and K-Ras proteins, once farnesylated, traffic differently in the cell and to potentially different microdomains at the plasma membrane, also suggests the possibility that, if location dictates function, the Ras isoforms are not really all alike. Together, these observations add yet another layer of potential complexity to the effect of FTIs and GGTIs on normal and transformed cells.

6. THE CONCEPT OF TARGET X ARISES

A second major complication of FTI development arose when the action of these drugs against human tumor cell lines was studied. In particular, Rosen, Sepp-Lorenzino, and colleagues *(77)* tested the ability of FTI treatment to impair the growth of a wide spectrum of human tumor cells. It was anticipated that tumor cells with mutated *ras* genes would be those most sensitive to FTI inhibition. Furthermore, in light of FTI-induced alternative prenylation, tumors that harbored K-*ras* or N-*ras* mutations would be expected to be insensitive. However, two unexpected results were seen. First, sensitivity did not correlate with *ras* mutation status and second, some tumor cells that harbored mutated K-*ras* were highly sensitive.

A partial explanation for these observations is the uncertain contribution of mutated *ras* to the transformed and tumorigenic phenotype of a particular tumor cell *(2)*. Mutated Ras may facilitate some aspect of tumor cell growth, such as invasion or metastasis, but not anchorage-independent proliferation or tumorigenicity. Hence, depending on the biological assay employed, FTI inhibition of oncogenic Ras function in a particular tumor cell may or may not be observed. Alternatively, mutated Ras may have served to facilitate an important step in tumor progression that is no longer relevant to the advanced tumor cell. Further, the *ras* mutation may have been secondary to the genomic instability characteristic of tumor cells, and may have no role in maintenance of the tumor cell

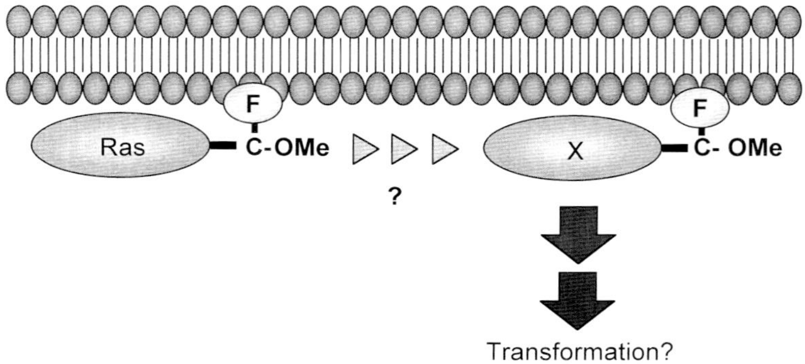

Fig. 3. Target X is a mediator of transformation: downstream or out of the stream of Ras signaling? Because FTI inhibition of cancer cell growth does not correlate with the presence of a mutated *ras* allele, FTIs may also inhibit the farnesylation of another protein(s), designated target X, that promotes cellular growth transformation.

growth phenotype. Finally, because Ras function may be aberrantly upregulated as a consequence of defects in other signaling components (e.g., tyrosine kinases), a *ras* mutation negative tumor may still be sensitive to FTIs via a Ras-dependent mechanism.

What determines the sensitivity of tumor cells to FTI inhibition? Is inhibition of Ras farnesylation important? Alternatively, are other farnesylated proteins the "real" targets of FTIs? At present, it is widely believed that, although inhibition of Ras farnesylation is a factor in at least some tumors, other farnesylated proteins (target X) targeted by FTIs are also likely to be important modulators of FTI responsiveness (Fig. 3). In addition to the lack of correlation between FTI susceptibility and *ras* mutation status, the ability of FTIs to inhibit the growth of K-Ras-transformed cells without inhibiting K-Ras prenylation or, apparently, K-Ras function also supports the existence of a target X for FTIs. A target X protein should either not become alternatively prenylated upon FTI treatment, or the function of X should be altered by alternative prenylation (Fig. 4). Whether this protein is a necessary component of Ras transformation or can facilitate transformation independent of Ras function are possibilities that remain to be established.

Increasing numbers of proteins have been identified as being modified by a farnesyl isoprenoid *(78)*, and several of these have been highlighted as potential target X proteins *(2,78)*. The candidates for target X that have attracted the most interest are members of the Rho family of proteins. Rho family proteins constitute one of the major branches of the Ras superfamily of Ras-related small GTPases *(62)*. Like Ras, Rho family proteins function as GDP/GTP-regulated switches involved in signal transduction that modulate a plethora of other cellular processes. Specific Rho family proteins have been implicated as necessary components for the transforming activity of Ras and other oncoproteins. Several Rho family proteins are substrates for FTase: RhoB, RhoE, Rho6, and Rho7. Where examined, the prenylation of Rho family proteins has also been shown to be critical for their function. Thus, some of these Rho family proteins possess the key properties expected of target X: a substrate for FTase, farnesylation-dependent function, and a mediator of cellular transformation.

Presently, experimental studies by Prendergast and colleagues *(79)* provide support for RhoB as one target X. RhoB is unusual in that it is normally modified by both FTase and GGTase I activity *(80,81)*. In the presence of FTIs, RhoB becomes exclusively geranylgeranylated and is growth-inhibitory *(82)*. Additional evidence in support of this

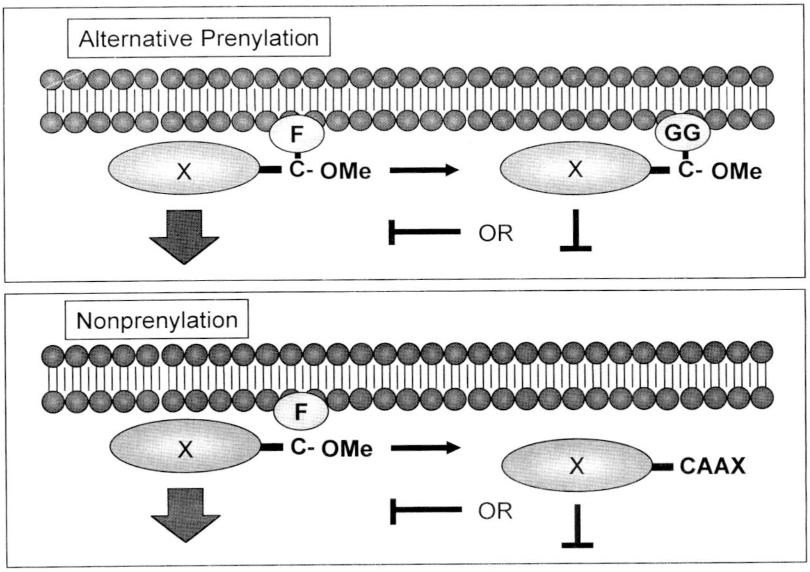

Fig. 4. FTase inhibition of target X function. This may occur either by promoting the formation of an alternatively prenylated, geranylgeranylated protein, or by promoting the accumulation of a nonprenylated protein. The GG-modified or nonprenylated target X may be a nonfunctional protein or may function as a dominant negative inhibitor of its farnesylated counterpart.

model includes the ability of an ectopically expressed, prenylation-independent, myristylated version of RhoB to render Ras-transformed cells resistant to FTIs *(83)*. However, other observations argue against RhoB as target X, or at least the sole target X. For example, because RhoB is already predominantly (>80%) GG-modified in the absence of FTIs, and RhoB-GG is growth-inhibitory *(82)*, it is hard to envision how the loss of the minor farnesylated, growth-stimulatory fraction of RhoB upon FTI treatment could result in the sometimes dramatic levels of growth inhibition seen in FTI-treated tumors. In addition, RhoB alone does not mediate all of the effects of FTI treatment, including apoptosis. The role of RhoB in cellular responses to FTIs is under intense and continuing investigation.

Other candidates for target X include RhoE *(84)*. This protein (also called Rho6 or Rnd1 *[85]*) causes disruption of actin stress fibers and focal adhesions *(85,86)*, which are actions also associated with Ras transformation. However, there is currently no evidence that RhoE exhibits growth-promoting activities or that RhoE function is required for Ras transformation. Furthermore, if RhoE were target X, then an exclusively GG-modified form of RhoE should render cells resistant to FTIs. We did not observe this to be the case (Cox and Der, unpublished data).

Two CAAX-containing phosphatases, PTP CAAX1 and PTP CAAX2, are also FTase substrates *(87)* and some evidence supports their candidacy for target X. Other interesting candidates are farnesylated centromere binding proteins (CENP-E, CENP-F) that control mitotic spindle formation and consequently cellular proliferation. A limited number of other farnesylated proteins have been identified but there is little or no evidence to date for their involvement in mediating the antitumor action of FTIs. Thus, it is entirely possible that target X is a protein that remains to be identified. Expression cloning to find additional FTase substrates or proteomic analysis of FTI-sensitive proteins may be required to identify this target. It is also entirely likely that there may be multiple target

X proteins, and that different cell types and/or tumor types may utilize different subsets of such proteins. These are topics of ongoing speculation.

7. ANIMAL MODELS: OF MICE AND MEN

The most widely used animal model in the evaluation of anti-tumor efficacy of potential cancer drugs involves tumor xenografts in the immune-compromised athymic nude mouse. This has been the starting point for animal-based studies of FTIs using Ras-transformed rodent fibroblasts or human tumor-derived cell lines *(11,12,14,15,49,64,71)*. Advantages of mouse xenografts include rapid tumor development, which is convenient for the experimental protocols of drug testing, and the opportunity to compare the activity of FTIs with a wealth of other previously characterized anticancer drugs. Disadvantages include tumor induction from homogenous cell populations, their growth as encapsulated subcutaneous tumors that do not accurately mimic invasive tumor cells that arise in specific organ locations, genetically different tumor and stromal tissues, and lack of a normal antitumor immune response. Clearly, these disadvantages may account, in part, for the poor clinical predictive value of antitumor efficacy observed in mouse xenografts.

The use of transgenic animals in which the tissue-specific expression of mutated *ras* alleles leads to stochastic development of tumors may overcome some of the limitations of mouse xenograft models. Published studies have assessed FTIs in H-*ras (13,49)* and N-*ras (88)* transgenic mouse models, and ongoing studies are employing K-*ras* transgenic mice. As described previously, FTIs have shown the ability to cause rapid tumor regression in H-*ras* transgenic mice *(13,49,89)*. However, the inhibition of tumor progression, rather than induction of tumor regression, seen in an N-*ras* transgenic model *(88)* prompts further debate regarding whether FTIs will be cytotoxic or cytostatic drugs against human cancers. Whether transgenic mouse models prove more predictive than the xenograft models for the testing of FTIs and other candidate drugs is an interesting question that will be answered only after the data from clinical trials become available.

An unavoidable limitation of these mouse models is the fact that the physiology of mice is distinct from that of humans. How great these limitations have been for the testing of FTIs will only be appreciated as FTIs are tested in clinical trials. Whether the remarkably nontoxic nature of FTIs seen in mice will also be observed in humans is one key aspect that awaits clarification.

In addition to concerns about the accuracy of mouse models in general, FTI animal studies have given rise to considerable caution regarding the usefulness of FTIs as anticancer drugs. First, the observations in xenografts that FTIs could block progression, but not reverse tumor growth of H-Ras transformed rodent fibroblasts, indicated that FTIs were cytostatic rather than cytotoxic drugs. Thus, continuous and long-term treatment with FTIs may be required for effective management of cancer patients. Second, in contrast to the model, the H-*ras* transgenic mouse tumors showed apparent complete and rapid regression when treated with FTIs. This was owing, in part, to induction of increased tumor cell apoptosis *(13,49,89)*. What animal model response would provide the best indication of how human tumors will respond? The hope is that human neoplasms will behave more like *ras* transgene-induced tumors than xenograft tumors in nude mice. However, even in transgenic mice, FTI withdrawal resulted in a rapid tumor regrowth, indicating that tumor cells can persist even with treatment. The prospects of long-term treatment argue that oral bioavailability of FTIs will be a necessity. Chronic treatment increases the prospects that tumor cell resistance or nonspecific toxicity may pose limitations for FTI treatment. Resistance to FTIs has been documented both in cell lines and in some animal studies after repeated cycles of drug treatment and withdrawal *(13,89,90)*.

8. QUESTIONS IN THE CLINIC

Like the cell culture and animal studies, Phase I and II clinical trials have also prompted many more questions than answers. One key question for the clinical trials has been the choice of the appropriate patient population. Tumors most frequently associated with *ras* mutations defined the initial patient population for FTI therapy (e.g., carcinomas of the pancreas, lung, and colon *[19]*). However, in light of the lack of correlation between FTI sensitivity and *ras* mutation status seen in cell culture *(77)*, it has become less obvious which patient population to target.

Perhaps other genetic markers, such as p53 mutation status, may define FTI-sensitive tumors. Some degree of correlation has been reported between FTI sensitivity and wild-type p53 function in human tumor cell lines *(91)*. The high FTI sensitivity of specific tissue-derived tumor cell lines (e.g., breast) may be another means to decide this question. However, a simple reliance on the sensitivity of specific tumor cell lines may be problematic, because very few established cell lines provide an accurate reflection of the more heterogeneous cell population that comprises a human tumor. Gathering histologic or genetic information may or may not be an effective method of predicting tumor response, absent identification of the primary protein target of FTI therapy.

A second important question in the design of clinical trials concerns the identification of endpoints to monitor the biologically effective dose of FTIs. In the majority of patients, routine monitoring of the tumor cells themselves will not be feasible. Thus, most trial designs involve the isolation and analysis of normal peripheral blood mononuclear (PBMN) leukocytes as a surrogate tissue in the hope that this will be predictive of the FTI activity in tumor tissue. Suggestions of changes to monitor include the ability of FTIs to block signaling activities associated with Ras activation, such as activation of the ERK mitogen-activated protein kinases. However, both published and unpublished studies indicate that ERK activation is not always a reliable marker for Ras activity *(92)*. Thus, a more direct measurement may be monitoring FTI inhibition of farnesylation in PBMNs, although the levels of H-Ras in such cells is low.

Because Ras may not be the effective target of FTIs, is inhibition of H-Ras farnesylation a useful measure to define the effective dosage for patients? Would surrogate marker proteins such as prelamin A *(93)* or DNAJ homologs *(94)* be more accurate? Because farnesylated proteins show different sensitivities to inhibition by FTIs, should total FTase activity be monitored instead? If so, is total inhibition necessary or will a 50% reduction be sufficient? These questions are under active debate.

There is concern that any chosen biologic endpoint in these trials may define a dosage of FTIs that is suboptimal for antitumor activity, particularly if FTIs have activities other than inhibiting FTase. An alternative is to use FTIs near the maximum tolerated dose. This approach would prevent the discarding of a viable anticancer agent because the wrong endpoint was targeted.

It is anticipated that treatment with FTIs alone will not be the most effective application for these drugs. Instead, combining FTIs with other conventional therapeutic approaches or with other novel target-based drugs has been proposed. Because FTIs showed cytostatic actions in xenograft tumor assays, combination approaches with conventional radiation or cytotoxic drugs were not anticipated to be useful. Current radiation and chemotherapy protocols are most effective when the growth fraction of a tumor is high, whereas FTI treatment would be expected to reduce this fraction. Surprisingly, additive and supra-additive effects of FTIs with both radiation and chemotherapeutic treatments have been described. The application of FTIs with conventional treatments may reduce the need for orally bioavailable forms of these drugs, because intramuscular or intravenous administration can be carried out concurrently with short-term treatments.

Oncogenic Ras can contribute to radiation resistance *(95)*, and FTI treatment has been shown to function as a radiosensitizer in some transformed and human tumor cells *(96, 97)*. The combination of taxol and FTIs has also shown additive or synergistic growth inhibition in different reports *(49,64,98)*. Additive or synergistic anti-tumor activity in vitro has also been reported for FTIs and other cytotoxic agents covering the spectrum of distinct mechanisms of action, including doxorubicin, cisplatin, 5-fluorouracil, vinblastine, vincristine, and cyclophosphamide *(49,64,98)*.

Another combination approach is the administration of other target-based drugs, such as inhibitors of the epidermal growth factor receptor, of the MEK1/2 kinases, or of angiogenesis. The rationale for such approaches is based on the likelihood that these drugs will inhibit signaling pathways that are functionally linked with Ras. However, combination of drugs with the least overlapping mechanisms of action has traditionally worked best. Nevertheless, because successful combination therapies have, for the most part, been developed empirically, all possible combination approaches should be considered.

Finally, there is concern with FTIs and other possibly cytostatic target-specific drugs about the appropriate measure of clinical efficacy. With conventional cytotoxic drugs, treatment-induced tumor regression defines efficacy. If FTIs do not prove to be cytotoxic agents against human tumors, then monitoring tumor regression would be an inappropriate clinical endpoint. Instead, the more difficult long-term assessment of inhibited tumor progression and prolonged patient survival will be required.

9. CONCLUSIONS AND FUTURE QUESTIONS

The long journey of FTIs from the bench to the clinic has been marked with the typical bumps in the road associated with all drug discovery efforts. However, the development of FTIs represents a radical departure from the rationale and approaches that have been employed to take more conventional cytotoxic drugs to the clinic. Consequently, new ground is being charted by these efforts.

At present, the precise mechanism of action of FTIs that accounts for their anti-tumor activity remains elusive. Although it is generally assumed that this mechanism is target-based and involves blocking FTase activity, we should remain cautious and open to the possibility that FTIs may alter more than FTase activity. The more simple idea of FTI mechanism may be analogous to the ideas that Ras simply activated Raf, and that all Ras proteins were functionally equivalent, concepts that once were embraced as dogma but have now been discarded for more complex scenarios.

If we assume that FTIs block the activity of one or more FTase substrates, then the identification of farnesylated target X proteins is certainly a vital step towards understanding the mechanism of FTI action. Identifying target X may also resolve key issues for clinical trials, such as the choice of a biologic endpoint that serves as an appropriate surrogate for clinical efficacy. Target X may also serve as a histologic or genetic marker for identifying the appropriate cancer patient population for FTI treatment. However, because every cell type utilizes a different array of intracellular signaling cascades, each of which may involve multiple farnesylated proteins, even the identification and functional characterization of every farnesylated protein may not permit predictable responses of particular cancers to FTIs.

The future role of GGTIs is even less clear, although equally interesting. The different roles of FTIs and GGTIs in inhibiting cell cycle and apoptosis suggest that their roles in treatment will not be overlapping. The identification of GGTIs as functional inhibitors that can cause apoptosis in desired cell populations when delivered appropriately is a potentially powerful tool for inhibiting local growth of unwanted normal cells in particular locations, such as cardiac arteries. The addition of GGTIs to FTIs as radiosensitizers has also yet to be tried in the clinic. Other potential uses in anticancer therapy remain to be determined, as essentially none of the studies critical to determining their utility in this disease have yet been performed. However, the increasing importance of known targets of GGTase I in cancer biology dictates that such studies are clearly merited.

In summary, together with inhibitors of protein kinases, FTIs, and potentially GGTIs represent the first wave of novel cancer therapeutics based on understanding the role of oncoproteins in cancer development and cellular proliferation. The effectiveness of these inhibitors will provide our first assessment of whether antagonists of signal transduction pathways will produce long-awaited and much-needed breakthroughs in cancer and cardiovascular treatment.

ACKNOWLEDGMENTS

Our research studies were supported by NIH grants to A.D.C. (CA76092) and C.J.D. (CA42978, CA69577) and a National Cooperative Drug Discovery Grant to A.D.C. and C.J.D. (CA67771). L.G.T. was supported an NIH Short Term Research Award. We thank our collaborators, Saïd Sebti and Andy Hamilton, and members of our laboratories for keeping this topic fresh, interesting, and enjoyable.

REFERENCES

1. Gibbs JB, Oliff A. The potential of farnesyltransferase inhibitors as cancer chemotherapeutics. *Annu Rev Pharmacol Toxicol* 1997; 37:143–166.
2. Cox AD, Der CJ. Farnesyltransferase inhibitors and cancer treatment: targeting simply Ras? *Biochim Biophys Acta* 1997; 1333:F51–F71.
3. Lerner EC, Hamilton AD, Sebti SM. Inhibition of Ras prenylation: a signaling target for novel anticancer drug design. *Anticancer Drug Des* 1997; 12:229–238.
4. Jackson JH, Cochrane CG, Bourne JR, Solski PA, Buss JE, Der CJ. Farnesol modification of Kirsten-ras exon 4B protein is essential for transformation. *Proc Natl Acad Sci USA* 1990; 87:3042–3046.
5. Kato K, Cox AD, Hisaka MM, Graham SM, Buss JE, Der CJ. Isoprenoid addition to Ras protein is the critical modification for its membrane association and transforming activity. *Proc Natl Acad Sci USA* 1992; 89:6403–6407.
6. Reiss Y, Goldstein JL, Seabra MC, Casey PJ, Brown MS. Inhibition of purified p21ras farnesyl:protein transferase by Cys-AAX tetrapeptides. *Cell* 1990; 62:81–88.
7. Reiss Y, Stradley SJ, Gierasch LM, Brown MS, Goldstein JL. Sequence requirement for peptide recognition by rat brain p21ras protein farnesyltransferase. *Proc Natl Acad Sci USA* 1991; 88:732–736.
8. James GL, Goldstein JL, Brown MS, Rawson TE, Somers TC, McDowell RS, et al. Benzodiazepine peptidomimetics: potent inhibitors of Ras farnesylation in animal cells. *Science* 1993; 260:1937–1942.
9. Kohl NE, Mosser SD, deSolms SJ, Giuliani EA, Pompliano DL, Graham SL, et al. Selective inhibition of ras-dependent transformation by a farnesyltransferase inhibitor. *Science* 1993; 260:1934–1937.

10. Garcia AM, Rowell C, Ackermann K, Kowalczyk JJ, Lewis MD. Peptidomimetic inhibitors of Ras farnesylation and function in whole cells. *J Biol Chem* 1993; 268:18,415–18,418.
11. Sun J, Qian Y, Hamilton AD, Sebti SM. Ras CAAX peptidomimetic FTI 276 selectively blocks tumor growth in nude mice of a human lung carcinoma with K-Ras mutation and p53 deletion. *Cancer Res* 1995; 55:4243–4247.
12. Nagasu T, Yoshimatsu K, Rowell C, Lewis MD, Garcia AM. Inhibition of human tumor xenograft growth by treatment with the farnesyl transferase inhibitor B956. *Cancer Res* 1995; 55:5310–5314.
13. Kohl NE, Omer CA, Conner MW, Anthony NJ, Davide JP, deSolms SJ, et al. Inhibition of farnesyltransferase induces regression of mammary and salivary carcinomas in ras transgenic mice. *Nat Med* 1995; 1:792–797.
14. Leftheris K, Kline T, Vite GD, Cho YH, Bhide RS, Patel DV, et al. Development of highly potent inhibitors of Ras farnesyltransferase possessing cellular and in vivo activity. *J Med Chem* 1996; 39:224–236.
15. McNamara DJ, Dobrusin E, Leonard DM, Shuler KR, Kaltenbronn JS, Quin J 3rd, et al. C-terminal modifications of histidyl-N-benzyl-glycinamides to give improved inhibition of Ras farnesyltransferase, cellular activity, and anticancer activity in mice. *J Med Chem* 1997; 40:3319–3322.
16. Njoroge FG, Vibulbhan B, Rane DF, Bishop WR, Petrin J, Patton R, et al. Structure-activity relationship of 3-substituted N-(pyridinylacetyl)-4-(8-chloro-5,6-dihydro-11H-benzo[5,6] cyclo-hepta[1,2-b] pyridin-11-ylidene)-piperidine inhibitors of farnesyl-protein transferase: design and synthesis of in vivo active antitumor compounds. *J Med Chem* 1997; 40:4290–4301.
17. Seabra MC, Reiss Y, Casey PJ, Brown MS, Goldstein JL. Protein farnesyltransferase and geranylgeranyltransferase share a common alpha subunit. *Cell* 1991; 65:429–434.
18. Zhang FL, Casey PJ. Protein prenylation: molecular mechanisms and functional consequences. *Annu Rev Biochem* 1996; 65:241–269.
19. Bos JL. *ras* oncogenes in human cancer: a review. *Cancer Res* 1989; 49:4682–4689.
20. Quilliam LA, Khosravi-Far R, Huff SY, Der CJ. Guanine nucleotide exchange factors: activators of the Ras superfamily of proteins. *Bioessays* 1995; 17:395–404.
21. Trahey M, McCormick F. A cytoplasmic protein stimulates normal N-ras p21 GTPase, but does not affect oncogenic mutants. *Science* 1987; 238:542–545.
22. Vojtek AB, Der CJ. Increasing complexity of the Ras signaling pathway. *J Biol Chem* 1998; 273:19,925–19,928.
23. Alessi DR, Cuenda A, Cohen P, Dudley DT, Saltiel AR. PD 098059 is a specific inhibitor of the activation of mitogen-activated protein kinase kinase in vitro and in vivo. *J Biol Chem* 1995; 270:27,489–27,494.
24. Weyman CM, Ramocki MB, Taparowsky EJ, Wolfman A. Distinct signaling pathways regulate transformation and inhibition of skeletal muscle differentiation by oncogenic Ras. *Oncogene* 1997; 14:697–704.
25. Oldham SM, Cox AD, Reynolds ER, Sizemore NS, Coffey RJ Jr, Der CJ. Ras, but not Src, transformation of RIE-1 epithelial cells is dependent on activation of the mitogen-activated protein kinase cascade. *Oncogene* 1998; 16:2565–2573.
26. Willumsen BM, Christensen A, Hubbert NL, Papageorge AG, Lowy DR. The p21 ras C-terminus is required for transformation and membrane association. *Nature* 1984; 310:583–586.
27. Willumsen BM, Norris K, Papageorge AG, Hubbert NL, Lowy DR. Harvey murine sarcoma virus p21 ras protein: biological and biochemical significance of the cysteine nearest the carboxy terminus. *EMBO J* 1984; 3:2581–2585.
28. Hancock JF, Magee AI, Childs JE, Marshall CJ. All ras proteins are polyisoprenylated but only some are palmitoylated. *Cell* 1989; 57:1167–1177.
29. Schafer WR, Kim R, Sterne R, Thorner J, Kim SH, Rine J. Genetic and pharmacological suppression of oncogenic mutations in ras genes of yeast and humans. *Science* 1989; 245:379–385.
30. Casey PJ, Solski PA, Der CJ, Buss JE. p21ras is modified by a farnesyl isoprenoid. *Proc Natl Acad Sci USA* 1989; 86:8323–8327.
31. Sattler I, Tamanoi F. Prenylation of Ras and inhibitors of prenyltransferases, in: *Regulation of the Ras Signaling Network* (Maruta H, Burgess AW, eds.), Landes, Austin, TX, 1996, pp. 95–137.
32. Finegold AA, Johnson DI, Farnsworth CC, Gelb MH, Judd SR, Glomset JA, Tamanoi F. Protein geranylgeranyltransferase of Saccharomyces cerevisiae is specific for Cys-Xaa-Xaa-Leu motif proteins and requires the CDC43 gene product but not the DPR1 gene product. *Proc Natl Acad Sci USA* 1991; 88: 4448–4452.
33. Del Villar K, Mitsuzawa H, Yang W, Sattler I, Tamanoi F. Amino acid substitutions that convert the protein substrate specificity of farnesyltransferase to that of geranylgeranyltransferase type I. *J Biol Chem* 1997; 272:680–687.

34. Dolence JM, Cassidy PB, Mathis JR, Poulter CD. Yeast protein farnesyltransferase: steady-state kinetic studies of substrate binding. *Biochemistry* 1995; 34:16,687–16,694.
35. Yokoyama K, McGeady P, Gelb MH. Mammalian protein geranylgeranyltransferase-I: substrate specificity, kinetic mechanism, metal requirements, and affinity labeling. *Biochemistry* 1995; 34:1344–1354.
36. Yokoyama K, Zimmerman K, Scholten J, Gelb MH. Differential prenyl pyrophosphate binding to mammalian protein geranylgeranyltransferase-I and protein farnesyltransferase and its consequence on the specificity of protein prenylation. *J Biol Chem* 1997; 272:3944–3952.
37. Huang CC, Casey PJ, Fierke CA. Evidence for a catalytic role of zinc in protein farnesyltransferase. Spectroscopy of Co^{2+}-farnesyltransferase indicates metal coordination of the substrate thiolate. *J Biol Chem* 1997; 272:20–23.
38. Hohl RJ, Lewis KA, Cermak DM, Wiemer DF. Stereochemistry-dependent inhibition of RAS farnesylation by farnesyl phosphonic acids. *Lipids* 1998; 33:39–46.
39. Park HW, Boduluri SR, Moomaw JF, Casey PJ, Beese LS. Crystal structure of protein farnesyltransferase at 2.25 angstrom resolution. *Science* 1997; 275:1800–1804.
40. Dunten P, Kammlott U, Crowther R, Weber D, Palermo R, Birktoft J. Protein farnesyltransferase: structure and implications for substrate binding. *Biochemistry* 1998; 37:7907–7912.
41. Strickland CL, Windsor WT, Syto R, Wang L, Bond R, Wu Z, et al. Crystal structure of farnesyl protein transferase complexed with a CaaX peptide and farnesyl diphosphate analogue. *Biochemistry* 1998; 37: 16,601–16,611.
42. Yonemoto M, Satoh T, Arakawa H, Suzuki-Takahashi I, Monden Y, Kodera T, et al. J-104,871, a novel farnesyltransferase inhibitor, blocks Ras farnesylation in vivo in a farnesyl pyrophosphate-competitive manner. *Mol Pharmacol* 1998; 54:1–7.
43. Vrignaud P, Bello A, Bissery MC, Jenkins R, Hasnain A, Mailliet P, Lavelle F. RPR 13401, a non-peptidomimetic farnesyltransferase inhibitor with in vivo activity. *Proc Amer Assoc Cancer Res* 1998; 39: 270, #1846.
44. Manne V, Yan N, Carboni JM, Tuomari AV, Ricca CS, Brown JG, et al. Bisubstrate inhibitors of farnesyltransferase: a novel class of specific inhibitors of ras transformed cells. *Oncogene* 1995; 10:1763–1779.
45. Ashby MN. CaaX converting enzymes. *Curr Opin Lipidol* 1998; 9:99–102.
46. Hancock JF, Paterson H, Marshall CJ. A polybasic domain or palmitoylation is required in addition to the CAAX motif to localize p21ras to the plasma membrane. *Cell* 1990; 63:133–139.
47. Boyartchuk VL, Ashby MN, Rine J. Modulation of Ras and a-factor function by carboxyl-terminal proteolysis. *Science* 1997; 275:1796–1800.
48. Dai Q, Choy E, Chiu V, Romano J, Slivka SR, Steitz SA, Michaelis S, Philips MR. Mammalian prenylcysteine carboxyl methyltransferase is in the endoplasmic reticulum. *J Biol Chem* 1998; 273: 15,030–15,034.
49. Liu M, Bryant MS, Chen J, Lee S, Yaremko B, Lipari P, et al.. Antitumor activity of SCH 66336, an orally bioavailable tricyclic inhibitor of farnesyl protein transferase, in human tumor xenograft models and wap-ras transgenic mice. *Cancer Res* 1998; 58:4947–4956.
50. Lebowitz PF, Sakamuro D, Prendergast GC. Farnesyl transferase inhibitors induce apoptosis of Ras-transformed cells denied substratum attachment. *Cancer Res* 1997; 57:708–713.
51. Suzuki N, Urano J, Tamanoi F. Farnesyltransferase inhibitors induce cytochrome c release and caspase 3 activation preferentially in transformed cells. *Proc Natl Acad Sci USA* 1998; 95:15,356–15,361.
52. Dalton MB, Fantle KS, Bechtold HA, DeMaio L, Evans RM, Krystosek A, Sinensky M. The farnesyl protein transferase inhibitor BZA-5B blocks farnesylation of nuclear lamins and p21ras but does not affect their function or localization. *Cancer* 1995; 55:3295–3304.
53. James G, Goldstein JL, Brown MS. Resistance of K-RasBV12 proteins to farnesyltransferase inhibitors in Rat1 cells. *Proc Natl Acad Sci USA* 1996; 93:4454–4458.
54. James GL, Goldstein JL, Brown MS. Polylysine and CVIM sequences of K-RasB dictate specificity of prenylation and confer resistance to benzodiazepine peptidomimetic in vitro. *J Biol Chem* 1995; 270: 6221–6226.
55. Lerner EC, Zhang TT, Knowles DB, Qian Y, Hamilton AD, Sebti SM. Inhibition of the prenylation of K-Ras, but not H- or N-Ras, is highly resistant to CAAX peptidomimetics and requires both a farnesyltransferase and a geranylgeranyltransferase I inhibitor in human tumor cell lines. *Oncogene* 1997; 15: 1283–1288.
56. Fiordalisi JJ, Chen Z, Johnson R, Toussaint LG, Casey PJ, Cox AD. Both CAAX and polylysine domains of K-Ras4B contribute independently to farnesyltransferase interaction. *J Biol Chem*, submitted.

57. Zhang FL, Kirschmeier P, Carr D, James L, Bond RW, Wang L, et al. Characterization of Ha-ras, N-ras, Ki-Ras4A, and Ki-Ras4B as in vitro substrates for farnesyl protein transferase and geranylgeranyl protein transferase type I. *J Biol Chem* 1997; 272:10,232–10,239.
58. Whyte DB, Kirschmeier P, Hockenberry TN, Nunez-Oliva I, James L, Catino JJ, Bishop WR, Pai JK. K- and N-Ras are geranylgeranylated in cells treated with farnesyl protein transferase inhibitors. *J Biol Chem* 1997; 272:14,459–14,464.
59. Rowell CA, Kowalczyk JJ, Lewis MD, Garcia AM. Direct demonstration of geranylgeranylation and farnesylation of Ki-Ras in vivo. *J Biol Chem* 1997; 272:14,093–14,097.
60. Hancock JF, Cadwallader K, Paterson H, Marshall CJ. A CAAX or a CAAL motif and a second signal are sufficient for plasma membrane targeting of ras proteins. *EMBO J* 1991; 10:4033–4039.
61. Cox AD, Hisaka MM, Buss JE, Der CJ. Specific isoprenoid modification is required for function of normal, but not oncogenic, Ras protein. *Mol Cell* 1992; 12:2606–2615.
62. Zohn IM, Campbell SL, Khosravi-Far R, Rossman KL, Der CJ. Rho family proteins and Ras transformation: the RHOad less traveled gets congested. *Oncogene* 1998; 17:1415–1438.
63. Qian Y, Vogt A, Vasudevan A, Sebti SM, Hamilton AD. Selective inhibition of type-I geranylgeranyltransferase in vitro and in whole cells by CAAL peptidomimetics. *Bioorg Med Chem* 1998; 6:293–299.
64. Sun J, Blaskovich MA, Knowles D, Qian Y, Ohkanda J, Bailey RD, et al. Antitumor efficacy of a novel class of non-thiol-containing peptidomimetic inhibitors of farnesyltransferase and geranylgeranyltransferase I: combination therapy with the cytotoxic agents cisplatin, taxol and gemcitabine. *Cancer Res* 1999; 59:4919–4926.
65. Vogt A, Qian Y, McGuire TF, Hamilton AD, Sebti SM. Protein geranylgeranylation, not farnesylation, is required for the G1 to S phase transition in mouse fibroblasts. *Oncogene* 1996; 13:1991–1999.
66. Miquel K, Pradines A, Sun J, Qian Y, Hamilton AD, Sebti SM, Favre G. GGTI-298 induces G0-G1 block and apoptosis whereas FTI-277 causes G2-M enrichment in A549 cells. *Cancer Res* 1997; 57:46–50.
67. Vogt A, Sun J, Qian Y, Hamilton AD, Sebti SM. The geranylgeranyltransferase-I inhibitor GGTI-298 arrests human tumor cells in G0/G1 and induces p21(WAF1/CIP1/SDI1) in a p53-independent manner. *J Biol Chem* 1997; 272:7224–7229.
68. Stark WW Jr, Blaskovich MA, Johnson BA, Qian Y, Vasudevan A, Pitt B, et al. Inhibiting geranylgeranylation blocks growth and promotes apoptosis in pulmonary vascular smooth muscle cells. *Am J Physiol* 1998; 275:L55–L63.
69. Olson MF, Paterson HF, Marshall CJ. Signals from Ras and Rho GTPases interact to regulate expression of p21Waf1/Cip1. *Nature* 1998; 394:295–299.
70. Adnane J, Bizouarn FA, Qian Y, Hamilton AD, Sebti SM. p21(WAF1/CIP1) is upregulated by the geranylgeranyltransferase I inhibitor GGTI-298 through a transforming growth factor beta- and Sp1-responsive element: involvement of the small GTPase rhoA. *Mol Cell Biol* 1998; 18:6962–6970.
71. Sun J, Qian Y, Hamilton AD, Sebti SM. Both farnesyltransferase and geranylgeranyltransferase I inhibitors are required for inhibition of oncogenic K-Ras prenylation but each alone is sufficient to suppress human tumor growth in nude mouse xenografts. *Oncogene* 1998; 16:1467–1473.
72. Umanoff H, Edelmann W, Pellicer A, Kucherlapati R. The murine N-ras gene is not essential for growth and development. *Proc Natl Acad Sci USA* 1995; 92:1709–1713.
73. Johnson L, Greenbaum D, Cichowski K, Mercer K, Murphy E, Schmitt E, et al. K-ras is an essential gene in the mouse with partial functional overlap with N-ras. *Genes Dev* 1997; 11:2468–2481.
74. Koera K, Nakamura K, Nakao K, Miyoshi J, Toyoshima K, Hatta T, et al. K-ras is essential for the development of the mouse embryo. *Oncogene* 1997; 15:1151–1159.
75. Jones MK, Jackson JH. Ras-GRF activates Ha-Ras, but not N-Ras or K-Ras 4B, protein in vivo. *J Biol Chem* 1998; 273:1782–1787.
76. Bollag G, McCormick F. Differential regulation of rasGAP and neurofibromatosis gene product activities. *Nature* 1991; 351:576–579.
77. Sepp-Lorenzino L, Ma Z, Rands E, Kohl NE, Gibbs JB, Oliff A, Rosen N. A peptidomimetic inhibitor of farnesyl:protein transferase blocks the anchorage-dependent and -independent growth of human tumor cell lines. *Cancer* 1995; 55:5302–5309.
78. Cox AD, Der CJ. Farnesyl transferase inhibitors: anti-Ras or anti-cancer drugs? in *Signaling Networks and Cell Cycle Control: Molecular Basis of Cancer and Other Diseases* (Gutkind JS, ed.), Humana Press, Totowa, NJ, 2000, pp. 501–518.
79. Lebowitz PF, Prendergast GC. Non-Ras targets of farnesyltransferase inhibitors: focus on Rho. *Oncogene* 1998; 17:439–445.
80. Adamson P, Marshall CJ, Hall A, Tilbrook PA. Post-translational modifications of p21rho proteins. *J Biol Chem* 1992; 267:20,033–20,038.

81. Armstrong SA, Hannah VC, Goldstein JL, Brown MS. CAAX geranylgeranyl transferase transfers farnesyl as efficiently as geranylgeranyl to RhoB. *J Biol Chem* 1995; 270:7864–7868.
82. Lebowitz PF, Casey PJ, Prendergast GC, Thissen JA. Farnesyltransferase inhibitors alter the prenylation and growth-stimulating function of RhoB. *J Biol Chem* 1997; 272:15,591–15,594.
83. Lebowitz PF, Davide JP, Prendergast GC. Evidence that farnesyltransferase inhibitors suppress Ras transformation by interfering with Rho activity. *Mol Cell Biol* 1995; 15:6613–6622.
84. Foster R, Hu KQ, Lu Y, Nolan KM, Thissen J, Settleman J. Identification of a novel human Rho protein with unusual properties: GTPase deficiency and in vivo farnesylation. *Mol Cell Biol* 1996; 16:2689–2699.
85. Nobes CD, Lauritzen I, Mattei MG, Paris S, Hall A, Chardin P. A new member of the Rho family, Rnd1, promotes disassembly of actin filament structures and loss of cell adhesion. *J Cell Biol* 1998; 141:187–197.
86. Guasch RM, Scambler P, Jones GE, Ridley AJ. RhoE regulates actin cytoskeleton organization and cell migration. *Mol Cell Biol* 1998; 18:4761–4771.
87. Cates CA, Michael RL, Stayrook KR, Harvey KA, Burke YD, Randall SK, et al. Prenylation of oncogenic human PTP(CAAX) protein tyrosine phosphatases. *Cancer Lett* 1996; 110:49–55.
88. Mangues R, Corral T, Kohl NE, Symmans WF, Lu S, Malumbres M, et al. Antitumor effect of a farnesyl protein transferase inhibitor in mammary and lymphoid tumors overexpressing N-ras in transgenic mice. *Cancer Res* 1998; 58:1253–1259.
89. Barrington RE, Subler MA, Rands E, Omer CA, Miller PJ, Hundley JE, et al. A farnesyltransferase inhibitor induces tumor regression in transgenic mice harboring multiple oncogenic mutations by mediating alterations in both cell cycle control and apoptosis. *Mol Cell Biol* 1998; 18:85–92.
90. Prendergast GC, Davide JP, Lebowitz PF, Wechsler-Reya R, Kohl NE. Resistance of a variant ras-transformed cell line to phenotypic reversion by farnesyl transferase inhibitors. *Cancer Res* 1996; 56: 2626–2632.
91. Sepp-Lorenzino L, Rosen N. A farnesyl-protein transferase inhibitor induces p21 expression and G1 block in p53 wild type tumor cells. *J Biol* 1998; 273:20,243–20,251.
92. James GL, Brown MS, Cobb MH, Goldstein JL. Benzodiazepine peptidomimetic BZA-5B interrupts the MAP kinase activation pathway in H-Ras-transformed Rat-1 cells, but not in untransformed cells. *J Biol* 1994; 269:27,705–27,714.
93. Sinensky M, Fantle K, Dalton M. An antibody which specifically recognizes prelamin A but not mature lamin A: application to detection of blocks in farnesylation-dependent protein processing. *Cancer Res* 1994; 54:3229–3232.
94. Andres DA, Shao H, Crick DC, Finlin BS. Expression cloning of a novel farnesylated protein, RDJ2, encoding a DnaJ protein homologue. *Arch Biochem Biophys* 1997; 346:113–124.
95. McKenna WG, Iliakis G, Weiss MC, Bernhard EJ, Muschel RJ. Increased G2 delay in radiation-resistant cells obtained by transformation of primary rat embryo cells with the oncogenes H-ras and v-myc. *Radiat Res* 1991; 125:283–287.
96. Bernhard EJ, Kao G, Cox AD, Sebti SM, Hamilton AD, Muschel RJ, McKenna WG. The farnesyltransferase inhibitor FTI-277 radiosensitizes H-ras-transformed rat embryo fibroblasts. *Cancer Res* 1996; 56:1727–1730.
97. Bernhard EJ, McKenna WG, Hamilton AD, Sebti SM, Qian Y, Wu JM, Muschel RJ. Inhibiting Ras prenylation increases the radiosensitivity of human tumor cell lines with activating mutations of ras oncogenes. *Cancer Res* 1998; 58:1754–1761.
98. Moasser MM, Sepp-Lorenzino L, Kohl NE, Oliff A, Balog A, Su DS, et al. Farnesyl transferase inhibitors cause enhanced mitotic sensitivity to taxol and epothilones. *Proc Natl Acad Sci USA* 1998; 95:1369–1374.

INDEX

R

S

T

U

V

W

X

Y

Z